Construction Contract

The Essentials

John Adriaanse
LLB, Dip Arb, FCI Arb, Barrister
Senior Lecturer in Construction Law
London South Bank University

palgrave
macmillan

First published 2005 by
PALGRAVE MACMILLAN
Houndmills, Basingstoke, Hampshire RG21 6XS and
175 Fifth Avenue, New York, N.Y. 10010
Companies and representatives throughout the world

PALGRAVE MACMILLAN is the global academic imprint of the Palgrave Macmillan division of St. Martin's Press, LLC and of Palgrave Macmillan Ltd. Macmillan® is a registered trademark in the United States, United Kingdom and other countries. Palgrave is a registered trademark in the European Union and other countries.

ISBN–13: 978–0–333–98087–3
ISBN–10: 0–333–98087–5

This book is printed on paper suitable for recycling and made from fully managed and sustained forest sources.

A catalogue record for this book is available from the British Library.

10   9   8   7   6   5   4   3   2
14   13   12   11   10   09   08   07   06   05

Printed in China

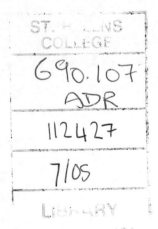

Contents

Table of Cases

Table of Statutes and Regulations

Statutes

Regulations

List of Abbreviations

ADR	Alternative Dispute Resolutions
A/E	architect/engineer
AIA	American Institute of Architects
ANBs	Adjudication Nominating Bodies
ARB	Architects Registration Board
CHA	Community Housing Association
CIA	Chartered Institute of Arbitrators
CIOB	Chartered Institute of Builders
CPR	Civil Procedure Rules
DOE	Department of the Environment
FASS	Federation of Associations of Specialists and Sub-contractors
GAIPEC	Groupe des Associations des Professionelles Européennes de la Construction
HGCRA 96	Housing Grants Construction and Regeneration Act 1996
ICE 7th	Institution of Civil Engineers, seventh edition
JCT 98	Joint Contract Tribunal, 1998 edition
JV	joint venture
LAD	liquidated and ascertained damages
LPA	local planning authority
M & E	mechanical and electrical
NEC	New Engineering Contract
NAM/T	Form of Tender and Agreement
NAM/SC	Named Sub-Contract Conditions (JCT 98)
NJCC	National Joint Consultative Committee
NPB	new parliamentary building
PFI	private finance initiative
PSW	Performance Specified Work
PWR	Public Works Contracts Regulations 1991
RIBA	Royal Institute of British Architects
RICS	Royal Institute of Chartered Surveyors
SOED	*Shorter Oxford English Dictionary*
UCTA	Unfair Contract Terms Act 1977
VAT	value added tax

Preface

Construction Law as a subject has grown rapidly in the last ten years. There is now a wide range of books for practitioners including Emden, Keating and Hudson, amongst others. There are also some highly detailed books on specialist topics (such as loss and expense, and adjudication) and some devoted to one particular standard form of contract. At student and construction professional level, the range of books is not as wide. Early in the 1990s the Faculty of the Built Environment at London South Bank University unitised its subject matter on all its courses. This allowed the development of 'advanced construction contract law' as an elective at level 3 (the final undergraduate year). This rapidly developed into a popular elective with students.

It soon became apparent that the difference in level between the student text-books and the practitioner works was difficult for all but the most able students to deal with. This book provides the materials to bridge the gap between those books. In addition, the development of a flourishing portfolio of postgraduate degrees in Building and Quantity Surveying, Construction Project Management and Construction Management has produced a need for a more comprehensive account of the subject than that offered by the student texts at undergraduate level. Such, however has been the development in the subject that a book, which reflects this, is also required at this stage of the development and the growth of the subject.

This books develops a number of themes:

- The interaction within construction contracts of the law of contract, tort and restitution
- The acknowledgement of the modern law of restitution
- The relationship between the common law and the standard forms of contract
- The introduction of the Housing Grants Construction and Regeneration Act 1996 part II.

This book is not intended as a comprehensive account of construction law. It is hoped that it will enable students at undergraduate, postgraduate and construction professional level to develop a deeper understanding of construction contract law. Recent legislation changes have meant that professionals require a much wider knowledge of construction law than the author needed when employed in the construction industry. What I have tried to do is provide a bridge for those who need to have a greater understanding of construction law. I hope this book will assist in that process.

As always it was difficult to decide what to do with the references to 'he' in the standard forms of contract. To avoid confusion, I have used 'he' throughout the book, though it should be read to include 'she'. I have also used neutral citations for the cases, thus indicating their easy availability on the Internet.

My thanks to all those students who choose to study advanced construction law and their contribution to my own understanding of the subject. My thanks to Keith Tweedy for his invaluable assistance in commenting on a number of chapters. All the mistakes and errors are of course mine. Lastly the contribution made by my wife Karen has been immense.

JOHN ADRIAANSE

Acknowledgements

Extracts from the Standard Form of Building Contract JCT 98 and the standard conditions of Nominated Sub-Contract (Conditions NSC/C) are reproduced with the kind permission of the Joint Contracts Tribunal Ltd.

Extracts from the ICE Form of Contract (7th edition 1999) is reproduced with the kind permission of the Institute of Civil Engineers.

1 The Nature of Construction Contracts

A variety of factors makes a construction contract different from most other types of contracts. These include the length of the project, its complexity, its size and the fact that the price agreed and the amount of work done may change as it proceeds.

Complex nature

The structure may be a new building on virgin ground. It may involve the demolition of an existing building and its full reconstruction. It could involve partial demolition and rebuilding, or the refurbishment and extension of an existing building or structure. This may be mostly below ground (in which case it is engineering) or above ground (in which case it is building). Building, however, includes foundations and other underground works. A building contract can consist of activities and services carried out both above and below ground.

The term 'construction contract' or 'construction law' is used throughout this book. The term includes contracts for building works and as well as engineering contracts. *Chitty on Contracts*, introduced for the first time in 1999, a chapter on the subject in its twenty-eight edition (see Ch. 37). Architects traditionally design and administer building contracts. The Architects Registration Board (or 'ARB', see www.Arb.org.uk) now regulates their academic and practical training. Consulting civil engineers, design civil engineering works and administer their construction. A large engineering contract will usually make provision for administration and supervision by site-based resident engineers.

What is a construction contract?

In *Modern Engineering (Bristol) Ltd* v. *Gilbert-Ash Northern* [1974] AC 689, Lord Diplock at 717B described a building contract as:

> 'an entire contract for the sale of goods and work and labour for a lump sum price payable by installments as the goods are delivered and the work done. Decisions have to be made from time to time about such essential matters as the making of variation orders, the expenditure of provisional and prime cost sums and extension of time for the carrying out of the work under the contract.'

It is important to realise that Lord Diplock was referring to a contract made using a standard form of building contract. Such contracts usually make provision for interim payments at regular intervals as the work proceeds, whereas a contract that is described as entire is a product of the common law. It may make provision for stage payments but, in essence, it requires the contractor to complete all its work before any entitlement to payment arises. A modern example of such an entire contract is *Discain Project Services Ltd* v. *Opecprime Developments Ltd* [2001]

EWHC Technology 450. The carrying out and completion of this contract (whether made using a standard form contract or entire) differs from other manufacturing processes. HHJ Newey OR in *Emson Eastern* v. *EME Developments* (1991) 55 BLR 114 within the context of practical completion of the work, described the differences at p.125:

> 'I think the most important background fact which I should keep in mind is that building construction is not like the manufacture of goods in a factory. The size of the project, site conditions, the use of many materials and the employment of various kinds of operatives make it virtually impossible to achieve the same degree of perfection that a manufacturer can. It must be a rare new building in which every screw and every brush of paint is absolutely correct.'

There is no special body of rules that applies to such contracts, whether described as building, engineering or construction contracts. Lord Reid said in *Modern Engineering (Bristol)* v. *Gilbert-Ash Northern* [1974] AC 689 that where the parties enter into detailed building contracts there were 'no overriding rules or principles covering their contractual relationships beyond those which generally apply'. This principle was supported by Lord Lloyd of Berwick in *Beaufort Developments (NI)* v. *Gilbert-Ash* [1998] UKHL, 2 All ER 788 where he stated that:

> 'Standard forms of building contracts have often been criticised by the courts for being unnecessary obscure and verbose. But in fairness one should add that it is sometimes the courts themselves who have added to the difficulty by treating building contracts as if they were subject to special rules of their own.'

The fact that the ordinary rules of the law of contract apply is subject to an important qualification. Legislation passed following the recommendations of the Latham Report has treated construction contracts as a special category requiring statutory intervention. The introduction of Housing Grants Construction and Regeneration Act 1996, part II (hereafter called the HGCRA 96) has also altered fundamentally the allocation of risks in construction contracts. All parties before entering into contracts have to consider how they will deal with the legislation. It also provides a much wider definition of what, for the purposes of the legislation, is a construction contract.

Section 104 (1) of the HGCRA 96 states that a 'construction contract' includes:

- the carrying out of construction operations;
- arranging for the carrying out of construction operations by others, whether under a sub-contract to him or otherwise;
- providing his own labour, or the labour of others, for carrying out construction operations.

Section 104 (2) extends the definition of a construction contract to any agreement to carry out architectural, design, or surveying work, or the provision of advice on building, engineering, interior or exterior decoration or the laying out of landscape. Note that a contract of employment is specifically excluded from the statutory definition. This definition is much wider than that given by Lord Diplock above. It includes the carrying out of design activities and the giving of advice, so widening the range of activities affected by the legislation. Construction operations are further defined by s.105 as including:

- all normal building and civil engineering works, including operations such as scaffolding, site clearance and painting and decorating as well as contracts for repair and maintenance
- consultants agreements on construction operations
- labour-only contracts
- contracts of any value

Certain contracts are excluded from the operation of the Act: see s.105 (2). The reason for this is that they did not suffer from the same ills identified by the Latham report. The petro-chemical and process industries are excluded and so are contracts concerning the supply and fixing of plant (including supporting steelwork). These activities are not classified as 'construction operations'. The off-site manufacture of components to be incorporated into the construction work is also excluded and so are contracts with residential occupiers' (see s.106). In a number of cases the meaning of construction operations has been considered. *Homer Burgess Ltd* v. *Chirex (Annan)* [2000] BLR 124 held that pipework was part of a pharmaceutical plant and not a construction operation. By contrast, in *Palmer Ltd* v. *ABB Power Construction* (1999) BLR 426 the sub-contractor work was held to come into the definition. This was so despite the main contract work being outside the definition. *Staveley Industries Plc* v. *Odebreght Oil & Gas Services* (2001) 98 (10) LSG 46 held that structures on the sea bed below low water mark are not part of the UK for the application of the Act.

There is a special requirement that the contract has to be in writing. 'Writing' is widely defined by s. 107, so word of mouth agreements referring to a written document or an exchange of letters is sufficient to bring it into the section. Whether the agreement is in writing has been an issue in a number of references to adjudication. In *Grovedeck Ltd* v. *Capital Demolition Ltd* [2000] BLR 181 an oral agreement was entered into. The parties went to adjudication under the Scheme for Construction Contracts despite the protests of one side that the Act did not apply. HHJ Bowsher QC refused to enforce the adjudicator's award finding that there was no written contract. The Court of Appeal in *RJT Consulting Engineers Ltd* v. *DM Engineering (Northern Ireland) Ltd* (2002) 18 Const LJ 425 had to consider the meaning of s.107. All three judges interpreted the section as requiring the terms of the agreement to be in writing, not the agreement itself (see chapter 16 below where the issue of writing is discussed further in relation to adjudication). Note also that private finance initiative contracts and finance agreements are excluded on policy grounds.

Risk allocation

Like all contracts, construction contracts are about the prior allocation of risk. Windward (1991) draws attention to the construction industry's need to make a profit on the employment of capital: 'If risk is an essential ingredient of the system which generates your profit, it is inevitable that there must be a structure for resolving disputes. It brings the relationship of the disputants back into balance so that life can resume its normal course.' Windward's reference was primarily to arbitration, the main method of dispute resolution prior to the introduction of adjudication. It is of course a matter of debate whether the introduction of statutory adjudication by the HGCRA 96 achieves that balance between disputing parties. Conceived by the Latham Report as essentially a 'quick fix' in that it was intended to resolve disputes as they arose rather than at the end of the project, this being the usual practice before its introduction. Standard forms of contract try to allocate risk equitably between the parties. In essence, the payment provisions of the HGCRA 96 can also be described as an attempt to introduce a measure of equity into the contractual relationship between contractor and subcontractor. Risks are varied in construction contracts, and include many factors that can affect the progress of the work.

1 The unforeseen:
    (a) unexpected ground conditions
    (b) unpredicted weather conditions
    (c) a shortage of material
    (d) a shortage of skilled labour
    (e) accidents, whether by fire, flood or carelessness
    (f) innovative design that does not work or proves impossible to construct.
2 The length of the contract. Projects vary in the time needed for completion, from days to years. During that time the risk allocation agreed at the time of contracting can change substantially. This is especially so with regard to the availability of materials and its costs. A contractor may have 'bought' the job because work was scarce at the time and the price of components low.
3 The number of participants, and parties in the project and the corresponding length of contractual chain cause their own problems. The risk of insolvency increases the longer the chain.
4 The particular relationship (often referred to as a triangular relationship of costs, time and quality) in which conflict is inherent. Contracting parties have different perceptions of how these factors of their relationship interact.
5 The interaction between liability for defective workmanship and for faults in design. The Latham Report in item 3.10 identified the lack of co-ordination between design and construction as a common source of dispute. Much of the innovation in procurement systems of recent years, stems from creating ways of minimising the effect of this clash.

Why use a standard form of contract?

There has been a proliferation of standard forms in the construction industry in recent years and there are many books available on specific forms of contract. It is not intended to compare those forms here or discuss their similarities or differences. There are, however, many advantages to be gained by using a standard form of contract.

Advantages

Some advantages are listed below:

1 The standard form is usually negotiated between the different bodies that make up the industry. As a result the risks are spread equitably.
2 Using a standard form avoids the cost and time of individually negotiated contracts.
3 Tender comparisons are made easier since the risk allocation is same for each tenderer. Parties are assumed to understand that risk allocation and their prices can be accurately compared.

Disadvantages

Some disadvantages are listed below:

1 The forms are cumbersome, complex and often difficult to understand.
2 Because the resulting contract is often a compromise, they are resistant to change. Much needed changes take a long time to bring into effect.

Participants

A construction contract is best described as a complex web of competing interests. A particular problem within construction contracts is that there is little interest in building long-term relationships. With the growth of partnering it is possible to argue that much has changed. Contracts such as the New Engineering Contract (NEC) now renamed the NEC Engineering and Construction contract, it is argued, provide flexibility and simplicity. Its success also depends on the parties building long-term relationships. In such a case the contract may well serve only as document to record the intentions of parties. Supply chain management is another much heralded innovation. When tested in a recession or an economic downturn, will these innovations may well provide proof that the construction industry has indeed changed.

The *traditional contract* is of great importance in understanding the problems and complexities of construction contracts. Only by analysing the relationship between the employer and the contractor is it possible to understand the problems that other forms of procurement try to resolve. At the heart of the traditional contract lies the conflict between design on the one hand and workmanship on the other. That conflict is complicated further by the need to allocate rights to third parties.

The traditional contract

In such a contract the employer contracts with an architect or engineer to carry out the design. The architect or engineer, acting as the agent of the employer, supervises the construction of that design. The contractor enters into a contract with the employer to build that design. There is no guarantee given by the employer to the contractor that the design can be built. In carrying out the work, the contractor employs both sub-contractors and suppliers of goods and equipment. These may in turn be classified as domestic or nominated. Out of the relationship between these parties arises the issue of privity of contract: the basic rule that only parties to a contract can enforce the contract. In summary:

(a) in the absence of a warranty there is no contractual relationship between the employer and subcontractors and suppliers;
(b) third parties have no contractual rights.

The use and spread of collateral warranties have resolved some of the problems caused by the doctrine in providing for the forming of contractual relations with third parties. The passing of the Contracts (Rights of Third Parties) Act 1999 provides an opportunity to simplify this area but at the moment, this piece of legislation has not been widely adopted by the construction industry. Instead, the standard form of building contracts, Private with Quantities 1998 edition issued by the Joint Contracts Tribunal (hereinafter called JCT 98) and the Institution of Civil Engineers Condition of Contract, Measurement Version, seventh edition (called ICE 7th) specifically excluded the application of the Act from these forms.

The tendering process

Whatever the method of procurement adopted, the tendering process in the UK is based on competitive bidding. To ensure transparency in this process the National Joint Consultative Committee (NJCC), an organisation consisting of the major professional bodies involved with construction has produced codes of procedure.

Open tenders

The first step in this type of tendering is an advertisement in the technical press calling for expressions of interest. Parties can obtain the documents needed from the body placing the advertisement or its agents. The advertisement usually contains a brief description of the location, the type of work being proposed, the scale of the project and the scope of the proposed work. Interested contractors are invited to apply for the details. The main disadvantage of this type of tendering is that it is indiscriminate in its approach, costly and likely to attract inexperienced tenderers. Local authorities have in the past tended to favour this method of procurement. Its use has been affected by European Directive 71/305/EEC 9 (as amended by directive 89/440/EEC), implemented in the UK by the Public Works Contracts Regulations 1991 (SI 1991/2680). These were followed by a number of regulations aimed at opening the European construction market. For more details see Emden (2001 p. 2009). In public procurement this is the main form of selection. The financial thresholds change every two years. Because it applies to all public and local government projects, health authorities, police, education authorities and so on, it is therefore applicable to a wide range of projects.

Single-stage selective tendering

The NJCC code considers that this procedure is suitable for both private and public sector works. This procedure restricts the number of tenderers by pre-selection from either an approved list or on an ad hoc basis. A limited number (up to six) are selected on the basis of general skill and experience, financial standing, integrity, proven competence with regard to statutory health and safety requirements and their approach to quality assurance systems. Thereafter, price alone is the criteria, the lowest tender being selected.

Two-stage selective tendering

The NJCC regards this as a suitable method where the early involvement of the main contractor is required before the scheme is fully designed. It enables the design team to make use of the contractor's expertise. The contractor also becomes involved in the planning of the project at an early stage. The first *stage* consists of the selection of the contractor on the basis of a competitively priced tender but with minimal information provided. The submission is on a basis of the layout and design of the works, clear pricing documents relating to the preliminary design and specification and the conditions of contract. In the second *stage* the employer's professional team collaborates with he selected contractor in the design and development of production drawings for the whole project. A Bill of Quantities (or it may simply be priced on drawings and a specification) is prepared and priced on the basis of the first *stage* tender. If an acceptable sum is produced the contract documents are then prepared. This method is considered to be useful for building works of a large or complex nature, where the brief is unlikely to change. It is recommended for projects where the design and construction phases may overlap and the contractor's design expertise can be utilized.

Selective tenders

Normally called design and build contracts, these are also called turnkey contracts which is a wider description of what the employer may expect (i.e., the employer puts the key into the door of the new factory and starts it up). The contractor's price is to carry out and complete the works in accordance with the conditions of contract. The tender includes the whole of contractor's proposal including price and design. The NJCC Code of Procedure for selective tendering for design and build describes its code as a procurement method that combines the design and construction responsibilities.

Other modes of selection

Negotiation

Contracts are seldom entered into on this basis alone. Parties may negotiate an extension to a contract, additional work outside the scope of the contract may be agreed or additional work may be carried out elsewhere for the same employer.

Joint venture

A joint venture is where two or more companies pool resources for a project beyond the resources of the single companies. It may be used on one project or the agreement may be for a specified period. The co-venturers accept joint and several liability for the project.

Other means of procurement

Various types of 'procurement systems' have evolved in recent years to deal with the difficulties perceived within the traditional contract.

(a) Management contracting.   Here the employer engages the management contractor to partake in the project at an early stage. Normally an experienced builder, the contractor is employed not to undertake the work but to manage the process. All the work is sub-contracted to works contractors who carry it out.

(b) Construction management.   This differs from management contracting in that the employer enters into a direct contract with each specialist. The employer engages the construction manager to act as a 'consultant' to co-ordinate these contractors.

(c) Project management.   The project manager is employed to co-ordinate all the work needed from design to procurement and construction on behalf of the client.

(d) Partnering.   The rise of partnering in UK construction can be seen as a response to the widely held view that the industry was inherently flawed. After the boom times of the 1980s and subsequent recession of the early 1990s it contracted sharply. A culture of conflict persisted in the industry with employers and contractors operating in a highly adversarial manner, with contractors taking on greater risks within a fiercely competitive market. In many cases, in order to secure contracts and survive, contractors had to tender at cost (or even under), and recover margins through building claims into the contracting process and withholding payments to sub-contractors (Critchlow 1998). The essence of any partnering agreement now involves a duty of good faith, mutual co-operation and trust between all parties involved in the construction process. This is discussed further in Chapter 2.

(e) The Private Finance Initiative (PFI).   The aim of the system known as PFI is to involve the private sector in the provision of public services. In essence the PFI contract is a concession granted by the public sector to the private sector. The private sector company provides the vehicle by which the project company secures the finance to provide services for running the asset provided. This may be a hospital, the provision of information technology projects and services or the running of the London Underground. The purpose of private and public partnerships is to share the risk of the project, and PFI is at the moment the dominant method. For further information reference should be made to www.pfi/ogc.gov.uk and www.doh.gov.uk/pfi. For a contractor's perspective see Martin Lenihan at www.scl.org.uk.

The incorporation of documents

The usual approach in books on contract law is to deal with this topic after discussing formation and before dealing with the terms of the contract. The approach taken here is to deal with it as part of the discussion of standard forms and the tendering process itself. Having commented earlier that parties use standard form contracts because they bring certain commercial benefits, the question of whether those documents are incorporated into the contract may prove crucial. Parties often make reference in contractual documents to the contract being 'subject to conditions available on request'. Such a reference, when brought to the notice of the other party, is sufficient to incorporate the current edition of those conditions of contract. This rule was decided in *Smith* v. *South Wales Switchgear Ltd* [1978] 1 All ER 18.

Smith overhauled South Wales Switchgear Ltd's electrical equipment for some years. The Company wrote to Smith asking him to carry out the overhaul of equipment. A purchase note requesting work which read 'subject to . . . our general Conditions of contract 2400, obtainable on request', was sent to Smith. He carried out the instructions as requested but did not request a copy. An unrequested copy of the 1969 conditions was sent to him. There were two other versions of the condition including the March 1970 revision.

Held: The reference in the purchase order incorporated the March 1970 revision. There were three reasons for the decision:

1 It was clear how Smith could have ascertained the terms.
2 It was common knowledge that conditions of contract change over time.
3 If he had asked for the conditions he would have received a copy of the current one.

Sometimes the terms on which the work is to be let, will be referred to in correspondence passing between the parties. Terms of a standard form of contract may be incorporated into the contract by such a reference. A contract was made by exchange of letters in *Killby and Gayford* v. *Selincourt* [1973] 3 BLR 104. On 11 February 1972 the architect wrote to the contractor seeking a price for alteration work. The letter concluded, 'subject to a satisfactory price between us, the general conditions and terms will be subject to the normal RIBA [Royal Institute of British Architects] contract'. The contractor provided a written estimate on 6 March 1972. The architect replied on behalf of the client accepting the estimate and also stated 'please accept this letter as formal instructions to start work'. No standard RIBA (now the JCT) contract was ever signed. It was held that the exchange of letters incorporated the current RIBA form of contract.

The parties may make a contract using a standard form of contract in circumstances where the use of the form is inappropriate. For example a main contractor may sub-let work proposing the use of a standard form of contract not intended for such contracts. Or, a fixed fee contract may use one which allow for variations. Where the parties do so, a court may try to make sense of the vague references to it. In *Modern Buildings (Wales)* v. *Limmer and Trinidad* (1975) 14 BLR 101 the sub-contract order read: 'to supply adequate labour, plant and machinery to carry out . . . [sub-contract] work . . . In accordance with the appropriate form for nominated sub-contractors (RIBA 1965) edition . . . All as per your quotation.' No RIBA sub-contract form existed and neither was there a RIBA 1965 edition. Held: the current green form was incorporated into the contract. The words in brackets would be disregarded to make the contract work. Buckley LJ observed that:

'Here it is not disputed that the written contract between the parties consisting of the quotation and the order contains all the essential terms in the contract and in my judgement, the "green" form of contract must be treated as forming part of the written contract, subject to any modifications that may be necessary to make the "green" form accord in all respects with the express terms agreed between the parties.'

Note here that the main contract was let under the JCT 63 standard form of building contract. Expert witnesses agreed that that the reference to the 'green' form would be understood in the trade as being a reference to the (former) Green Form of nominated sub-contract.

In *Brightside Kilpatrick Engineering Services* v. *Mitchell Construction* (1973) *Ltd* (1975) 1 BLR 64, the printed references to the green form had been deleted. The Court of Appeal had to construe references to the main contract condition contained in the sub-contract order. Buckley LJ said at 65:

'The contention for the defendants has been that the statement in the document of 24 March 1972 that "the conditions applicable to the sub-contract with you shall be those embodied in the RIBA as above" is a clear reference to the main contract and the effect of that is to import into the contractual relationship between the contractor and the sub-contractor all the terms of the head contract . . . substituting for references to the building owner references to the contractor, and substituting for references to contractor references to the sub-contractor'.

In agreeing with this approach the court held that the main contract provisions had to be modified to reflect the contractor – sub-contractor relationship. Note that the FASS 'green' form was the one approved by the Federation of Associations of Specialist and Sub-contractors (FASS).

The construction of contracts

Amongst the issues that often arise in construction contracts is the meaning of words used by the parties in their written contracts. The process by which courts arrive at this meaning is called construing the contract. The resulting meaning as determined by the court is called the construction of the contract. Lord Diplock in *Pioneer Shipping* v. *BTP Toxide* [1982] AC 724 at 726 said that the object in construing any commercial contract is to ascertain: 'What each [party] would have led the other to reasonably assume were the acts he was promising to do or refrain from doing by the words in which the promise were expressed'.

Expressed intention

In construing a contract the court applies the rule of law that while it seeks to give effect to the intentions of the parties it must give effect to the actual words used. It must decide what the parties actually meant by using those words. Intention does not, however, equate to 'motive, purpose, desire or state of mind'. Those are subjective states. The common law adopts an objective standard of construction by excluding general evidence of the actual intention of the parties. The meaning of the phrase 'expressed intention', has been considered by Lord Hoffman in a number of recent cases. In *Mannai Investment Co Ltd* v. *Eagle Star Life Assurance Co Ltd* [1997] AC 749, he reiterated the well-known principle that the law was not concerned with the subjective intention of the speaker. He went on to comment that contained in the expression, 'the meaning of the words', was an ambiguity that could not be ignored. This ambiguity was itself twofold. One concerned the meaning of the actual words themselves, whether contained in dictionaries or in the effect of their syntactical arrangement in grammar. The second was what the person who used them understood them to mean with regard to the factual background in which they were used. He went on to say:

'It is the background which enables us, not only to choose the intending meaning when the word has more than one dictionary meaning but also to understand a speaker's meaning, often without ambiguity, when he has used the wrong words.'

Bowdrey and Cottam conclude that Lord Hoffman's analyses of what he describes as all the 'old intellectual baggage' of legal interpretation has been discarded. A view shared by McKendrick (2000 p.194). In *Investors Compensation Scheme Ltd* v. *West Bromwich BS* [1998] 1 WLR 912 Lord Hoffman provided a summary of the principles involved:

1   Interpretation is the ascertainment of the meaning which the document would convey to a reasonable person, having all the background knowledge, which would reasonably be available to the parties in a situation in which they were at the time of the contract.
2   The background was famously referred to by Lord Wilberforce as the 'matrix of fact' but this phrase is, if anything, an understated description of what the background may include. Subject to the requirement that it should have been reasonably available to the parties and to the exception to be mentioned next, it includes absolutely anything which would have affected the way in which the language of the document would have been understood by a reasonable person.
3   The law excludes from the admissible background the previous negotiations of the parties and their subjective intent. They are admissible only in an action for rectification. The law makes this distinction for reasons of practical policy and, in this respect only; legal interpretation differs from the way utterances

are generally interpreted in ordinary life. The boundaries of the exception are in some respects unclear, but this is not the occasion to explore them.

4   The meaning which a document (or any other utterance) would convey to a reasonable man or woman is not the same thing as the meaning of the words. The meaning of words is a matter of dictionaries and grammars; the meaning of documents is what the parties using those words against the relevant background would reasonably have been understood to mean. The background may not merely enable the reasonable person to choose between the possible meaning of the words which are ambiguous but even (as occasionally happens in ordinary life) to conclude that the parties must, for whatever reason, have used the wrong words or syntax; see *Mannai Investment*.

5   The 'rule' that words should be given their 'natural and ordinary meaning' reflects the common sense proposition that we do not easily accept that people make linguistic mistakes, particularly in formal documents. On the other hand if one nevertheless concludes from the background that something must have gone wrong with the language, the law does not require judges to attribute to the parties an intention to which they plainly could not have had. Lord Diplock made this point vigorously when he said in *Antaios Compania Naviera S.A.* v. *Salem Rederierna A.B.* [1985] AC 191 at 201: 'if detailed semantic and syntactical analyses of words in commercial contracts is going to lead to a conclusion that flouts business common sense, it must be made to yield to business common sense'.

Extrinsic evidence

The general principle for written contracts is that extrinsic evidence, outside of the document itself, is not admissible by a court. No extrinsic evidence may normally be adduced to contradict, vary, add or subtract from the written terms. There are a number of exceptions.

Blanks

Where a complete blank is left in a material part of a contract, evidence is not admissible to fill those blanks: see *Kemp* v. *Rose* [1858] 32 LT, (OS) 51, where the date of completion was omitted from the contract. To insert it would have resulted in the imposition of an onerous liquidated and ascertained damages clause. The court refused to admit evidence that each party had been told of the date. See also *Temloc Ltd* v. *Errill Properties* (1988) 39 BLR 30 where 'Nil £' was held to equal a blank.

Preliminary negotiations

Where the parties have reduced their negotiations to a final written contract, preliminary negotiations such as letters cannot be referred to for the purpose of explaining the parties' intentions. In *Davis Contractors* v. *Fareham Urban District Council* [1956] 2 All ER 145, a covering letter was sent subject to certain terms. After negotiations, a formal contract was signed which defined the contract documents but without including the letter. It was held that the term contained in the letter was not part of the contract.

Deletions from the printed documents

A question that often arises is whether it is permissible to look at deletions from printed documents. If it is permissible, for what purpose can it be used? Keating (2001 p. 81) says there are two schools. One school, an old weighty authority, states that it is not permissible to look at deletions at all. The rule was followed in a number of recent cases: see for example *Wates Construction* v. *Franthom Property* (1991) 53 BLR 23.The other school says that where the parties use a printed form and delete part of it, the deleted part can be looked at to decide what they have agreed to leave in. This was the approach adopted in *Motram Consultants* v. *Bernard Sunley* [1975] 2 LLR 197, where the House of Lords referred to the deletion from an existing contract of a clause giving the employer the right to set-off monies owed to the contractor. It decided that the parties had directed their minds to the question of set-off and decided that it should not be allowed. The second school was preferred. Where there is an ambiguity it is possible to look at deletions in order to throw light on the subject.

Where intrinsic evidence is permissible

Two occasions can be identified:

1 Where foreign or technical words can be proven. If there is no dispute about technical words, the court can inform itself by reliable means. Counsel can provide an explanation, dictionaries can be consulted or an assessor appointed.
2 Custom or usage. Evidence is permissible to show that the words were used according to a special custom or usage known to the trade or locality applicable to the contract. Everyone must know about it. In *Symonds* v. *Lloyd* (1895) 6 CB 691 it was accepted by the court that 'reduced brickwork' meant brickwork nine inches thick.

Rules of construction

This is only applied where there is some ambiguity or inconsistency. Where the meaning of words is plain, the court gives effect to them.

Ordinary meaning

The court decides as a matter of fact whether the case is within the ordinary meaning or not. The grammatical and ordinary sense of a word is to be adhered to. In doing so it also adopts the reasonable meaning. Where terms are ambiguous and two meanings are possible, the court adopts the reasonable meaning.

Written words prevail

Where there are printed contracts with clauses inserted or filled which are then inconsistent with the printed words, the written words prevail over the printed. Written words reflect the selection made by the parties.

Ejusdem generis rule

The phrase means 'of the same type'. This means that where words of a particular class of words, are followed by general words, the general words are treated as referring to matters in the same class. Thus in a clause dealing with non-delivery, where the port was unsafe 'in consequence of war, disturbance, or other causes', ice was held not be within the class. Another example is the presence of a clause that entitled the contractor to an extension of time if the works were 'delayed by reason of any alteration or condition . . . or in the case of a combination of work-men, or strikes, or by default of sub-contractors . . . or any other cause beyond the contractor's control'. The clauses were held to be limited to those ejusdem generis within the clause described. It did not therefore, include the employer's default in failing to give the contractor possession of the site.

Contra proferentem

This simply means that where the words are ambiguous they are construed against the profferer, the person who drafted the document. As it is their own document, they must know what they meant by the words they used. Where the representative bodies or committees have drafted a standard form of contract such as the JCT 98 or the ICE 7th, the rule has no application.

Summary

1   All contracts for work and materials are subject to the provisions of the HGCRA 96.
2   It introduces a statutory definition of a construction contractor that is much wider than the usual one at common law.
3   Contracts are about the allocation of risks which are varied in such contracts.
4   The length of the chain in construction contracts increases the nature of the risks.
5   Standard forms of contract attempt to allocate risk fairly.
6   Modern innovations in procurement methods try to resolve the basic conflict between the design and workmanship, which is at the heart of the traditional contract.
7   Competitive tendering and the various procurement systems allocate risk in different ways.
8   Standard forms may be incorporated into the contract by an exchange of letters. Whether they are can also depend on how easily the documents could have been obtained by the party arguing that they were not part of the contract.
9   Courts will attempt to use vague references to such contracts in order to make a contract work.
10 The meaning of expressed words used is subject to the rules of construction. The result of the process is called construing the contract.

2 Outline of the Law of Obligations

Is there a law of obligations?

The purpose of this chapter is to provide an introduction to the 'law of obliga-tions' that forms one of the essential themes of this book. Its inter-relationship and application are demonstrated in *J. Jarvis & Sons Ltd* v. *Castle Wharf Developments Ltd and ors* [2001] Lloyds LR, PN 308. For completeness, an analysis of the evolving area of *good faith* is included. Put shortly, should the law of oblig-ations be underpinned by an overriding principle of good faith? The adoption of such a concept was proposed for all construction contracts by the Latham Report, and subsequently incorporated in some standard forms.

Unlike the civil law of continental Europe, English law does not generally recognise a law of obligations. Samuel and Rinkes (1996, p.2) suggests that the term is being used with increasing frequency to describe the law of *contract*, *tort* and *restitution*. He argues that as a term of convenience it causes few problems. As a term of art (or science) it lacks the precision of the civilian (continental) legal thought for a number of reasons:

(a) common law lawyers are more concerned with *remedies* than with rights;
(b) no clear distinction is drawn between rights *in personam* and in rights *in rem*;
(c) the distinction between substantive law and legal procedure is not clearly drawn.

The interaction between the three obligations was considered by Lord Goff in *Henderson* v. *Merret Syndicates Ltd* [1995] 2 AC at 145. There he observed at p.184 that:

> 'The situation in common law countries, including of course England is excep-tional, in that the common law grew up within a procedural framework unin-fluenced by Roman law. The law was categorized by reference to the forms of action, and it was not until the abolition of the forms of action by the Common Law Procedure Act 1852 that it became necessary to *reclassify the law* in substantive terms. The result was that common lawyers did at last separate our law of obligations into contract and tort, though in doing so they relegated *quasi-contractual* claims to the status of an appendix to the law of contract, thereby *postponing* by a century or so the development of *a law of restitution*.
> Even so there was no systematic reconsideration of the problem of *concur-rent claims* in *contract* and *tort*. We can see the courts rather grappling with unpromising material drawn from the old cases in which liability in negligence derived largely from categories based on status of the defendant. In a sense, we must not be surprised; for no significant law faculties were established at our universities until the late 19th century, and so there was no academic opinion available to guide or stimulate the judges.' (emphasis added)

The case of *Henderson* v. *Merret Syndicates Ltd* ended *the dual debate*, that 'systematic reconsideration of the problem of *concurrent claims* in *contract* and *tort*' referred to above. This debate accelerated following the decision in *Murphy* v. *Brentwood District Council* [1991] AC 398. The House of Lords confirmed in *Henderson* that professionals could be sued in both *contract* and *tort*.

An obligation is defined in the *Shorter Oxford English Dictionary* (*SOED*) as 'an agreement, enforceable by law whereby a person or persons become bound to the payment of a sum of money or other performance'. The word also has moral or religious overtones as, for example, in a 'day of obligation', which is a day when it is an obligation for the faithful to abstain from work and attend divine service· (Burke quoted in *SOED*, p.1427 under obligation).

By contrast Zimmerman (1996, p.5) traces the word to its Roman origin 'obligatio'. Its root, *lig*, indicates that something or someone is bound; just as we are all bound (to God) by virtue of our 're-*ligio*'. In the *private law* of the European countries (often called civilian law) the term has a *technical* meaning. It refers to a two-ended relationship that appears from one end as a personal right to claim and from the other as a duty to render performance. As far as Roman terminology was concerned the *obligatio* or *legal bond* refers to both the debtor (the person bound to make performance) and the creditor (the person who has put his confidence in this specific debtor's will and capacity to perform). In other words the *legal bond* is looked at from either end of the two-ended relationship. It could refer to the creditor's rights as well as the debtor's duty.

The Roman concept is difficult to equate with the English obligation. An obligation in English law is merely oriented towards the person bound, not to the person entitled. My obligation refers only to my *duties*, not my rights. For this reason the adoption of a law of obligation is considered by many commentators to be premature at this stage in the development of English law. However, it does provide a convenient and useful overview of the interplay of obligation within a construction contract.

Traditionally the law of contract and the law of tort are treated as different and distinct subjects in the writing and teaching of English Law. Rogers argues persuasively in *Winfield and Jolowitz on Tort* (2002) that a general book on the law of obligations would become unmanageably large. However, he also states that where a particular factual area of legal liability study (such as 'professional liability') is involved this is *essential*.

The House of Lords in two recent cases, namely *D & F Estates* v. *Church Commissioners for England* [1988] 3 WLR 368 and *Murphy* v. *Brentwood District Council* [1991] 1 AC398 414, *reinforced* this distinction. Within the *context* of the liability of construction professionals, whether engineers, architects, surveyors or project managers, they co-exist as related duties. Contractors also owe duties in contract and tort but their tortious duties have been *restricted* in recent years as the result of those recent cases mentioned above.

In *Tai Hing Cotton Mill Ltd* v. *Lui Chong Hing Bank Ltd* [1986] AC 80, Lord Scarman summarised the issue of concurrent liability at p. 107. First, he commented that 'Their lordships do not believe that there is anything to the advantage to the law's development in searching for liability in tort where the parties are in a contractual relationship'. This was particularly so where the parties were in a commercial relationship. Second, by sticking to the contractual principle that the parties have, subject to few exceptions, the right to determine their own contractual obligations. To add tortious rights merely makes the quantification of duties more difficult. Adding it into the analyses introduces uncertainty in the relationship by using such concepts of proximity and character of the relationship into contractual relations. Third, he observed different consequences flowing from characterising the liability as arising in contract or tort. One example he gave of this was the existence of different limitation periods.

Keating on Building Contracts suggested, in the preface to its fifth edition, that in the light of *Murphy* the law on professional negligence should be reconsidered. It questioned whether the existence of duties to a client in contract should also attract duties in tort. *Henderson* rejected this argument emphatically, stressing that professionals were free to *exclude* duties in tort when entering into contracts (see the further discussion in Chapter 13 below).

The recognition of the modern law of restitution

In the 1980s, whenever a claim for damages was made in contract, it was the usual practice to make a claim in tort (especially in negligence) as an alternative head of claim. By the 1990s a claimant might make instead a claim for *unjust enrichment* based on a law of restitution. McMeel, in *The Modern Law of Restitution* (2000, p.5), cites Dawson's comment that:

> 'It seems clear that the Anglo-Saxon legal system has come late to the problems of restitution. This tardiness is only relative. Therefore it appears that the prevention of unjust enrichment as a distinct and independent motive for judicial action is apt to be recognised late in any legal system after provision has been made for primary institutions through which societies organise.'

McMeel states that it is now generally accepted that three of House of Lords cases in the 1990s finally established the *modern law of restitution*.

1 *Lipkin Gorman* v. *Karpnale Ltd* [1991] 2 AC 548 *recognised* the principle of *unjust enrichment* as the foundation of the law of restitution.
2 *Westdeutsche Landesbank Girozentrale* v. *Islington London Borough Council* [1996] AC 669 rejected the implied contract fiction (or quasi-contract, as it is also called).
3 *Banque Financière de la Cité* v. *Parc (Battersea) Ltd* [1999] 1 AC 221 *adopted the analytical methodology* to the structure of unjust enrichment enquiry adopted by the leading scholarly works. (see p. 22 below)

The law of restitution (in construction contracts) based on unjust enrichment features largely in relation to *benefits arising out of contracts that fail* for some particular reason to come into existence. Such claims are common in the construction industry. They often arise out of the contrary demands of getting work started whist contract terms are being negotiated.

The differences between contract and tort

These are listed below.

1   A contract can be described as an *expectation* created by *binding promises*. Both parties expect to gain financially from their bargain. The aim of damages for breach of contract is to put the innocent party in the position they would have been in had the contract been carried out (in other words to protect their *expectation interest*).
2   Tort's function is the maintenance of the *status quo*. The aim of damages is to restore the victim to the position they would have been in had the tort not been committed (i.e., to protect their *reliance interest*).
3   All legal systems have experienced in recent years a number of problems caused by developments in technology. This has tended to blur the distinction between contract and tort.
4   English law has attempted to solve this problem by reasserting the primary distinction between contract and tort. It has done this by restating that *damage in tort is dependent on physical injury* and by refusing a remedy for *economic loss*. Such losses can only be recovered in contract. The only exception is *liability of professionals*. They can be sued in either contract or tort and the plaintiff can choose the most advantageous basis.

From a practical lawyer's viewpoint there may be considerable overlap between contract and tort in any factual situation (see Winfield & Jolowicz 2002 p.5 for some examples). The recent case of *Albright & Wilson UK Ltd* v. *Biachem Ltd and ors* [2002] ULHL 143 provide a clear illustration of this overlap. Two separate contracts, made with different companies to deliver tanker loads of chemicals to a plant, were scheduled for the same day. By chance, the same haulage contractor was employed to deliver both loads. As a result of a muddle, the delivery notes were swapped and the wrong chemical was delivered. A tanker load of sodium chloride was emptied into a storage tank containing epichlorohydrin, resulting in an explosion. The respondent sued both companies for the resulting damage. It could have brought its claim in either contract or in tort: in contract, by alleging that the companies had failed to deliver the right chemical and their breach had resulted in the damage; or in tort of negligence, alleging they had failed their duty of care. Had it done so, it was open to the defendants to make a claim for contributory negligence, since the respondent should have checked before emptying the chemical into the tank, that it was the expected one. To avoid this possibility it chose to frame its action in contract only.

Despite the fact that there can be concurrent liability there is no double recovery of damages. Winfield & Jolowicz produce the following classification:

1   Tortious duty arises by law and is not based on agreement or consent. This distinguishes it from contract (and bailment), which has no existence outside agreement or consent. Even so, tortious duties can arise out of consent: see, for

example, the duty of care owed to a lawful visitor under the Occupiers Liability Act 1957 and liability for gratuitous statements at common law.

2 The law fixes the content of the duty in tort whereas the content of the duties is fixed by the contract itself. However, a clearer basis for distinguishing between the two is to consider the 'aims of the two heads of liability'. The core of contract is the enforcing of promises and tort the prevention of or compensation for harm. This has two principal consequences:

   (a) mere failure to act is not actionable in tort;
   (b) there can be no damages awarded in tort for the 'loss of an expectation'.

Further differences: the measure of damages

In *Copthorne Hotel (Newcastle)* v. *Ove Arup Assoc.* (1996) Const LJ 402, HHJ Hicks QC adopted the following formulation (based on *South Australia Asset Management Corporation* v. *York Montague Ltd* [1996] 3 WLR 87) to describe the difference:

1 Warranty (contract): A warrant's that the house is in good repair. If it is not, A must pay must pay the cost of putting it in good repair.

2 Negligent advice (tort): A negligently advises that the same house is in good repair. If the advice were sound B would not have got a sound house but a warning that it was unsound. B's damages are not the cost of repair. B must show that the position would have been different from that which ensued as a result of the advice.

The warranty (contract) measure of damages is the extent B would have been better off had the information been right. In contrast, the negligent measure is the extent that B is worse off because the information was wrong.

Contracts enable us to deal with things that may happen and make allowances for these eventualities. Torts are essentially accidents and cannot be prepared for in advance. That is why insurance cover is taken to guard against the unforeseen. This difference can be clearly illustrated by examining the celebrated case of *Beswick* v. *Beswick* [1968] AC 58 HL.

Peter Beswick made an agreement with his nephew John, to assign to him his business in return for a weekly sum of £6.50 for his life, and in the event of his death, pay the sum of £5 per week to his widow Ruth for her life. John paid Peter promptly during the next 11 years until Peter died intestate. John made one payment to Peter's widow and then refused further payments on the basis that he owed her no legal liability. Ruth Beswick (aged 74 and in poor health) sued John in her capacity as administratix for her late husband's estate and in her personal capacity for performance of the agreement. Held (HL): that the widow as administratix of a party to the contract was entitled to an order for specific performance of the promise made by the nephew.

The principle illustrated here is that Peter could have arranged things differently to get the same result, by simply arranging for his wife to be a party to the contract. This would have achieved the same result arrived at in the House of Lords, without the considerable costs or the legal distortion involved.

The difference in limitation periods

On the surface the limitation period is the same as contract: 6 years from the date of the tort. Time in tort starts to run when the damage occurs: see *Pirelli General Cable Works* v. *Oscar Faber and Partners* [1983] 2 WLR 6. Prior to *Pirelli* time started to run when the damage was discovered. *Pirelli* created a problem since it was difficult to identify when time started to run in a building that had been damaged, since time could have started to run long before the damage was discovered. This resulted in the claimant (*Pirelli*) being out of time when they discovered the damage to the chimney of their factory. What is important is to realise is that the limitation period in tort could be significantly longer than in contract. Thus parties with a contract could sue in tort if the contractual limitation period was over. A third party could also sue a party responsible for the damage but with whom it did not have a contract.

The Latent Damage Act 1986

The Act was passed to deal with the problems created by the decision in *Pirelli* and introduced new statutory provisions into the Limitation Act 1980. Broadly speaking it effects were as follows:

1 It kept the rule in *Pirelli*: the limitation period started 6 years from the date damage occurred, or
2 3 years from the date the damage was discovered, whichever happened first.
3 Both were subject to a long-stop of 15 years, which ended all claims.
4 The Act confirmed that third parties had the right to sue under it. This right was effectively ended by the combination of the case of *D & F Estates* and two years later *Murphy*. The effect of these cases were the economic cost of buildings defects could no longer be recovered in using the law of tort (see also chapter 14 below).

Note that the Latent Damage Act 1986 now applies primarily to professionals, since contractors and sub-contractors cannot be sued in tort for economic loss. Only where there is injury to persons or damage to other property are they liable in tort.

Outline of the law of restitution

As mentioned above, Lord Goff observed in *Henderson* that reclassifying the law of obligations into contract and tort only became necessary after the passing of the Common Law Procedure Act 1852. In doing so, 'quasi-contractual claims were relegated to the status of an appendix to the law of contract, thereby postponing by a century or so the development of a law of restitution'. This position was itself only altered recently by the three House of Lords cases mentioned at the beginning of this chapter. Professor Burke claims this late evolution was due to Blackstone's *Commentaries on the Laws of England*, since it 'taught everyone from the 18th century onwards, that restitutionary obligations depended on an "implied contract" ' (Introduction to the Law of Restitution, 1996, p.4). The effect was to relegate non-contractual obligations to the appendix of the textbooks on contract. Equitable restitution was also treated in much the same way since equity always follows the law. Rescission (an equitable remedy) became attached to contract and obligations were imposed on trustees by calling them constructive trustees. Burke's definition is as follows: 'Restitution is the response which consists in causing one person to give up to another an enrichment received at his expense or its value in money' (1996, p.13).

Restitution was recognised as a vital component of a mature civil law long before its acceptance in the 1990's. Lord Wright observed in the *Fibrosa Spolka Akcyjna* v. *Fairbairn Lawson Combe Barbour Ltd* [1943] AC 32 at 61 that:

> 'It is clear that a civilized system of law is bound to provide remedies for cases of what has been called unjust enrichment or unjust benefit, that is to prevent a man from retaining the money or some benefit derived from another which it is against conscience that he should keep. Such remedies are generically different from remedies in contract and tort, and are now recognised to fall within a third category which has been called quasi-contract or restitution.'

In considering the relationship of the law of restitution to other legal concepts, it is important to realise that other legal concepts establish their own limits. McMeel uses the following classification to illustrate this:

1 Property law establishes the interest that persons can have in things. It defines and constitutes what counts as *wealth* in an advanced society.
2 The law of wrongs (tort) provides remedies for interference with and wrongfully causing of harm to a person's interests, most importantly physical harm, proprietary rights, but also to some extent their economic interests.
3 Contracts and the law of transactions generally provide the dynamic component. The exchange of value via the medium of contracts takes the form of money. Not all transfers of wealth and not all transactions turn out as planned.
4 The law of restitution governs transactions that go wrong and undoes shifts in wealth that require reversal (*Casebook on Restitution*, 1996, p.1).

Unjust enrichment

In *Lipkin Gorman* v. *Karpnale Ltd* [1991] 2 AC 548 the House of Lords recognised the principle of unjust enrichment as the foundation of the law of restitution. Birks (1996) considers the word restitution as implying that the person who has to make restitution, as having received something. It is concerned with gains to be given up, not with losses to be made good. Restitution is therefore the response which consists in causing one person to give back something to another. What has to be given back is the benefit received. The value of the benefit arises in a number of ways:

- money received
- the receipt of property
- a benefit obtained from the receipt of services
- the discharge of a valid obligation which was owed
- profits made from wrongdoing.

Money received normally presents no difficulty in valuing the benefit received. Property is more difficult since the claimant may argue for an objective valuation, whilst the defendant would argue that the benefit was of little value to him. The valuation of the benefit from the receipt of services presents many difficulties in showing a benefit accruing to the recipient.

'At the expense of' identifies the person on whose behalf payment was made and to whom payment is due (per Evans LJ, in *Kleinwort Benson Ltd* v. *Birmingham City Council* [1997] QB 380). There are two ways that the requirement that the unjust enrichment should be at the expense of the claimant could be expressed. First, the claimant can demonstrate a corresponding plus and minus – I mistakenly gave you £100; your gain exactly matches my loss. Second, the claimant can demonstrate that the defendant benefited from committing a wrong against him and is therefore in breach of a legal duty. The first category is termed autonomous or subjective unjust enrichment, and the second termed restitution for wrongs. Only the first category is dealt with in this book.

The structure of the enrichment enquiry

Banque Financière de la Cité v. *Parc (Battersea) Ltd* [1999] 1 AC 221 adopted the analytical methodology to the structure of unjust enrichment enquiry, adopted by the leading scholarly works. Lord Steyn (at p.227) said that four questions arise.

(a) Has [the defendant] benefited or been enriched?
(b) Was the enrichment at the expense of [the claimant]?
(c) Was the enrichment unjust?
(d) Are there any defences?

Enrichment

Has the defendant benefitted by the receipt of money, property, services or profited by wrongdoing? Money claims are well developed whereas non-money claims are not well served by authority. In *BP Exploration (Libya) Ltd* v. *Hunt (No. 2)* [1979] 1 WLR 783, Goff J observed at p.799 that in money claims:

> 'by receipt, the recipient is inevitably benefited; and (subject to problems arising from such matters as inflation, change of position and the time value of money) the loss suffered by the plaintiff is generally equal to the defendants gain, so no difficulty arises concerning the amount to be paid.'

He went on to consider other benefits, such as goods and services. Goods may not be capable of restoration; they may be consumed or transferred to another. Services by their nature cannot be restored. The value of the benefit resulting to the recipient may be debatable. 'From the very nature of things, therefore, the problem of restitution in respect of such benefits is more complex than in cases where the benefit takes the form of money.'

At the expense of?

Here the plaintiff must demonstrate that it has a right to sue. It can demonstrate this in two ways. First the plaintiff can show a corresponding plus and a minus (i.e., the defendant has a benefit which matches exactly the benefit passing from the plaintiff to the defendant). Second, the plaintiff can assert that the defendant benefited by committing a wrong. As a result of that breach a duty is owed to the plaintiff.

Unjust

Should the defendant return (disgorge) the benefit received? To decide this English law asks a number of concrete questions. Was the plaintiff mistaken in transferring the benefit or was it compelled to do so? Did the basis of the transfer collapse? Did the defendant receive the benefit by reason of a wrong done to the plaintiff? The main reasons for restitution in English law are mistake, compulsion, failure of consideration, wrongdoing and claims in *quantum meruit*.

Defences

Defences to a claim in restitution are good faith in purchase or change of position. Sometimes public policy may preclude a claim in restitution. In *Lipkin Gorman* v. *Karpnale* [1991] 2 AC 548 Lord Goff stressed at p. 578:

> 'The recovery of money in restitution is not, as a general rule, a matter of discretion for the court. A claim to recovery money is made as a matter of right; and even though the underlying principle of recovery is the principle of unjust enrichment, nevertheless, where recovery is denied, it is denied on legal principle.'

A claim for restitution can thus be defeated or diminished, by one of the defences mentioned above.

Good faith in purchase

This defence rests on the principle that it would unfair to compel the recipient to disgorge in full where there has been an intermediate expenditure on the faith of receipt. If available and established, the defendant has a complete defence. The difficulty that this defence poses is that there are clear tensions between two conflicting legal principles of law with this defence:

(a) Both ownership and wealth has to be protected;
(b) In a market economy consensual transactions have to be promoted and facilitated.

What this means is that wealth passing under a consensual transaction (whether by conveyance in the case of land or by contract) is governed by terms of that transaction. It is only where that transaction is legally ineffective that unjust enrichment has a role to play. Where there is a valid and enforceable contract there is no place for restitution. Only when the transaction has failed for some reason and no contract ever came into existence, is there a role for the law of restitution.

Change of position

This defence was first recognised by the House of Lords in *Lipkin Gorman* v. *Karpnale* [1991] 2 AC 548. In doing so. it left open the scope of the defence to be decided on a case by case basis. Lord Goff made some suggestion as to the circumstances in which it may apply:

> 'Why do we feel it would be unjust to allow restitution in cases such as these? The answer must be that, where the innocent defendant's position is so changed that he will suffer an injustice if called upon to repay or to repay in full, the injustice of requiring him to repay far outweighs the injustice of denying the plaintiff restitution. If the plaintiff pays the money to the defendant under a mistake of fact, and the defendant then, acting in good faith, pays the money or part of it to charity, it is unjust to require the defendant to make restitution to the extent that he has changed his position. Likewise, on the facts as those in the present case, if a thief steals my money and pays it to a third party who gives it away to charity, that third party should have a good defence to an action for money had and received. In other word bona fide change of position should in itself be a good defence in cases such as these.'

Public policy

The limitation defence is an example of public policy providing a defence. The rules are complex since the Limitation Act 1980 treats civil obligations as arising out of either contract or tort. Note, however, the recommendations of the Law Commission that the time limits should be the same for all obligations.

Westdeutsche Landesbank Girozentrale v. *Islington London Borough Council* [1996] AC 669 rejected the implied contract fiction (or quasi-contract, as it is also called). This is discussed further in Chapter 3. That chapter also considers the complex subject of the *valuation* of a *quantum meruit* claim and restitution for services provided.

Obligations in action

As acknowledged earlier in this chapter, there are three kinds of obligation that arise in English law. One of the themes of this book is that they are *reflected* in construction contracts. The case of *J. Jarvis and Sons Ltd* v. *Castle Wharf Developments and ors* [2001] CA is *unusual* in bringing together actions in *contract*, *tort* and *restitution*. It does not say anything new about the law, but it demonstrates the obligations that are the theme of this book coming together through rather unusual circumstances.

This was an appeal against the judgment of his HHJ Cyril Newman OC (deceased). On a trial of preliminary issues he held the following parties liable to J. Jarvis and Sons Ltd (Jarvis). Jarvis was the design and build contractor for a development in Nottingham.

1 Castle Wharf Developments (Castle) was liable to compensate Jarvis for the value of work carried out on a *quantum meruit* basis or in *restitution*.
2 Castle and Gleeds Management Services (Gleeds) were jointly and severally liable to Jarvis in *tort* for losses due to their negligence.
3 Franklin Ellis Architects Ltd (Ellis) were liable for damages in *contract*.

Judge Newman died before completing his order or fully revising his judgment. The three parties were given leave after a two-day hearing by his HHJ Havery QC to challenge the conclusions of the judge.

Castle was a company created to develop a site in Nottingham, Gleeds were the quantity surveyors and Ellis the architects. Later the contract with Ellis was novated to Jarvis. Novation is a three-party agreement to transfer an existing contract between two parties to a third party. One of the original parties then drops out and has no further liability. All three must, however, agree for novation to be effective.

As the site was at a sensitive and prominent location in Nottingham, the local planning authority (LPA) took a close interest in the development; in fact, it was in a conservation area. The first phase of the development comprised the demolition of existing buildings and the erection of an office block (office A), a pub/diner, access roads and external works including a pedestrian bridge over a canal. Office A was the proposed office of the *Nottingham Evening Post*, a newspaper of the Northcliffe group which Castle was determined to keep on site. Planning permission (subject to conditions) for the scheme was obtained on the basis of drawings submitted by Franklin. The scheme when put out to tender proved to be too costly. Castle and Ellis then started negotiation with the planners to revise the approved scheme without having to make a new application for planning consent.

Subsequently Gleeds, on behalf of Castle, invited building contractors to tender for the design and construction of the first phase of the development. The employer's requirements included the document (which was not properly updated) being submitted to tenderers in the first round. The contract was to be the JCT standard form of Building Contract with Contractors Design 1981 edition. The contract with the architect for the development was to be novated to the successful tenderer. Amongst other things the contractors were to be responsible for ensuring compliance with all statutory requirements and for obtaining all necessary approvals.

Jarvis submitted its Contractor's Proposal. This differed from the approved scheme contained in the drawings submitted to the planners by Ellis. At a meeting on 30 October 1996 between Gleeds (acting as agent for Castle) and Jarvis the proposals were discussed. Jarvis were told that 15 items in its proposal were unacceptable to Castle, to Northcliffe and also to the LPA. Subsequently Jarvis produced a 'menu of options' relating to those 15 items. These were priced and discussed at a further meeting and agreement reached on the Contractor's Proposal as modified by the 15 items.

On 1 November, Castle wrote Jarvis a letter of intent, subject to contract, to enter into a contract with Jarvis for a price to exceed £4.26 million and to be 'subject to final mutual agreement'. It continued: 'The contract is intended to include provisions that will comply with the detailed planning approval and you will obtain all necessary additional statutory consents.' Possession of the site was to be given on 4 November with a completion date of 31 October 1997. In the event of a contract not being entered into Jarvis was to be paid on a basis of *quantum meruit*.

Jarvis (who by now employed Ellis) had a number of meetings with the LPA. The LPA was unenthusiastic and Jarvis subsequently submitted their revisions through Ellis. On 18 November, despite not having received the comments of the LPA, Jarvis started work on-site. At a meeting with the LPA four days earlier Jarvis had been warned that since planning permission had not been granted work carried out was at their own risk. By 18 February 1997 the building had reached the stage of steel erection. On 22 April the LPA served an enforcement notice to take effect on 27 June 1997. This was for breach of planning control in carrying out development without planning permission. As a result work came to a stop. Planning permission was granted for a revised scheme on 17 July 1997. Jarvis and Castle then entered into a contract for a scheme for a sum not to exceed £6 million. Office A was later completed. In May 1998 Jarvis commenced proceedings against Castle with Gleeds and Ellis added as additional defendants later on.

Jarvis alleged:

(a) That Castle and/or Gleeds induced them to tender for the carrying out of the design and building work on the basis of a *misrepresentation* made by Gleeds (that planning permission had been granted for the scheme they tendered on);

(b) By reason of the negligence of Gleeds, and the negligence and breach of *contract* of Ellis, it had suffered loss through the absence of planning permission and the intervention of the LPA.

Because the judge had not revised his judgment the Court of Appeal granted all parties permission to appeal. It stressed its reluctance to order a new trial. The difficulty that faces an appellate court is that it cannot interfere with the *findings of facts* made by the trial judge. The only time it can interfere with those findings of fact by the judge, who has heard and seen the witnesses, is where the judge misused that advantage (*Watt* v. *Thomas* [1947] AC 484 at 488 per Lord Thankerton).

Jarvis's Case against Castle

This was based on the alleged *misrepresentation* made by Gleeds. As principals they were *vicariously liable* for the tort committed by their agent. Gleeds, as the provider of information to another, when it knew the information would be *relied on* by that person, owed a duty of care under the principle of *Hedley Byrne & Co.*

Ltd v. *Heller & Partners* [1964] AC 465 not to cause them economic loss. (Note: the extent of the principles arising from *Hedley Byrne* is dealt with in Chapter 13). See also the examination of *juridical nature* of misrepresentation by Goff LJ in *Whittaker* v. *Campbell* [1983] 3 Al ER 582, cited by McMeel (2000, p.74).

> 'Looked at realistically, a misrepresentation, whether fraudulent or innocent, induces a party to enter into a contract in circumstances where it may be unjust that the representor should be permitted to retain the benefit (the chose in action) so acquired by him. The remedy of rescission by which the unjust enrichment of the representor is prevented, though for historical and practical reasons treated in the books on the law of contract, is a straightforward remedy in restitution subject to the limits which are characteristic of that branch of the law.'

The difficulty facing Jarvis (as Gibson LJ recognised at para. 51) was that English law has not recognised a duty of care arising where a professional agent, acting within the scope of his authority on behalf of a known principal, provides information to a third party. Castle argued that authority was against imposing a duty of care to a contractor on the members of professional team rather than on the employer agent: see *Pacific Associates Inc and anor* v. *Baxter and ors* (1988) 44 BLR 33. In support it used the arguments made in *Hudson* (1995 at p. 161):

> 'An essential consideration in rejecting a proposed generalised or affirmative duty of care owed by an owner's construction professional team to a contractor within the framework of a typical construction project is the element of conflict of interest, calculated to detract from the wholehearted performance of a professional's duty to his client, which the incorporation of such a duty into the parties' relationship must invariably create, it is submitted.'

The argument was rejected because in these circumstances there was *no contract* between the contractor and the employer at the time of the alleged misstatements. Gibson LJ said at para. 53:

> 'The position, as it appears to us, is this. There is no reason in principle why the professional agent of the employer cannot become liable to the contractor for negligent misstatement made by the agent to the contractor to induce a contractor to tender, if the contractor relies on those misstatements.'

Being reluctant to order a retrial, the court accepted that a duty of care arose in these circumstances. If Jarvis relied upon the *misstatements* to make a tender, certain facts stood in its way.

1 No contract was entered into, since the letter of intent was issued subject to contract; once it knew the truth it was free to walk away from the development.
2 Jarvis was aware that it was proceeding with the construction work at its own risk.
3 Jarvis was a very experienced contractor with particular expertise in design and build projects.

The court decided that there was *no reliance* by Jarvis on the misstatement made by Gleeds beyond the beginning of November. The case against Castle (other than the reasonable costs of the work stated in their letter of intent) was dependent on its agent being held liable. The decision that Gleeds was not liable meant that there was no cause of independent liability to be found against it.

The Case against Ellis for Breach of Contract and Negligence

Jarvis's claim against Ellis was based on three matters:

1 They failed in their duty to inform or warn them that planning permission was for a different scheme.
2 They had erroneously told them that the planners had approved certain matters.
3 They failed to produce detailed drawings to meet the conditions attached to planning permission.

The court decided that *no damage* flowed from the information in point 2 above. In any case, where the client has *relevant experience* in the relevant area (and Jarvis was a very experienced contractor with particular expertise in design and build), the duty is to only give advice if it is sought (*Carradine Properties Ltd* v. *D J Freeman & Co.* [1999] LLR Practice Note 483 applied). Since Jarvis had started work when it was well aware that no agreement had been reached on their proposals, any advice Ellis might have given would have been superfluous.

Good faith and fair dealing

Chitty states that the notion of good faith has not been generally accepted in English law. It is, it argues, mostly a 'doctrinal dispute' of little practical application (1999, p.17). In the light of recent developments it is instructive to consider whether, contrary to that argument when applied to construction contracts, the doctrine is of some value. The Latham Report recommended that all construction contracts should contain '*a duty to trade fairly*' and that tenders should be evaluated on quality *and* price.

The starting point for any discussion of the issue must be Duncan-Wallace (1986, p.459). There he argued that any transfer of risk from contractor to employer should be balanced by a reduction in price. This had nothing at all to do with *fairness*, *justice* or *morality*. In England in any case, such a rule has only been implied into contracts of insurance. Such a contract is based on *the utmost good faith*, the doctrine of *uberrima fides*. This means that there must be *full disclosure* by the person seeking insurance cover of all the *material circumstances* that could affect the risk the insurer is being asked to provide cover for. A failure to disclose means that when the failure to disclose all material facts is discovered, the doctrine allows the insurer to *disclaim* the policy. See *St Paul Fire and Marine Insurance Ltd* v. *McConnel Dowell Construction Ltd and Ors* [1994] 1 All ER 96. The insurers were able to avoid the policy because the contractor failed to tell them that shallow spread foundations were being used, instead of driven piles as originally proposed. The contractor found it had no insurance cover when spread foundations failed.

Duncan-Wallace considers the doctrine of good faith has limited application (1995, p.102). He questions whether it adds anything to:

(a) the notion that promises to should be honoured (see for instance, *Williams* v. *Roffey Brothers & Nicholls (Contractors) Ltd* [1990] 1 All ER (1990) 48 BLR 69, discussed further in chapter 4).
(b) the development of economic duress as a means of discharging contracts.
(c) conspiracy.

Hudson agrees with *Chitty* that in England the rule overlaps with economic duress. The case of *D&C Builders* v. *Rees* [1965] 2 All ER 837, it is suggested would now probably be regarded as an example of *economic duress*. The county court decided it on the classical grounds of the absence of consideration. The Court of Appeal held itself bound by *Foakes* v. *Beer* [1884]. An agreement to accept payment of a lesser sum was held not to be binding; Denning LJ argued that the defendants had simply acted inequitably by taking advantage of the builder's financial circumstances, thus demonstrating the fine line that exists between *legitimate competitiveness* and *unacceptable coercion*.

The arguments raised *against good faith* are put cogently by Brownsword (2000, p.100) as comprising five essentially negative themes:

1 The English law of contracts is essentially individualistic. Requirements to take into account the legitimate interests or expectation of each other is incompatible with the adversarial approach underpinning it.
2 Good faith is a restriction on the pursuit of self-interest. It presents a moral dimension with unclear standards for judging it.
3 How is a court to decide when parties have broken off negotiations for their own economic interests or simply because they could not see a way of closing the bargaining distance between the parties?
4 It restricts the doctrine of freedom of contract and challenges self-regulation in private matters.
5 Contractual contexts differ and the law should reflect those differences.

Cornes (ICLR, 1996, p.97) described how the second edition of the New Engineering Contract (NEC) had been 'Lathamised'. He criticises the use of simple language and novel drafting as not always being a proper basis for effective contractual provisions. The danger with using non-traditional words and phrases is that they have not been tested judicially. Precedents set by previous cases are of value since they create a basis for the proper construction of contracts. The NEC has adopted the following clause: 'The Employer, Contractor, Project Manager and Supervisor shall act in good faith as stated in this contract *in a spirit of mutual trust & co-operation.*' Mutual trust and co-operation is called a duty of good faith in the USA and the civil systems of the European Union.

The USA position

Hudson confirms (1995, p.102) that the implied covenant of good faith and fair dealing has been accepted by nearly all American states. A requirement not to do anything to *deprive the other of the benefit of that agreement*, it includes the following:

(a) big companies not using a bad faith defence against small companies with limited means;
(b) regulatory bodies making decisions which will excuse further performance by itself;
(c) insurance companies using delaying tactics to pay justified claims.

Every contract governed by the USA Uniform Commercial Code requires '*honesty in fact in the conduct of the transaction concerned*' (clause 1–203). This was changed in 1981 to 'every contract imposes on each party *a duty of good faith and fair dealing* in its performance and enforcement'.

Odams considers that the standard form of the American Institute of Architects (AIA) introduces good faith and fair dealing into the contract. In its standard form of agreement between owner and contractor (AIA Document A101), the following clauses achieve this objective:

1 Clause 3.2.1 is based on *good faith*. It imposes on the contractor an active duty to search for discrepancies, errors, omissions (and so on) between various contractual documents and to report them at once. Furthermore, the contractor is not liable for any damage resulting from such discrepancies, errors and so on, 'unless the contractor recognized such discrepancies, errors . . . etc. and knowingly failed to report it'.
2 Clause 3.7.3 is based on fair dealing. The contractor has no duty to ascertain whether the contract documents are in accordance with the applicable laws, statutes, codes, and so on. If, however, the contractor observes 'that portions of the contract documents are at variance therewith, the contractor shall promptly notify the architect in writing' (Odhams, 1995, p.169).

European Union

The term civil law is used here to describe the private law of civil law countries. It includes the law of obligations. In France, article 1134 of the Civil Code states that all 'contracts *must be executed in good faith*', whilst in Germany the Burgelikje Gewets Book in clause 242 requires that 'everyone must perform his contract in a manner required by *good faith*'. This gave the courts great freedom and is fundamental to German law. A party to a contract must take into account the *legitimate interests of the other*.

Commonwealth

In *Renard Construction* v. *Ministry of Public Works* [1992] 26 NSW LR 234, the court considered the scope of the following clause.

'If the Contractor defaulted in the performance of *any* stipulation etc, in the contract or *refused* or *neglected* to comply with *any* direction of the Principal or Superintendent *given in writing* the Principal might suspend payment under the contract, and by *giving notice in writing* call on the Contractor to *show why* the work should not be taken over and the contractor expelled from site, or the contract cancelled.'

The 'show cause' notice had to (i) specify the default, etc., and (ii) the notice period within which the contractor had to show cause '*to the satisfaction of* the Principal'. If it failed the principal might exercise power to take over the work or cancel the contract.

The contract was for the construction of a large sewage works. A 'show cause' notice was issued and, four months after the deadline, the work was still not completed. The contractor had accelerated the work significantly and had completed the most difficult and unprofitable part of the work. The principal cancelled the contract, unaware that:

(a) the delays were due to non-supply of materials by the employer;
(b) an extension of time had been granted;
(c) defects when identified had been promptly fixed by the contractor;
(d) the number of men had been increased since the 'show cause' notice and a new foreman appointed;
(e) the new contractor would be unable to finish any earlier.

The arbitrator decided (a) that the determination was unreasonable; (b) that the contract was wrongly determined; and (c) that he should award *reasonable remuneration* greater than the contract price. The government appealed. The High Court held that it was sufficient for the principal to act honestly in good faith, which he had done. There was no requirement in contract to act reasonably and the contract was validly determined.

Held by the NSW Court of Appeal: The principal's consideration of the contractor's case and subsequent determination must be exercised reasonably. This was not done. Because the principal's decision was grounded in misleading, incomplete and prejudicial information, he could not 'be satisfied' and hence the determination had been wrongful. The initial finding by the arbitrator was upheld.

Hudson considers that there are two views of this decision.

1 The narrow view. The harsh wording plus 'show cause' requires a reasonable approach to make the contract work. The ruling only applies to this particular case.
2 The broad view. The law implies a term in construction contracts containing termination clauses of reasonableness. This ruling applies to other clauses as well. Moreover, the requirement of reasonableness implies good faith and fairness. Support for this contention comes from Priestly J's alternative basis for the decision at 268: 'People . . . have grown used to courts applying standards

of fairness which are consistent with the duty of good faith existing in all contracts.'

Sir Thomas Bingham emphatically rejected such a duty in *Interfoto Picture Library Ltd* v. *Stiletto Visual Programmes Ltd* [1989] QB 433 where he said:

'In many civil law systems outside the common law, the law of obligations recognises and enforces an overriding principle that in making and carrying out contracts parties should act in good faith. This does not mean that they should not deceive each other . . . its effect is perhaps mostly aptly conveyed by such metaphorical colloquialisms as "playing fair," "coming clean" or "putting one's cards upwards on the table". It is a principle of fair and open dealing. English law has, characteristically, committed itself to such an over-riding principle but has developed piecemeal solutions in response to demon-strated problems of unfairness.'

For an example of a case where the Court of Appeal applied an element of good faith, see *D Mooney* v. *Henry Boot Ltd* (1996) BLR. The groundwork sub-contractor encountered very hard and dense formations of gypsum that made its work more difficult than anticipated. The main contractor refused extra payment for the difficult ground conditions. A number of disputes relating to non-payment were referred to arbitration. During the arbitration the sub-contractor became aware that the main contractor and the employer had agreed a settlement that included extra money for delay and disruption to the drainage work.

The sub-contract conditions were under the FCEC blue form which provided in clause 10(2) 'that the main contractor shall pass such proportion of the benefit as may be fair and reasonable'. The Court of Appeal held that any money claimed under the main contract in good faith and received by the contractor had to be dealt with by clause 10(2).

Other approaches

Harrison and Jansen Const LJ 1999 present two different arguments or approaches to good faith. Harrison suggests that though it is possible to infer from the old cases that such a duty was present in earlier English law, on closer analysis it boils down to the simple principle that if you prevent a man from carrying out his contract work, he is excused from carrying it out. Jansen focuses on the civil law experience. Good faith in a construction contract is about co-operation, fitness for purpose and duties to warn. These are implied terms in construction contracts (see Chapters 6 and 9). His analysis of the European posi-tion suggests that English law has already dealt with good faith in a different way in construction contracts.

Consumer Contracts

The Unfair Terms in Consumer Contracts Regulations 1999 requires both parties to act in *good faith*. The Regulations apply to consumer contracts and only where standard forms of contract are used. The major interpretation of the act has come in *Director of Fair Trading* v. *First National Bank* [2001] UKHL 52. The relevance of this case from the point of view of construction contracts was the conclusion of the House of Lords on the *nature of good faith* in the *Regulations*. Lord Bingham of Cornhill said in para. 17:

> 'The requirement of good faith in this context is one of open and fair dealing. Openness requires that the terms should be expressed fully, clearly and legibly, containing no concealed pitfalls or traps. Appropriate prominence should be given to terms which might operate to the disadvantageously to the customer. Fair dealing requires that the supplier should not, deliberately or unconsciously, take advantage of the consumer's necessity, indigence, lack of experience, unfamiliarity with the subject matter of the contract, and weak bargaining position.'

Note the observations made in *Pratt* in December 2003, which are discussed further in Ch. 3 below. Giving the opinion of the Privy Council, Lord Hoffmann observed that the 'the nature of the implied duty to act fairly and in good faith has been subject to a great deal of discussion in Commonwealth authorities.' He went on to say that the question of whether there was a 'more general obligation to perform a contract fairly and in good faith' was a 'somewhat controversial question into which [in the circumstances of the case] it is 'unnecessary for their Lordships to enter.' One effect of the introduction of adjudication has been the use of the process in contracts with consumers. In a number of cases where the winning party has sought to enforce the adjudicator's award, the losing party has challenged the procedure as being incompatible with the Regulations. This is discussed further in the context of adjudication: see Ch. 16 below.

Summary

1   English law does not recognise a general law of obligation but it has become convenient to treat contract, tort and restitution as part of a general law of obligations.
2   These obligations are reflected in construction contracts.
3   Contract is based on agreement and is concerned with the expectation interests of the parties.
4   Tortious duties are imposed by law and protect the reliance interest of the innocent victim.
5   English law has reaffirmed the distinction between contract and tort by refusing a remedy for economic loss in tort.
6   The restitution of benefits received based on unjust enrichment comes into play when contracts fail for some reason to come into existence.
7   The case of *Jarvis* reflects the law of obligations in a construction context.
8   The profile of the concept of good faith was raised by publication of the Latham Report.
9   There are arguments against the imposition of such a duty, and in particular that it adds nothing to English law.

10 The case of *Director of Fair Trading* v. *First National Bank* [2001] suggests that the doctrine will continue to be considered even though the regulations only relates to consumer contracts.

11 However, the introduction of adjudication into consumer contracts, mean that adjudicators and constructions professionals and parties to construction contracts, need to be aware of the Regulations.

3 The Formation of Construction Contracts

In *VHE Construction Plc* v. *Alfred McAlpine Construction Ltd* [1997] EWHC Technology 370, HHJ Bowsher QC remarked that the issue in this case was:

'[Whether] there is a contract between the parties, and if so what were the terms? [then added]. It is remarkable how this question is probably the most frequent issue raised in the construction industry. On projects involving thousands and sometimes millions of pounds, when a dispute arises about payment, the first issue very often is to decide whether there was a contract and if so what were the terms of the contract, if any.'

Now this may appear to be a surprising question to pose since English law adopts an *objective theory* of contract formation. Its distinguishing feature is the reasonable expectation of businessmen and women. Courts are therefore *loath* to hold that no contract exists especially where the transaction is of a *commercial character*.

In most cases the formation of a contract is demonstrated by the *mechanism* of *offer* and *acceptance*. Given that the three basic essentials for the formation of a valid contract are agreement, the intention to create legal relationships and consideration, a further question arises: what is so special, complex or different about construction contracts that makes their formation so difficult a problem that it often requires the help of the courts?

Agreement is proven in most cases by showing that there was an offer followed by its acceptance. The offer must be clear and unambiguous and capable of being accepted as it stands. Any attempt to tinker with the offer destroys it and amounts to a counter-offer. An offer must be distinguished from an invitation to treat. This is only an offer to negotiate and its acceptance does not create a valid contract. An invitation to a contractor to submit a tender is only an invitation to treat or an invitation to make an offer. The return of the completed tender by the contractor is the offer. It is the acceptance by the employer of that offer, whether made orally or in writing or by conduct, that creates a valid contract.

Responses to tenders

The responses by the employer to the submission tenders are various and quite complicated (in this context the employer may be the sponsor of the project, its agent or a main contractor negotiating with prospective sub-contractors or suppliers).

1 Specific tendering procedures contained in the invitation to tender may not have been followed.

2 Further work may be requested to explain the tender.
3 The submitted tender may have been qualified.
4 A battle of forms may result.
5 A letter of intent is issued.
6 The tender may be accepted by conduct.

The submission of a tender

The rules about the withdrawal of tenders (or offers) are surprisingly relaxed in English law compared to civil law jurisdictions. In France, for example, (depending on the complexity of the tender) it has to be kept open for a reasonable time where none has been specified. Withdrawing it before that time lays the offerer open to a claim in delict (tort). In England, however, a tender may be withdrawn (revoked) at any time before its acceptance, and the withdrawal or revocation must be brought to the attention of the offeror. Tenderers sometimes qualify their tenders. They may restrict the time it stays open by stating 'this tender shall be open for acceptance for six months from the date of submission', or 'our tender is valid for six months from the date of receipt'. Employers counter this risk by requesting *tender bonds*. A tenderer may be required to provide a financial bond to cover the costs of re-tendering, should the tenderer withdraw its tender. If it does so, it will be forfeit value of the bond.

Submitting a tender

Substantial projects require contractors to follow *complex procedures* when submitting tenders. They may have to pre-qualify, a process by which unsuitable or inexperienced tenderers are eliminated, or they may have to submit formal accounts, current certificates of insurance, lists of key personnel and projects successfully carried out. They have to invest money and time in site visits, the preparation of schemes of work, assessment of the design process and they must also produce prices.

> 'The purpose of the [tender] system is to provide competition, and therefore reduce costs, although it by no means follows that the lowest tender will necessarily result in the cheapest job. Many a "low" bidder has found his prices have been too low and has ended up in financial difficulties, which has inevitably resulted in additional costs to the owner, whose right to recover from the defaulting contractor is usually academic'

per Iancoucci J in *MJB Enterprises Limited* v. *Defence Construction (1955) Limited and others* (1999) 15 Const. LJ 455. He went on to say that a prudent owner would also take into consideration the capability and experience of the contractor and how realistic the tender price was by comparison with others. Bingham LJ in *Blackpool and Fylde Aero Club* v. *Blackpool BC* [1990] 1 WLR 1195 described this process 'as heavily weighted in favour of the invitor'. The invitor does not:

(a) commit himself to proceed with the project;
(b) accept the highest (or lowest) tender;

(c) need to accept any tender;
(d) need to give reasons for his acceptance or refusal of a tender.

The question that then arises is whether there is any requirement for *fairness in the tendering process*. The type of tender being considered is very different from what would be described as an *uncomplicated request* for a bid, which is one that requires the potential bidder to calculate a few prices, to make some telephone calls checking availability of material and to establish when the work could be done. Such a bid is normally no more than an *invitation to treat*, whereas a tender not complying with those procedures mentioned before will be automatically rejected. The leading English case on the status of such a conforming tender is *Blackpool and Fylde Aero Club* v. *Blackpool BC*, although it must be stressed that there was nothing overcomplicated about that particular tender. Bingham LJ described what a tenderer is entitled to at p.1,202: 'not as a matter of mere expectation but of *contractual right*, to be sure that his tender will after the deadline be opened and considered in conjunction with all other conforming tenders, or at least that his tender will be considered if others are'.

In *Blackpool* the plaintiff flying club had a concession to fly pleasure flights from the local aerodrome. The concession came up for renewal and the plaintiff placed its tender in the council's delivery box an hour before the expiry of the 12 o'clock deadline. The council's staff failed to open the post box that day, and returned the tender some days later marked 'late'. It was not considered and the council awarded the contract to another bidder. This was in fact a lower bid than that submitted by the plaintiff, but the council had made it clear that they would *not be bound to accept the lowest tender*. The plaintiff responded by starting proceedings in the High Court, arguing that the express request for tenders gave rise to an implied obligation on the part of the council to consider all tenders duly received. At first instance the judge agreed. The council appealed. The Court of Appeal dismissed their appeal against the decision of the trial judge. HHJ Lloyd QC in *Harmon CFEM Facades (UK) Ltd* v. *The Corporate Officer of the House of Commons* [1999] EWCA Technology 199 described *Blackpool* as 'perhaps no more than authority for the proposition that a contracting authority undertakes to consider all tenders received with all other conforming tenders, or at least that his tender will be considered if others are'.

The Court of Appeal held that there was an *implied undertaking* by the employer, and that tenders submitted in accordance with *published conditions* would *at least be properly considered*.

Blackpool was distinguished by the Court of Appeal in *Fairclough Building Ltd* v. *Port Talbot Borough Council* (1993) 62 BLR 82. The issue there was whether the defendant was in breach of an implied contract in removing the plaintiff from the tender list. At the time the project was first advertised its local director was married to a senior assistant architect employed by the council. By the time the firm was short-listed with five others, the wife was the council's principal architect. The Court of Appeal held that there was an implied contract arising out of the conduct of the council to consider their tender. In the circumstances they had two choices, either to remove the tenderer from the list or remove the principal architect. In removing both the council had acted properly and was not in breach of contract. It stressed that no one had acted improperly in any way. It held further that the defendant council was under a duty to 'honestly consider the tenders of those whom they had placed on the short list, unless there were reasonable grounds for not doing so'. Note also the observation by Parker LJ that even if there had been a breach of the implied contract, the loss of a chance of remaining on the select list of tenders would have been worth 'precious little'.

Assessing damages for breach

Failure to consider a conforming tender creates a further problem. What has the disappointed tenderer lost beyond the possibility of winning the contract, which was a mere chance? After all, in any competition such as this there is always going to be only one winner but many losers. If, however, a breach of the implied contract entitles the disappointed tenderer to damages, *how are these to be assessed*? Is the tenderer entitled to (i) its *reliance* losses, i.e., the costs of tendering) or (ii) its *expectation* losses, based on the profit it would have made had it won the contract? In short, how is the loss of a chance to be quantified?

The Canadian Supreme Court in *MJB Enterprises Limited* v. *Defence Construction (1955) Limited and Others* (1999) considered the matter. The defendant invited tenders and awarded the contract to the lowest tenderer, despite the fact that the bid did not comply with the tender specification. The tender document included the usual disclaimer that the lowest, or any, tender would not necessarily be accepted. The winning tender included a hand-written note outlining a schedule of final costs, even though amendments to the tender documents required tenderers to submit one price only. The plaintiff brought an action alleging breach of an implied contract. They argued that the winning tender should have been disqualified and their tender, being the next lowest, should have been accepted. The project was for the construction of a pump house, the installation of a water distributing system and the dismantling of a water tank on the Canadian forces base in Sunfield, Alberta. Four tenders were received and the contract was awarded to a company called Sorochan Enterprises Limited, which submitted the lowest tender.

Held: The *submission* of a tender in response to an invitation to tender may give rise to contractual obligations (*contract A*). This arises quite separately from the contract arising out of the *acceptance* of the tender (*contract B*). In this case a contract *A* arose. Implied into that contract was an obligation to accept only *compliant* tenders. This need not be the lowest compliant tender. Below are given the reasons why the case is relevant.

1 The Supreme Court accepted that English, Australian and New Zealand law had adopted the principle that contractual obligations can arise on the submission of a tender.
2 The practical problem in the case was that there were three different fill materials specified for backfilling the water system trenches. Their prices varied

significantly. The choice of fill was to be made at the time of construction. That being so, the *financial risk* of the cost of materials was placed on the tenderer, as *only one price was required*. Sorochan, the winning tenderer, gave one price but added a rider that the price would increase by $\$x$ if another fill material was used.

3 The parties agreed damages before trial of \$398,121.27 subject to liability. Contrary to *Blackpool*, which specified the damages as relating to a *loss of a chance*, the Canadian courts have been bolder and in this case awarded the contractor the *profit* it would have made if it had been awarded contract B.

The application of *Blackpool*

The basis of the implied contract arising on the submission of a conforming tender was accepted in *Harmon CFEM Facades (UK) Ltd* v. *The Corporate Officer of the House of Commons* [1999] EWCA Technology 199. In this action the plaintiff, Harmon, claimed (amongst other things) that the defendant:

(a) had failed to observe the Public Works Contracts Regulations 1991 (PWR);
(b) was in breach of obligations under Articles, 6, 30, 59 and 65 of the Treaty of Rome (as amended at Maastricht);
(c) breached the obligations stemming from certain European Directives concerning the procurement of public works (e.g., 71/305 and 89/665);
(d) was under an *implied contract* in the treatment of its tenders.

The case arose out of the tendering process which led to the award of a contract for fenestration for the new Parliamentary building ('HNPB') at Westminster. The preliminary issues for decision were whether Harmon had been treated equally, openly and fairly; whether there had been discrimination of a 'buy British' nature in favour of the successful contractor; and whether the damages or compensation recoverable should include exemplary or aggravated damages (although this aspect of the case will not be considered further here).

The NPB was to provide offices for some 210 Members of Parliament and their staff, together with committee rooms and other accommodation. The brief to the architects was to design a building with a planned life of 200 years and to create a 'showpiece of British design'. It was to be known as 'Portcullis House'. When its foundation stone was unveiled on 3 February 1998, the cost for the building was estimated to be £250 million. This included allowances for inflation, furnishings, equipment, and professional fees, it was one of the most expensive buildings to be built in London.

Harmon (the subsidiary of a French company) tendered for the works contract for the fenestration. The contract was, however, awarded to a company called Seele Alvis Fenestration Ltd. It was specially formed to take over the tender submitted by a consortium comprising a German company, Glasbau Seele GmbH (Seele for short), and a British company, Alvis Vehicles Ltd (Alvis), a subsidiary of Alvis Plc. The fenestration was not an ordinary or orthodox one: it was designed to provide a façade that would be proof against terrorist attack and against bomb blast, and hence the choice of the material considered necessary by the architects and engineers to achieve these objectives.

Harmon based its implied contract argument on *Blackpool*. HHJ Lloyd QC found as a fact that the situation differed from *Blackpool*. In saying so he adopted the remarks of the appeal judges in *Fairclough Building*. He went on to say that the facts here were very different. The House of Commons was *bound to proceed* with the project; it *had* to accept some tender; and it had to give reasons if it rejected a tender.

In his detailed judgment HHJ Lloyd QC considered that by repeating the *offer to consider alternatives*, on 11 September and on 2 and 30 October 1995, it was to be implied in that offer that by submitting a tender any alternatives would be equivalent to the schemes for which revised tenders were being sought; these were options only in *terms of refinements* of design detail which would reduce cost, although these were confidential to the tenderer, and these alternatives would fall short of *different proposals* which were more than matters of detail but related to changes of design, about which tenderers were not informed. Therefore tenderers were entitled to assume the changes were not major ones.

In his judgment HHJ Lloyd QC said at 1.8–1.20:

> 'Even though all tenderers accepted that they would not be entitled to see alternatives of detail which were considered to be commercially confidential to a given tenderer, H of C [the House of Commons] in soliciting new or revised tenders under the European public works regime (to which effect is given by the PWR) impliedly undertook towards any tenderer which submitted a tender that its submission would be treated as an acceptance of that offer or undertaking and:
> (a) that the alternative submitted by any tenderer would be considered alongside a compliant revised tender from that tenderer; and
> (b) that any alternative would be one of detail and not design;
> (c) that tenderers who responded to that invitation would be treated equally and fairly.'

These contractual obligations, the judge said, derived from a contract to be implied from the procurement regime required by the European directives, as interpreted by the European Court, whereby the principles of fairness and equality form part of a preliminary contract of the kind that he had indicated. *Emery Construction* v. *St John's (City) RCSB* (1996) 28 CLR (2nd) 1 (a Canadian case), showed that such a contract may exist at common law against a statutory background that might otherwise provide the exclusive remedy. In his opinion he considered that 'it is now *clear in English law* that in the *public sector* where competitive tenders are sought and responded to, a contract comes into existence whereby the prospective employer *impliedly agrees to consider all tenderers fairly*: see *Blackpool* and *Fairclough*'.

The House of Commons was in breach of all these obligations. In his judgment it *also broke the implied duty to treat all tenderers fairly and equally* by considering an alternative design without giving any other tenderer the opportunity of competing with it on its terms.

The damages for loss of a chance

In *Harmon* v. *The Corporate Officer of the House of Commons* [2000] EWCA Technology 84, HHJ Lloyd QC gave judgment in application by Harmon for an interim payment of damages. The following awards were made:

(a) wasted costs of tendering amounting to £210,054;
(b) the loss of gross margins of £3,696,930.

The final costs later were settled out of court by the parties.

The Canadian experience

In *English law* it is clear that in the *public sector* where competitive tenders are sought and responded to, a contract comes into existence whereby the prospective employer *impliedly agrees to consider all tenderers fairly*. It is therefore of interest to examine the Commonwealth cases in order to consider whether they have, as demonstrated by the Canadian Supreme Court in *MJB Enterprises*, gone much further than the English cases. The Canadian cases illustrate the various factors that may make the bidding process unfair. Factors which Courts have taken into account include the following:

1   The fact that the bid was irrevocable. In the *Queen in the Right of Ontario* v. *Ron Engineering & Construction* [1981] SCR 111, a tender bond accompanied it.
2   Reliance has been placed on the sub-contractor's bid. In *Northern Construction* v. *Gloge Heating & Plumbing* (1986) 27 DLR (4th) 265, the sub-contractor was unable to withdraw its tender. The main contractor had, with the knowledge of the sub-contractor, included it when it submitted its own tender to an employer.
3   Discussions with only one tenderer even though the contract was finally awarded on the original basis on which all parties tendered: see *Best Cleaners & Contractors* v. *R in Right of Canada* [1985] 2 FCR 241.
4   Not informing the bidders of its own policies. Any local bidder, within 10 per cent of the lowest price, would be awarded the contract thus giving one party an unfair advantage. The defendant was in *breach of its duty to treat all tenderers fairly*; see *Chinook Aggregates Ltd* v. *Abbotsford (Municipal Districts)* (1989) 35 CLR 241.
5   Statutory provisions. In *Emery Construction* v. *St John's (City) RCSB* (1996) 28 CLR (2nd) 1, the local policy was to be found in a statute, which required preference to be given to local content. The plaintiff's tender was based upon the price of structural steel from a New Brunswick company, whereas its competitor had used the price from a local Newfoundland company and was awarded the contract. If the plaintiff had used a Newfoundland sub-contractor its tender would have been higher than its competitor. The plaintiff lost. The Newfoundland Court of Appeal held that the statutory obligations had also become contractual obligations in the process of seeking tenders.

HHJ Lloyd QC in *Harmon* found that the decision in *Chinook* and *Emery* provided useful guidance. Note: these extracts are taken from the very detailed and comprehensive review given in his judgment in *Harmon*.

The implied contract and tendering procedures

The nature and effect of this type of contract was considered by the Privy Council in *Pratt Contractors Ltd* v. *Transit New Zealand (New Zealand)* [2003] UKPC 83. This was an appeal by an unsuccessful contractor in two successive rounds of tendering for the same project. The first round was aborted after the employer had been wrongly advised that all the tenderers had to agree to an extension of the time limits in the first round. The basis of the argument for the contractor was that the request for, and submission of tenders, gave rise to a preliminary (or implied) contract. This contract contained express and implied terms by which the tenderer was to be selected. In the light of the modern authorities, the employer did not dispute this. It accepted this proposition and also that it had an implied duty to act fairly and in good faith. In the course of giving the opinion of the Privy Council, Lord Hoffmann observed at para 45:

> 'The nature of the implied duty to act fairly and in good faith has been subject to a great deal of discussion in Commonwealth authorities. In *Pratt Contractors Ltd* v. *Palmerston North City Council* [1995] NZLR 469, Gallen J said that fairness was a "rather indefinable term". In *Hughes Aircraft Systems International* v. *Airservices Australia* [1997] 146 ALR, 36–37, Finn J said the duty in cases of preliminary procedural contracts for dealing with tenders is a manifestation of a more *general obligation* to perform a contract fairly and in good faith. That is a somewhat *controversial question* into which it is unnecessary for their Lordships to enter because it is accepted that, in *general terms*, such a *duty existed* in this case.' (emphasis added)

He went on to say that it was the specific content of that duty on the employers that was the issue here. What were the particular acts that, that duty, imposed on the employer when evaluating the tenders submitted? In dismissing the contractor's appeal the Privy Council held:

1 The evaluation of a tender should express the views held honestly by the body responsible for recommending its acceptance.
2 All tenders had to be treated equally. Where tender acceptance was based on price plus other attributes (which may be subjective), two observations can be made:
   (i)   where the tenderers had different attributes (for example in their professional qualifications, experience and track record) it would be expected that different marks would be awarded;
   (ii) where the evaluation was be 'subjective' it did not mean that persons with 'views' could not be appointed – especially where the employer was paying for their expertise.
3 There was no need to act judicially (i.e., like a judge or official carrying out statutory duties). For example, the inspector carrying out a planning inquiry

cannot visit the site accompanied only by the officials of the planning authority. After the award of the tender there is no need to debate with the losing tenderer about the basis of the award.

4 Bad faith means taking steps to avoid receiving information that might prove a party making the decision or recommendation wrong.

In conclusion Lord Hoffmann said at para. 49:

'It follows that their Lordships do not think that the findings of fact justify a conclusion that there was a breach of the express and implied terms of the preliminary procedural contract at either of the tender rounds. They also agree with the Court of Appeal that even if there were such a breach in the first round, it would have had no causative effect on Pratt's failure to obtain the contract. They therefore humbly advise Her Majesty that the appeal should be dismissed. The appellants must pay the costs before their Lordships' Board.'

Implications

The PC accepted there was an 'implied contract' to treat tenders fairly and in good faith. It refused to be drawn into the controversy as to whether this was a general duty in contracts generally. In evaluating a tender, the employer was entitled to employ persons on the team who had experience of the contractor's past performance. In the summary of the issues Lord Hoffmann at para 3 of the case commented that parties who invite tenders want to choose in 'what they consider to be in their best commercial interests.' This is so whether they are a private company or a public authority. They do not want to be hobbled by 'quasi-judicial rules.' The terms of any implied contract is to be found in the actual terms of the call for tenders and the procedures described there.

Employer responses to the submission of tenders

Pre-contract work

Normally a tenderer cannot recover its tender costs. These are the speculative costs incurred in tendering for work. On occasions the tenderer may be asked to do work in anticipation of being awarded the contract. *William Lacey (Hounslow) Ltd* v. *Davis* [1957] 1 WLR 932 is the leading case. The defendant owned war-damaged premises, which he wished to develop. Lacey submitted the lowest tender for the contract and was led to believe that it would receive the contract. At the request of the defendant it carried out estimates which the defendant needed to further his claim for compensation from the War Damage Commission. He was also required to submit further estimates that the plaintiff provided for the suggested alterations to the premises, made by the Commission. The defendant later sold the premises and no contract was ever entered into.

Held: The work carried out by the schedule fell outside the normal work of a builder. The defendant was required to pay a reasonable sum for the work carried out. Barry J said 'On the evidence I am quite sure that the whole of the work covered by the schedule fell outside the normal work carried out by a builder, by custom and usage, gratuitously, when invited to tender for the erection of a building.'

In *Marston Ltd* v. *Kigrass Ltd* (1989) 15 Con LR 116 [QB] the claim was for the cost of additional work above the tender costs for a design and build contract. A factory had been burnt down and the proposed work was to replace it. The contract was never placed because the insurance monies did not cover the costs of a new factory complying with local authority standards.

Held: the preliminary work was above that normally done when preparing a tender. The result was a benefit to the defendants and they had to pay a reasonable sum for it.

Regalian Properties Plc v. *London Docklands Development Corporation* (1995) 11 Const. LJ 127 demonstrates the risks in undertaking work *without a contract*. The claimant tendered for the development of land in London Docklands. The defendant accepted its tender *'subject to contract'*. It incurred costs in the expectation that it would be awarded the contract. The parties failed to reach agreement and the development was made uneconomic by the crash of the property market. Regalian brought an action to recover its costs of £2.3 million spent on the proposed development.

Held: that the parties entered into negotiations expressly *'subject to contract'* and thus on terms that *each party was free to withdraw from negotiations at any time*. On that basis, each party to such negotiations must be taken to know that pending the conclusion of a binding contract, any costs incurred by him in preparation for the intended contract will be incurred at his own risk. He will be unable to recover any money spent in preparation if no contract results. *Regalian* was distinguished in *Easat Antennas Ltd* v. *Racal Defence Electronics Ltd* (2000) unreported. In anticipation that the sub-contract would be placed with them if the contractor was successful in its bid, the sub-contractor expended money on the design and testing of the specified equipment. The bid was successful but the sub-contract was awarded to another company. The sub-contractor brought a claim in restitution against the contractor. It was awarded the costs it had agreed to absorb if the promised sub-contract was awarded to it. The difference between the cases was that in:

(a) *Regalian* the work was carried out as part of conditions that needed be satisfied to be awarded the contract.
(b) In *Easat* work was carried out in anticipation of being awarded the contract if the defendant succeeded in its own bid. As that bid was successful the situation was different to *Regalian*.

The battle of the forms

Ever since *Hyde* v. *Wrench* (1840) 3 Beauv 334, the rules concerning the battle of the forms have been known. They were confirmed by Megaw J in *Trollope & Colls*

Ltd v. *Atomic Power Constructions Ltd* [1963] 1 WLR 333 where he observed that those rules still apply. What then is the battle of the forms? It usually starts when one party makes an offer to the other based on its own terms. Or where the other party specified that any offer should be made based on the terms contained in the invitation to tender. Thereafter the parties may try to reach agreement by the exchange of various kinds of communications. They may do this by the exchange of letters, faxes, emails, estimates, orders, and acknowledgements, each of which claims to be on the usual terms of the party producing it. They may also have meetings which one later dispute as having reached the agreement claimed by other. Uncertainty then results as to what terms and conditions apply to any contract later formed. Different views have been expressed as to whom wins this battle. Lawton LJ in *Butler Machine Tool Co Ltd* v. *Ex-Cell-O Corporation (England)* [1979] 1 All ER 963 thought it should be conducted on classical 18th century lines. That it to say, on the basis of offer and acceptance. With any new offer being regarded as counter-offer that destroys the original offer. The winner then is the one whose last terms have been accepted, either by words or conduct. Where there is no such acceptance, there is probably no contract at all. *Butler* is generally regarded as a leading example of the battle of the forms.

On May 23, 1969 the sellers offered to deliver a machine tool for the price of £75,535 with delivery in 10 months. Any orders were to be in accordance with the seller's terms and these would prevail over any terms of the buyer's order. The seller's terms included a price variation clause stating that the goods would be charged at the price ruling at the date of delivery. On 27 May the buyers replied with an order containing their own terms and conditions and containing no price variation clause. The order had a tear-off acknowledgement for signature and return which accepted the order 'on the terms and conditions thereon'

On June 5, the sellers, after acknowledging receipt of the order on June 4, returned the acknowledgement form duly completed with a covering letter stating that delivery was to be 'in accordance with our revised quotation of May 25 for delivery in . . . March/April. The machine was ready in September but the buyers could not accept delivery until November. The sellers claimed an extra £2,292 for the increase in costs between the actual date of delivery and the agreed one.

Held (Court of Appeal): The buyer's order of 27 May was not an acceptance of the seller's quotation of 23 May but a counter-offer. This meant that the sellers had accepted it by their letter of 5 June. The contract was completed on that date without a price variation clause.

Out of this case arises the perception that Lord Denning is the leading authority on the battle of the forms. Lord Denning attempted to adopt a European approach to the battle of the forms. By examining the documents that pass between the parties, it is possible to draw out all the issues that the parties are agreed on. Conflicting terms are then scrapped and replaced by *implied terms* from the general law. By adopting this approach the parties would more often have a contract than not. LJJ Lawton and Bridges rejected this solution, preferring the classical approach of offer and acceptance to decide the issue. In *Gibson* v.

Manchester City Council [1978] 1 WLR 520 Lord Denning returned to the views he had broached in *Butler*. There was, he said, no need for a strict offer and acceptance in deciding whether a contract had come into being. By examining the conduct of the parties and providing everything material had been agreed between them, a contract would come into existence.

This approach was emphatically rejected by the House of Lords in *Gibson* v. *Manchester City Council* [1979] 1 WLR 294. It stressed that the *classical approach* of *offer* and *acceptance* was the correct one.

For a further example of the battle of the forms see also *Drake & Sculley Engineering Ltd* v. *J. Jarvis and Sons* [1996] EWCA Civ 1242. The dispute concerned a claim by the sub-contractor for monies due. The defendant counter-claimed for delays alleging the terms of the sub-contract allowed it an unlimited right of set-off against sums due. This claim was seven times greater than the claim for payment. They were particularly aggrieved because their employer had under their main contract deducted sums arising out of alleged delays. The sub-contractor in turn claimed that the terms of the DOM/1 standard sub-contract had been incorporated and this restricted the right of set-off.

Below is a summary of events:

1 On 1 March the defendants invited tenders for mechanical and electrical work.
2 On 4 April the sub-contractors submitted a tender.
3 On 3 August they submitted a revised tender.
4 On 31 October the defendants sent an order form on the reverse side of which were their standard conditions. A letter accompanied the order. This was agreed by the parties to be a counter-offer.
5 The sub-contractors accepted the counter-offer by starting work.

The Court of Appeal had to decide whether the DOM/1 conditions contained in the invitation to tender had been superseded by the conditions contained in the order. It decided that in the words of Philmore LJ, with whom the other Lord Justices agreed, that:

> 'In my judgment the DOM/1 terms and conditions are specifically incorporated into the counter-offer and shall be applicable . . . to the extent that they [the conditions of the order] are in conflict with the DOM/1 terms and conditions, the standard conditions on the reverse of the form are ineffective.'

Agreements to negotiate

It is commonplace in the construction industry for letters of intent to be issued. Very often and for a variety of reasons, the response of the employer (and main contractors) to the submission of a completed tender is the issue of a *letter of intent*. The English approach to the issue of such letters is quite cavalier compared to civil law countries. For example, in Germany a letter of intent must be headed *freibleibend*, which translates roughly as 'not binding' or its English equivalent 'subject to contract'. For the confusion caused by this see *Hok Sport* v. *Geoffrey King and Gernot Frauenstein* [2000] EWHC Technology 64, a letter of intent saga between a proposed German company and an American one based in London. Its American director stated during the hearing that he had not understood its significance of the phrase.

A great puzzle is the alacrity with which contractors, sub-contractors and construction professionals respond to such letters. It may well be that their legal consequences are not always appreciated or that they are so commonplace they are taken for granted. *Letters of intent* are only indications only of future intent. No contract is formed unless the wording of the letter indicates otherwise. What has to be remembered is that such letters are part and parcel of the *negotiating process*, pending agreement being reached between the parties.

Whether or not a contract has been formed is of great importance to the parties. Judges are then called upon to consider whether the facts and the intention of the parties are compatible. In *Hescorp Italia Spa* v. *Morrison Construction Ltd and Impregilo UK Ltd* [2000] EWCA Technology 143, HHJ Hicks QC adopted the following extract from the judgment of Lloyd LJ in *Pegnan SpA* v. *Feed Products Ltd* [1997] 2 Lloyds LR 601 (with which Stocker and O'Connor LJJ agreed).

'1 In order to determine whether a contract has been concluded in the course of correspondence, one must first look to the correspondence as a whole.

2 Even if the parties have reached agreement on all the terms of the proposed contract, nevertheless they may intend that the contract shall not become binding until some further condition has been fulfilled. That is the ordinary "subject to contract" case.

3 [The parties] may intend that the contract shall not become binding until some further term or terms have been agreed.

4 Conversely, the parties may intend to be bound forthwith even though there are further terms still to be agreed or some further formality to be fulfilled . . .

5 If the parties fail to reach agreement on such terms, the existing contract is not invalidated unless the failure to reach agreement on such further terms renders the contract as a whole unworkable or void for uncertainty.

6 It is sometimes said that the parties must agree on essential terms and that it is only matters of detail that can be left over. This may be misleading since the "essential" in that context is ambiguous. If by "essential" one means a term without which the contract cannot be enforced then the statement is true, since the law cannot enforce an incomplete contract. If by "essential" one means a term that the parties have agreed to be essential for the formation of a binding contract then the statement is tautologous. If by "essential" one means only a term that the Court regards as important as opposed to a term which the Court regards as less important or a matter of detail, the statement is untrue. *It is for the parties to decide whether they wish to be bound and, if so by what terms, whether important or not* (emphasis added). It is the parties who are, in the memorable phrase coined by the Judge, "masters of their contractual fate". Of course the more important the term the less likely it is that the parties will have left it for future decision. But there is no legal

obstacle which stands in the way of the parties agreeing to be bound now while deferring important decisions to agreed later. It happens every day when parties enter into so called "heads of agreement".' (p. 619)

In *Hescorp Italia Spa*, HHJ Hicks QC found that the parties were in dispute about three essential matters: (a) the completion date; (b) the amount of liquidated damages payable for late completion; and (c) the requirement that the contract be executed as a deed. In those circumstances he found that no contract had been entered into.

Essential terms

In *Courtney & Fairburn Ltd* v. *Tolaini Brothers* [1975] 1 WLR 297, Tolaini wished to develop a site in Hertfordshire known as the Thatchet Barn hotel. Courtney agreed to introduce Tolaini to a developer who could provide the finance for the deal, but one condition was that his company would be appointed the building contractors. In letters between the parties Courtney wrote that if the introductory meeting led to 'a financial arrangement acceptable to both parties you will be prepared to instruct your quantity surveyor to negotiate fair and reasonable contract sums in respect of the project . . . as they arise'. Tolaini agreed. The meeting and the finance were arranged. Negotiations over the price started with the quantity surveyors but the parties could not agree the price of the building works. Tolaini appointed other contractors to complete the work. Courtney sued for breach of contract.

Held: There was no binding contract between the parties.

1 An agreement to negotiate was no substitute for an agreement as to price. If there was no agreement as to price, and no agreed method of ascertaining the price (except by negotiations between the parties), there was no binding contract.
2 A 'contract to negotiate' was too vague to be a contract in law.

Lord Denning said:

'Now *price* in a building contract is of *fundamental importance*. It is so *essential* a term that *there is no contract unless the price is agreed* or unless there is an agreed method of ascertaining it, not dependent on the negotiations of the parties themselves. In a building contract both parties must know at the *outset*, before the work is started, what the price is to be or, at all events, what agreed estimates are. No builder and no employer would ever dream of entering into a building contract for over £200,000 without there being an estimate of cost and an agreed means of ascertaining the price.' (emphasis added)

Lord Denning also observed that the law did not recognise a contract to negotiate. 'I think we must apply the general principle that when there is a fundamental matter left undecided and the subject of negotiations, there is no contract.'
    A more recent Court of Appeal judgment concerned the extent to which the sub-contractor engaged to carry out the design and installation of mechanical and

electrical work had agreed all the essential terms. Work started on request and a letter of intent was subsequently issued. Disputes arose at the end of the work concerning (i) the value of the work and (ii) the extent to which the employer's requirements had been incorporated in the original price of the work. No formal contract was ever entered into. *Carillion Construction Ltd (trading as Crown House Engineering)* v. *Ballast Plc (Formerly Ballast Wiltshier Plc)* [2001] EWCA Civ 1098 was an appeal against the judgment of HHJ Faulks QC. On a trial of preliminary issues he decided the following:

1 A contract had been concluded when the claimant first wrote to the defendant accepting its offer made by letter.
2 The sub-contractors only had to comply with those elements of the employer's requirements for which they had tendered.
3 The terms of the main contract had not been incorporated into the sub-contract.

The main contractors appealed against these findings, contending that the sub-contract had been formed at a later date and that the employer's requirements had been incorporated. The nub of the argument was whether there was a fixed price covering all of the employer's requirements or a fixed price with exclusions identified in the offer. In carrying out the work excluded from the fixed price, the sub-contractor claimed payment for doing it. The main contractor argued that agreement had been reached in negotiations between the parties at a number of meetings.

LJ Aldous, in giving the main judgment, concluded that a contract was entered into but at a later date than that held by the judge. He found that there was no evidence that the sub-contractor had warranted that the fixed price included all the employer's requirements. Therefore the appeal failed.

Hale LJ agreed with Aldous LJ that Crown House had not agreed to comply with the employer's requirements in every respect and limited their fixed price to works included in their specification. She commented at para. 54:

'This case is however a good example of isolating such issues from the underlying issues in the case: it became apparent only in the course of the hearing that these relate principally to the scope of the works covered by that fixed price rather than to their quality. Even on a preliminary issue, some examples of what the case is in fact about can be most helpful.'

This case was presented as being about when the contract was formed, whether all the essential terms were agreed and what terms were incorporated. In fact the real issue was the price to be paid for the completed work and whether the extra work was included in that price.

Letters of intent

It is not being suggested here, that letters of intent are not extremely useful devices in 'kicking off' commercial relationships. They are frequently used in the initial stages of very large contracts awarded by major oil companies and cause very little dispute or litigation. There are many sound practical reasons for using them. It speeds up work, parties may have to make arrangements with third parties or finance has not been finalised, and so on. As an interim measure it works very well.

A simple type of letter could read:

(Heading)

Dear Sirs,

With reference to your tender dated . . . it is our intention to award the contract to you, subject to agreement being reached in further negotiations.

However, as the work is urgent, we authorise you to begin design work. We will pay you for such design work at £x per hour for each of your design staff, payable monthly in arrears, subject to a maximum limit of £10,000.

We reserve the right to require the termination of such design work at any time.

Yours faithfully
(Client's signature)

The purpose of a letter of intent is to express an intention to *enter* into a *contract* at a *future date*. It is purely a term of *convenience*. Unlike expressions such as *'subject to contract'* which have a defined legal meaning or the phrase *'without prejudice'*, it has no substantive legal meaning. Windward suggests that there are *four categories* of letters of intent:

(a) an expression of intention to enter into a contract at a future date which does not give rise to any legal obligation; or

(b) an expression of intention to enter into a contract at a future date which does not give rise to any legal obligation but does not exclude other claims to payment, or

(c) the creation of a conditional obligation which will be subsumed by a wider contractual obligation when formal contracts are executed; or

(d) a legally binding contract in that the letter of intent is an offer capable of being accepted.

The leading construction case on the status of a letter of intent is *British Steel Corporation* v. *Cleveland Bridge & Engineering Co. Ltd* (1981) 24 BLR 94. The defendants were involved in the construction of a bank in Saudi Arabia. They negotiated with the claimants for the supply of steel nodes. Work started on the receipt of the following letter of intent.

'We are pleased to advise you that it is the intention of Cleveland to enter into a sub-contract with your company for the supply of steel casings which will form the roof nodes on this project. The price will be as quoted in your telex dated 9th February 1979 . . . The form of sub-contact to be entered into will be our standard form of sub-contract for use with the ICE General Conditions, a copy of which is enclosed for your consideration . . . We request you to proceed immediately with the works pending the preparations and issue to you of the official form of sub-contract.'

No contract was ever signed because the parties could not reach agreement on the sequence of the delivery of the nodes. The claimants brought an action for £229,832, the price of the nodes delivered. The defendants counter-claimed for losses of £868,735, alleging that the nodes were delivered late and out of sequence.

Held: No contract was ever formed between the parties. British Steel was, however, entitled to be paid upon a *quantum meruit*.

A case usually set opposite the conclusion in British *Steel* is *Turriff Construction & Turriff Ltd* v. *Regalia Knitting Mills Ltd* (1971) 9 BLR 20. There the court came to a different conclusion. It was held that the letter of intent created an *ancillary contract* under which Turriff was entitled to its interim costs. The letter of intent had been given as an assurance to Turriff that it would be paid for the preparatory work even if the project were abandoned.

Whether or not a letter of intent creates a contract depends *entirely* on its *wording*. An analysis of cases in the last ten years suggest that letters of intent fail to end in agreement for the following *reasons*:

1 The primary reason is that in many cases the parties *do not reach agreement* in their negotiations (*British Steel*). Evidence before the Court of Appeal in *Jarvis Interiors Ltd* v. *Gilliard Homes Ltd* [2000] BLR 33 suggested that in 85 per cent of cases work started before contracts were concluded. The Editors of Building Law Reports, (BLR) suggest in their commentary on that case that practitioners do not find it rare for letters of intent not to lead to completed contracts. Giaquinto (*Construction Law*, October 1999, p.32) states that a high percentage of works started with a letter of intent are completed without a formal contract ever being concluded.

2 The *formal paperwork* is *never completed* for some reason. In *Hall & Tawse South Ltd* v. *Ivory Gate Ltd* (1998) EWHC Technology 358, extensive refurbishment and redevelopment of office premises was carried out at 34–40 Jermyn Street, London. Near completion the parties were in dispute as to the contract terms that governed their contract. The contract was nine months late and the parties were £2 million apart on payment of additional preliminaries for this period. Work started on 24 October on receipt of a letter of intent sent the previous day.

> 'We confirm that it is our intention to enter into a formal contract in the form of the Standard form of Building Contract 1980 Edition Private with Qualities (including amendments 1–14) to appoint your Company as building Contractors . . . in the sum set out in your tender dated . . . The final form of Building Contract together with the terms of the *escrow* agreement are those to be agreed between our solicitors . . . and your solicitors . . . and confirmed by the quantity surveyors . . .

Pending the execution of the Building Contract and the escrow agreement we are now instructing you to begin the Works . . .

In the event that we decide not to appoint you or proceed with the works for any reason, then we will reimburse you after making due allowance for all previous payments made to you which you have made and all reasonable costs properly incurred by you together with a fair allowance for overheads and profit . . .

Will you please signify your acceptance of the above terms.

Yours faithfully'

Escrow is a deed, bond or other engagement delivered to a third party to take effect upon a future condition and not till then to be delivered to the grantee (SOED).

Held: A contract was *formed by conduct*. Hall & Tawse had accepted the offer by starting work the day after the receipt of the letter of intent. Where no express terms had been agreed a reasonable sum was payable. In this a case a *reasonable* sum would be computed *using the agreed bills of quantities*. All other issues would be dealt with using the provisions of the JCT 80 – the standard form of building contract current at the time.

3   This is a purely tactical one. Defective work surfaces (*Trentham / Sculley*) and one party prefers not to arbitrate (or now perhaps be taken to adjudication). In *Jarvis* the main argument was whether the arbitration clause had been incorporated.

4   The contract rates are unattractive compared to payment on a basis of *quantum meruit*: see *VHE Ltd* v. *Sir Alfred McAlpine*. Confusion arose between the independent quantity surveyor employed by VHE to prepare its application for payment. He used what he thought was the agreed contractual rates whereas VHE argued that no agreement was reached. The judge summed it up by saying that the plaintiffs had done the work. Should they be paid pursuant to the contract, and if so what contract, or are they to be paid a *quantum meruit*? In money terms there is a considerable difference between the parties

5   The proposed formation of the company collapses because the proposed client is unable to procure the finance: see *Hok Sport* v. *Geoffrey King and Gernot Frauenstein*.

6   This is demonstrated by *J. Jarvis and Sons Ltd* v. *Castle Wharf Developments Ltd and others* [2001] CA. Castle wrote Jarvis a letter of intent, subject to contract, to enter into a contract with Jarvis for a price to exceed £4.26 million and to be 'subject to final mutual agreement'. Although the contract was subsequently signed the work started without planning permission being obtained (see Chapter 2 for a fuller discussion).

A condition precedent

This is where one party has to perform a condition before the other's promise to perform arises. *CJ Sims Ltd* v. *Shaftesbury PLC* (1991) 60 BLR 94 illustrate this. A property company invited tenders for the construction of offices at 7 Tanner

Street, London. The tender documents included the terms of the standard contract, JCT 80. The claimant submitted a tender in late August and on 6 September 1988, the defendant sent a letter of intent confirming that they would appoint them on the terms of the standard form and asking them to start work at once. The letter stated:

> 'in the unlikely event of the contract not proceeding, CJ Sims will be reimbursed their reasonable costs which have been and will be incurred and costs for which they are liable including those of their sub-contractors and suppliers, such costs to include loss of profit and contribution to overheads, all which must be *substantiated in full to the reasonable satisfaction of our quantity surveyor.'*

Negotiations took place between the parties, but no agreement was ever reached. The claimant carried on working and completed the project. They received payment from time to time. In an action against the developers they claimed £1,090,038 plus value added tax (VAT) as their reasonable costs of carrying out the work.

Held: The letter of Intent created a *condition precedent* so that the claimants were not entitled to make a claim until they had substantiated their claim to the reasonable satisfaction of the defendant's quantity surveyor. The result was that until the condition was satisfied, i.e., the quantity surveyor certified the amount owed), *no payment* was due to the contractor.

Note that the recipient of the specimen letter of intent is entitled to interim costs if no contract is ever made; in other words, an *ancillary contract* is created. This can create liability for quality and suitability of the work. Whether such liability arises *depends* on the *words* used in the letter. See *Twintec Ltd* v. *GSE Building and Civil Engineering Ltd* [2002] EWHC 605 (TSC). The letter of intent was Issued to secure the claimant's services quickly. All parties proceeded on the basis that the letter would constitute commitment by the defendant. It amounted to an acceptance of the claimant's offer to do the work.

Held: A contract came into existence.

The use of the phrase 'subject to contract'

Regalian Properties above is an example of the meaning of the phrase. Each party to the negotiations is free to walk away at any time before a contract is concluded.

In *Jarvis Interiors Ltd* v. *Gilliard Homes Ltd* [2000] BLR 33 CA, the Court treated (i) the issue of the letter of intent and (ii) the negotiations between the parties on the use of the JCT 80 standard form of building contract (together with amendments to it) as creating a result that it was in fact 'subject to contract'. This prevented a contract from coming into being. Counsel's argument, that this formulation would send tremors through the construction industry, was dismissed by the Court of Appeal. The misunderstanding that can be created is demonstrated by *Comyn Ching Limited and Others* v. *Radius Plc* [1997] 15-CLD-08-28.

Comyn Ching Ltd were architectural ironmongery suppliers within a group of companies that wished to computerise their stock control and accounting operations. They approached Radius Plc, who supplied computer hardware and software. Ching did not employ consultants and they had a limited knowledge of computers. Radius offered to make an investigation into Ching's requirements for a fee of £6,930, which was rejected. The parties met on 25 November 1988 to negotiate. Ching agreed to pay a deposit of £15,000 and handed Radius a cheque for that amount. A letter of intent was then signed by Radius that was headed 'subject to contract'.

Ching issued a writ in September 1994, shortly before the limitation period expired. Ching argued that the above facts gave rise to a contract under which Radius had agreed to provide advice and to investigate its requirements. Ching argued further that by responding to its invitation to tender, Radius had impliedly offered to ascertained the requirements of Ching and propose equipment and software to satisfy its requirements.

Held:

1 The letter of intent did not give rise to a binding contract particularly as it was headed 'subject to contract'.
2 Radius did not enter into a contract by accepting the deposit cheque. It only agreed to carry out certain works asked for by Ching.
3 Ching's subsequent conduct was inconsistent with their claims that there was a contract.
4 The proposition that, by responding to Ching's invitation to tender, Radius impliedly offered to ascertain its requirements and advise was extraordinary, totally unacceptable and a further example of Ching demanding something for nothing.

No coincidence of offer and acceptance

Although the classic test for the presence of an agreement is a clear offer and an unconditional acceptance, the courts are reluctant to conclude that where the work is executed no contract exists. Where the transaction was fully performed (i.e., the work was carried out by one party and payment made by the other), it is *highly unlikely* that no contract was ever formed: see *Percy Trentham Ltd* v. *Archital Luxfer Ltd* (1992) 63 BLR 44. In this case the Court of Appeal decided that the traditional analysis of offer and acceptance was not always the test needed to decide whether a contract had been formed. The sub-contractor had done the work following the issue of a letter of intent and the main contractor had made regular monthly payments. In arbitration the main contractor was held liable to pay damages to the employer for defective work. The sub-contractor had carried out some of this work. It refused to accept responsibility for the defects or to go to arbitration since it argued that the parties had no contract. There had been no offer that had been accepted. The Court of Appeal held that a contract had come into existence even though there was no coincidence of offer and acceptance since '*no reasonable business person would conclude that no contract existed*'. Steyn LJ said that at p.52: 'before I turn to the facts it is important to consider briefly the approach to be adopted to the issue of contract formation in this case. It seems to me that there are four matters of importance.'

1 English law adopts an objective theory of contract formation, which in the present case was 'the reasonable expectation of sensible businessmen'.
2 In the vast majority of cases the coincidence of offer and acceptance will be the mechanism of contract formation.
3 Where both parties have performed the transaction it was unrealistic to argue that there was no intention to create legal relationship.
4 It is possible to hold that a contract that only comes into existence during and as a result of performance covers the pre-contractual performance.

For an example of a case where the work was executed and paid for, but no contract was found to exist, see *Drake and Sculley Engineering* v. *Higgs and Hill* (1995) 11 Const. LJ 214. The court found that though all *essential terms* had been agreed, the parties had also agreed that no contract would come into existence until a *formal contract* was signed.

The issue of letters of intent and whether a contract was formed out of subsequent negotiations arose in the case of *Jarvis Interiors Ltd* v. *Gilliard Homes Ltd* [2000] BLR 33. Gilliard was the developer of a scheme for 36 flats in London and approached Jarvis for a price for doing the fitting-out work. As a result of their dealings Gilliard issued a letter of intent. Jarvis started work and Gilliard made interim payments as the work progressed. The parties carried on negotiations in the meanwhile. Gilliard wanted a fixed price (a guaranteed maximum price) contract using the JCT 80 Standard Form of Building Contract. This meant the deletion of its variation clauses, amongst other things. Jarvis complained that the architects on site, were changing the agreed specifications and demanded to be paid for the extras. Gillard terminated both the negotiations and the work.

The judgment of the Court of Appeal was as shown below:

1 Although the contract was executed *Trentham* did not apply. This was because the parties had agreed that no contract would exist until a formal contract was signed as a deed.
2 No contract came into existence on 1 December when the parties shook hands on their agreement that the contract price would be £1,325,000. Evans LJ at p.47 said:

> 'When two experienced businessmen shake hands on what they regard as a deal, I am loath, and in my view courts are always loath, to hold that no legally binding contract came into existence . . . This is for two reasons. First it is not possible to say what the terms were . . . Secondly the agreement on the 1 December was subject to contract meaning no legal contract would come into existence until a formal contract, meaning a contract under seal, was signed.'

3 The Court treated the issue of the letter of intent and the negotiations between the parties on the use of JCT 80, together with amendments to it, as meaning the agreement was 'subject to contract'. This prevented a contract from ever coming into being.

4 A further reason was that the letter of intent said 'In the event that we do not enter into a formal contract with you through no fault of Jarvis International, you will be reimbursed all fair and reasonable costs and these will be assessed on a *quantum meruit* basis.' This suggested that the parties had considered the possibility that no contract might be formed.

Held: No contract was ever entered into and no arbitration clause was incorporated. Jarvis was entitled to a reasonable price for the work carried out on the basis of a *quantum meruit*.

Jarvis was not followed in *Hescorp Italia Spa* v. *Morrison Construction Ltd* [2000] EWHC Technology 123. HHJ Hicks QC decided it was a decision based on its own facts. Counsel for the claimant had drawn attention to it, in support of a contention that the execution of a contract as a deed was in issue. Although not a letter of intent case, the parties failed to reach agreement on the price, starting date or the rate for liquidated damages. Work had started on request and Judge Hicks found no contract existed between the parties.

The Court of Appeal considered both *Trentham* and *Jarvis* in *Stent Foundations Ltd* v. *Carillon Construction (Contracts) Ltd* (2000 unreported). Work was completed following the issue of a letter of intent. Dyson J, sitting in the Technology and Construction Court, found a contract existed between the parties. All the essential terms had been agreed. The respondents challenged that finding.

The argument

Carillon (formerly Wimpey Construction Management, or WCM) accepted before the Court of Appeal that all the *essential terms* had been agreed. That being so, they argued instead that the agreement was in effect 'subject to contract'. In support of their argument that formality was a *condition precedent* to a contract between the parties, they made two points. First they argued that no negotiation could take place before the management contract was in place with their client, Wiggins Waterside Limited ('Wiggin') and Carillon. Second, Stent understood that the sub-contract would be embodied in formal documentation. This meant than the agreement was in effect one that was 'subject to contract'. LJ Hale in response to that argument disagreed, saying: 'It also seems to me that everyone behaved as if a contract was in place. Payment was made under the contract. The developing dispute about ground conditions was handled by WCM as if it was under contract.'

Counsel for WCM accepted that *Trentham* stood in the way of any argument that, once the transaction been concluded, it was still possible to argue that no contract existed. However, in *Jarvis* the existence of a condition precedent precluded the existence of a contract although the transaction was concluded. LJ Hale dealt with those arguments by saying that the history of the letter of intent supported rather than undermined the claimant's case. It also contemplated a works contract being made with reimbursement of costs but no profit element. Once Stent had started work, that position was no longer acceptable. As it was, they were paid in accordance with the agreed tender sums, not the letter of intent. She also distinguished *Jarvis* as a case where too many substantial matters remained in active dispute. Brook and Swinton Thomas LJJ supported the dismissal of the appeal.

Other informal agreements

In *Clarke & Sons* v. *ACT Construction Ltd* [2002] EWCA Civ 972, the Court of Appeal was faced with the familiar issue of whether there was a contract between the parties and, if so, what were the terms? The agreement had been made informally. The appellant Clarke had wanted to turn the site of a disused cold store into a coach depot for its expanding business. This required, amongst other things:

(a) obtaining planning permission for refurbishment and change of use;
(b) putting together a professional team to prepare the documents for the application for planning consent;
(c) carrying out structural demolition and reconstruction.

It approached the respondent ACT, whom it knew from previous dealings, to assist with the planning application. ACT went on site in April 1993. A great deal of demolition work was required, including the drilling out of concrete and paving. Also required was the construction of structural steel framing, cladding, extensive roofing and paving, the installation of equipment purchased by Clarke and the partial demolition, building, rebuilding and refurbishment of office accommodation.

ACT made applications for 'interim payment' and from time to time payments were made on account. In May 1994, ACT was asked to discontinue work and leave the site. In 1997 ACT submitted its final reconciliation of the account which amounted to £208,608. It started proceedings to recover the amount. Clarke counter-claimed, alleging that the account had been overpaid and claiming damages for failure to complete within a reasonable time and for defective work.

HHJ Thornton QC decided that ACT started work on an informal basis without any contractual framework. His conclusion is quoted at para. 26 by LJ Ward:

> 'I conclude that there was no contract between the parties. The parties' relationship was not a contractual one, with the consequence that the value of the work carried out by the claimant could be recovered and paid for, but on a basis of a *quantum meruit*, a reasonable price, a restitutionary basis in fact.'

In the Court of Appeal Ward LJ disagreed with these findings, a conclusion shared by his fellow Lord Justices. At para. 27 he stated that there were three ways in which a contract might arise.

1 Where there is an entire contract. A recent example of a large commercial contract where the rule was applied is *Discain Project Services Ltd* v. *Opecprime Developments Ltd* [2001] EWHC Technology 450.

2   Where the parties have entered into a 'formal contract'.
3   A contract can arise in circumstances where no price has been agreed and the scope of work is not yet finalised. Provided that there is an instruction to carry out work and that instruction is accepted the parties will have a contract. In such a contract the law implies an obligation to pay a reasonable sum for that work.

He held that the third type of contract arose here. The reasonable sum he classified as a 'contractual *quantum meruit*'. For a statutory provision for the payment of a reasonable sum see s.15 of the Supply of Goods and Services Act 1982. In an otherwise complete contract where no price is specified, a reasonable sum is payable for the work or service provided.

Calculating the reasonable sum

The judge held this to be *cost plus 15 per cent*. Much of the appeal consisted of Clarke challenging the validity of the 15 per cent. The expert witness for ACT was a quantity surveyor. He gave evidence that the mark-ups at the time were typically 8–12 per cent. The submissions of Clarke's expert were rejected because an engineer rather than a quantity surveyor gave them. An engineer could not properly deal with within the expertise of quantity surveyors and as there was no support from the pricing books, there was no other evidence the judge could rely on.

All three judges agreed that the mark-up of 15 per cent should stand. It allowed Clarke's appeal to the extent that it would be applied to *all costs*, even those costs the judge held could not be re-opened. How defective work and delays in completion would have affected the calculation of the reasonable sum was not considered since it was unclear whether there were any.

The implied contract and *quantum meruit*

McMeel (2000, p.27), observes that by the mid-eighteenth century much of the modern law of restitution was in place. It was, however, contained in a body of common law doctrine called *quasi-contract*, the word 'quasi' indicating that it is 'just like' or 'almost' a contract. The modern approach is to regard the 'implied contract theory' as a historical fiction based on the old-fashioned system of pleadings abolished in 1852. McMeel describes the theory as claiming that a thief impliedly promises to repay the victim of his theft. *Chitty* (*Beale*, 1999, p.1,470) states that the theory of the implied contract was 'unequivocally and finally' rejected by the House of Lords in *Westdeutsche Landesbank Girozentrale* v. *Islington London Borough Council* [1996] AC 669, 710. See also at 718, 720, and 738. Lord Brown Wilkinson stressed that:

'Subsequent developments in the law of restitution demonstrate that this line of reasoning is no longer sound. A restitutionary claim at common law is based not on an implied contract but on *unjust enrichment*: in the circumstances the law imposes an obligation to repay rather than implying a fictitious agreement to repay . . . In my judgement your lordship should now *unequivocally and finally reject* the concept that [an action for] money had and received is based on an implied contract.'

The term is best defined nowadays by reference to the principle of *unjust enrichment* whereby one party is enriched by receiving a benefit at the expense of the other party. At common law *quasi-contract* now forms a part of the law *of restitution*. The debate about quasi-contract is not entirely settled. The leading construction law textbooks of Keating, Emden and Hudson still use the expression. It is suggested that the continuation of this terminology may derive from the often quoted statement of Goff J (as he was then) in *British Steel Corporation* v. *Cleveland Bridge & Engineering Co. Ltd* (1981) 24 BLR 94 where he said:

'In most cases where work is done in pursuant to a request contained in a letter of intent, it will not matter whether a contract did or did not come into existence; because if the party who has acted on the request is simply claiming payment, his claim will usually be based on a *quantum meruit*, and it will make no difference whether that claim is contractual or quasi-contractual. Of course, a *quantum meruit* claim (like the old actions for money had and received, and for money paid) straddles the boundaries of what we call now contract and restitution, so the mere framing of a claim as a *quantum meruit* claim, or a claim for a reasonable sum, does not assist in classifying the claim as contractual or quasi-contractual.'

The essence of the claim

Some benefit has been conferred on A by B but B never intended that A should have that benefit. In the circumstances it would be unjust for A to retain the benefit or retain it without paying for it. Such a claim in relation to construction contracts may be for 'money had and received', or a claim in *quantum meruit*.

'Money had and received'

This is illustrated by *Furguson* v. *Sohl* (1992) 62 BLR 94. The contractor stopped work before completion of the contract. By then it had been paid £26,000 out of a contract value of £32,000. This sum was found by the judge as £4,600 in excess of what was due under the contract. The Court of Appeal, affirming the decision of the judge held that the owner was entitled to recover the money 'as money had and received'. Tucker J in striking out the pleadings in *Ocean Mutual* v. *FAI General Insurance* [1998] 16-CLD-05-30 observed that 'quasi-contract is no longer the favoured legal analysis or a claim for money had and received'. Hence a claim under the then Order 11 rule 1(d) could not proceed since the words covered only contractual claims.

Quantum meruit

This is not a case of returning money paid, but of recovering the value of work and services performed. Such a claim arises where a benefit has been conferred on A by B which justice requires should be reimbursed by A.

Two legal contexts are identified

1 Where an 'implied' contract exists

One the parties have requested work outside a contract and an intention to pay is implied but no price has been fixed, the promise to pay a reasonable sum then arises. An example of this is *William Lacey (Hounslow)* v. *Davis* [1957] 2 All ER 712. The plaintiff was held to be entitled to a reasonable payment for work carried out during tendering. Keating (2001) merely states of this case that the modern legal analyses may be that the obligation to pay 'sounds in quasi-contract or, as we now say, restitution' (p. 19).

2 No agreement exists or subsequently comes into existence

It would be inequitable to allow one party to receive the benefit of work or services provided to the other. Hudson describes this as the true quasi-contractual remedy of *quantum meruit* or restitution. Keating further uses the term quasi-contract in describing a claim for the costs of work carried out where contractual negotiations are never completed. By contrast the High Court of Australia in *Pavey & Matthews Pty Ltd* v. *Paul* (1987) 61 ALJ 151 accepted that 'the true foundations of a *quantum meruit* claim does not depend on the existence of an implied contract'.

From a practical point of view, restitution and a reasonable sum would in most cases amount to the same thing. However, restitution is about the benefit received and if the employer gets no benefit the amount due to the contractor may be nothing. See for instance, the case of *Regalian*, where a claim for expenditure spent on preliminary design, failed amongst other reasons because no benefit was conferred. Whereas *Easat* can be described as a case where it succeeded because a benefit was obtained from the services provided.

Quantum meruit and variations

While it is true to say that a claim for a *quantum meruit* cannot be made where there is an existing contract, a variation to the contract may be made *outside* the existing framework. In such a case a claim for a reasonable sum may be made. *Costain Civil Engineering Ltd and Tarmac Construction Ltd* v. *Zanen Dredging and Contracting Co. Ltd* (1996) 85 BLR 85 illustrates this situation. The two companies formed a joint venture (the JV) for the construction of the A55 Conway bypass and river crossing for the Welsh Office. Part of the work was the construction of a tunnel under the River Conway. Six pre-fabricated tunnel elements cast from reinforced concrete were to be floated out into the river and sunk in a trench dredged from the bed of the estuary. The six sections, when immersed, connected and drained, would form the tunnel. Casting the tunnels took place in a dry basin excavated next to the river. Once completed a channel would be cut into the river, connecting it with the basin so that the work could take place.

Instead of backfilling the basin, the JV entered into a contract to provide a marina for the Crown Estates Commissioners. As a consequence they instructed the sub-contractor by way of a *variation* to execute the work necessary to accommodate the changed plan. They did not inform them of the supplemental agreement. In a reference to arbitration, the arbitrator made an interim award that Zanen was entitled to a payment for the work by way of a q*uantum meruit* for work done outside the contract. Zanen was awarded the sum of £370,756 plus on costs of the works themselves and a share of the profits made, amounting to £386,000. The JV appealed.

Held: Since a variation clause in a sub-contract does not require the sub-contractor to carry out a main contractor's obligation outside the contract, the sub-contractor was entitled to a *quantum meruit* payment. Such a payment could include a sum in respect of the competitive advantage the provider of the service contributed to enable the JV to bid profitably for the work. It resulted in them being able to strike a much better bargain.

Assessing a reasonable sum

The courts have laid down no guidelines. Where *quantum meruit* is recoverable for work done outside the contract it is *wrong* to regard the work as though it was performed under the contract.

The contractor should be paid a *fair and reasonable rate* for the work done. Useful evidence could be abortive negotiations as to price, a calculation based on the net cost of labour and materials used plus a sum for overheads and profit, measurement of work done and materials supplied and the opinion of quantity surveyors, experienced builders or other experts as to a reasonable sum (Keating, 2001, p.102).

The award of a reasonable sum

In assessing a reasonable sum can the employer take into account the following?

1 Defective work carried out by the contractor.
2 Delays caused by the contractor to other contractors employed on the site.
3 Tardy or badly planned work.
4 Abortive negotiations as to price.

This matter arose in *Crown House Engineering* v. *Amec Projects* (1989) 48 BLR 32. The claimant alleged in its amended statement of claim that it was entitled to a reasonable sum for the work carried out. Crown were sub-contractors for mechanical and electrical work (work package 16) carried out for Plessey Properties at a site in Plymouth in Devon. The defendants were the contractors for the design and construction of the works. A dispute arose during which doubts surfaced as to whether a contract had ever been entered into. Amec counter-claimed for damages alleging the following complaints:

(a) late completion of the work;
(b) failure to properly integrate and arrange their work with work carried out by others;

(c) failing to devote adequate resources to the work;
(d) failing to prepare drawings properly and to execute quality assurance;
(e) claims for disruption to other contractors;
(f) the cost to the defendants of having to carry out certain works themselves;
(g) failing to keep their working areas clean and tidy and free from debris result-
    ing in damage to the works of others.

In dismissing the respondents' appeal the Court of Appeal made the following observations: Slade LJ at p. 54:

> 'I am not convinced that either that learned work [Goff and Jones on the Law of Restitution] and any of the reported cases cited to us affords a clear answer to the crucial question of law. On the assessments of a claim for services rendered based on *quantum meruit*, may in some circumstances (and if so, what circumstances) be open to the defendant to assert that the value of the services falls to be reduced because of their tardy performance or because the unsatisfactory manner of their performance had exposed him to extra expense or claims by third parties?'

Stocker and Bingham LJJ both agreed that there was *no clear answer* to the question posed. Both accepted that within the concept of restitution the extent to which cross-claims could be entertained in a claim for a *quantum meruit* required full argument in a later case.

Part of the difficulty for the respondents stemmed from the decision in *British Steel Corporation* v. *Cleveland Bridge & Engineering Co Ltd* (1981) 24 BLR 94. The plaintiffs succeeded in a claim for *quantum meruit*. The counter-claim by the defendant was based on alleged delays (the nodes being delivered late and out of sequence). Their claim was for damages for breach of contract and Goff J (as he was then) found no contract was ever made. The claim therefore failed and it was not necessary to consider whether the expense incurred by the defendant should have been reflected in the *quantum meruit* awarded.

Hudson's observes that 'the resulting obligation of the defendant is not to pay a reasonable sum based on costs incurred by the plaintiff, but to reimburse him to the *value* of the *advantage*, if any received by the defendant' (1995, p.144).

Colin Reese QC, sitting as recorder in *Birse Ltd* v. *St David Ltd* (2000) BLR 57, commented on the flexibility of the law of restitution to achieve justice in a situation where work had been carried out and no contract formed. Adopting the observations of LLJ Slade and Bingham in *Crown House* he observed that in his opinion the *builders' price* might well serve as a cap on the level of recovery. In addition tardy performance and the level of liquidated damages agreed as part of contract negotiations might well form part of the calculation of the *quantum meruit*.

This issue was touched on in *Serck Controls Limited* v. *Drake & Scull Engineering Ltd* (2000) CILL 1643, where a letter of intent was issued instructing the subcontractor to proceed. The parties had agreed the price and the scope of works but were unable to agree the programme and the conditions. The letter provided an assurance that they would be reimbursed their reasonable costs. HHJ Hicks QC had to decide whether the reasonable sum payable was (i) the value of the work was the contractor's reasonable cost of executing it or (ii) the value of the work to the employer.

While accepting that *quantum meruit* could cover a spectrum from costs at one end to value at the other, he decided that those reasonable costs included an element for profit and overheads. In addition the tender price was *not* a determining factor in assessing a reasonable sum. The defendant was under no duty to comply with the main contractor's programme. It could *not* therefore be charged for out-of-sequence working. The judge observed that a contractor seeking a *quantum meruit* payment on a complex construction site could not wholly ignore the desirability of co-operation with others working there.

The question of a limit on the amount claimed under a *quantum meruit* was considered by the Court of Appeal in *Rover International Ltd* v. *Cannon Film Ltd* [1989] 1 WLR 912. It held that breaches under a contract void *ab initio* (void from the very beginning) could not be taken into account in determining the appropriate *quantum meruit*.

The claim in restitution

In *Discain Project Services Ltd* v. *Opecprime Developments Ltd* [2001] EWHC Technology 450 the judge found that no contract had been made for the fabrication and erection of a deck. As a result the defendant was entitled to the payment of a *reasonable sum* on a *restitutionary* basis. (The defendant it should be noted, made no claim for damages for beach of contract.) He then added:

'However, it is convenient from a practical point of view, to calculate what sum was due to Discain by calculating what would be a reasonable price for the deck, if it were not defective in the respects admitted on behalf of Descain, and then to deduct the reasonable cost of remedying the admitted defects.'

Note *Clarke & Sons* v. *ACT Construction Ltd* [2002] where a reasonable sum was held to be costs plus 15 per cent in the circumstances of the case. This was a *quantum meruit* due under a contract; it did not include a claim for defective work or delay in completion. The Court of Appeal remitted the case to the Technology and Construction Court for recalculating the sum based on costs plus 15 per cent. It left the question of deduction for defective work or delay in completion to the judge. Most contractors would be delighted to settle for a *quantum meruit* based on these figures.

A.L. Barnes Ltd v. *Time Talk (UK) Ltd* [2003] EWCA Civ 402 was an appeal against the award of a contractual *quantum meruit* in circumstances where the parties had agreed to conspire to defraud the client. The issue on appeal was against the decision of the judge to award the contractor the value of the work done. The appellant claimed that the dishonest assistance of the contractor's director, in the breach of trust by the client's director and its project manager, prevented any recovery of monies claimed. For this it relied upon *Taylor* v. *Bhaill* [1996] CLC 377. The headmaster of a private school entered into an agreement with a builder. In return for awarding him the contract to repair a storm-damaged wall, the builder would inflate the estimate to the insurance company by the sum of £1000 to be paid to the headmaster. On completion of the work, he refused to pay the builder any money and the builder sued for payment. The Court of Appeal decided that it would not enforce a contract to defraud a third party. In *A.L. Barnes Ltd*, Longmore LJ described *Bhaill* as 'an example of a contract to commit a crime' (conspiracy to defraud the insurers or at least to obtain money from them by deception): see para 8 (1). Since this was not alleged, the case was not of application. The second argument was that although the performance of the contract was not illegal, its purpose was: see for example *Pearce* v. *Brooks* (1866) LR 1 Ex. 213. There the claimant knew that the carriage was to be used for an illegal purpose. In contrast, this was a contract for the supply and fitting out of outlets for the sale of mobile phones and there was no illegal purpose in that. Hence the contracts were, and Barnes entitled to, a *quantum meruit* for the work carried out. The monies claimed for project management fees were not enforceable since they formed no part of the *quantum meruit* claim.

Summary

1 The issue of whether there is a contract and what the terms are arises frequently in construction contracts.
2 The submission of a completed tender for a complex project may lead to the forming of an implied contract. The damages for the loss of a chance where the tender is not considered may be the costs of tendering plus the profit the contractor could have made.
3 Pre-contract work may trigger a claim in restitution where the work produced is of value to the employer.
4 The battle of the forms is based on the classic approach of offer and acceptance. It arises often out of negotiations on the terms of a particular contract.
5 Where the work has been carried out and payments made, courts are reluctant to find no contract exists.
6 Agreements to negotiate are not legally enforceable as contracts.
7 Letters of intent fail to create contracts for a number of reasons. No agreement can be reached, the paperwork is not completed, the contract rates prove unattractive or finance is not obtained. The parties may simply be unable to agree the allocation of risks.
8 Agreement may be 'subject to contract' for a number of reasons even where the parties have not used the actual expression. Usually it means simply that the parties have no contract and are free to walk away from negotiations at any time.
9 *Quantum meruit* arises out of a reversal of unjust enrichment, not an implied or quasi-contract.
10 The valuation of a *quantum meruit* is of great practical importance to the parties.
11 *British Steel* did not address the question of whether deductions for defective performance can be made.
12 A number of cases suggest this might be possible.

13 *Rover International Ltd* v. *Cannon Film Ltd* [1989] suggests this is not possible. The contract ceiling could not be taken into account in calculating the *quantum meruit.*

14 The theoretical basis for rejecting such a possibility is that it undermines the contractual framework by suggesting equivalent remedies are available in restitution.

4 Consideration

At first consideration may appear to be an unlikely topic to require detailed discussion within construction contracts. Such contracts, whether made between commercial companies or householder and builder, usually involve the outlay of significant sums of money. This would suggest that consideration and the *intention to create legal relations* are always likely to be present. This perception is not, however, entirely accurate. The doctrine of consideration is of *major importance* in construction contracts. This is the inevitable consequence of *a number of factors* likely to be present, such as the length of the contractual chain stretching beyond the parties to their sub-contractors and suppliers. In addition, the system of competitive bidding means that one party may well under- price the cost of the work, or the price of the work escalates beyond their control, thus raising the possibility that one party may wish to renegotiate the agreement. Such contracts are often made for the benefit of third parties, and thus it is inevitable that the rights of third parties will be involved at some stage.

Lord Dunedin approved, in *Dunlop Pneumatic Tyre Co. Ltd* v. *Selfridges and Co. Ltd* [1915] AC 847, the definition provided by Pollock (1936, p.133). Consideration is 'an act of forbearance of the one party, or the promise thereof, is the price for which the promise of the other is bought, and the promise thus given for value is enforceable'. *Chitty* considers that the doctrine is there to provide legal limits on the enforceability of agreements, even when they are meant to be legally binding. Such a valid agreement can only be vitiated by mistake, misrepresentation, duress or illegality (1999, p.167).

In English law, a promise is not as a general rule, binding as a contract unless (i) it is made under seal or (ii) it is supported by consideration. In this context consideration means *price*, the price paid for the other party's act or promise.

Why consideration?

Simpson (1996), discussing the origin of the doctrine in the sixteenth century, states that it was meant to be the *factors* which *motivated* the promisor when he made his promise. He suggests that *motive* might be a more approximate word. He adds that the 'essence of the doctrine of consideration, then, is the adoption by the common law of the idea that the legal effect of a promise should be depend upon the factor or factors which motivated the promise' (see p.321). Later on, he states that 'in modern terms one can see the plausibility of the theory – a promise which lacks any adequate motive cannot have been serious, and therefore ought not to be taken seriously'. Milsom (1981) commenting on the same topic, thinks that in looking to the circumstances in which agreements were enforceable, the common law judges would have looked at the practice of the local courts. These practices were later formalised in a doctrine of consideration. (In short, it is suggested that consideration may well have grown out of the need to *police informal agreements*.)

There has been considerable academic debate on the reasons *why* consideration is needed to enforce contracts and the literature is copious. See Furmston (2001, p.79) for some of the authorities. McKendrick (2000) discusses at length the differences of approach between professors Atiyah and Treitel – two leading commentators on the law of contract. He cites the following comments by Atiyah:

> 'in truth the courts never set out to create a doctrine of consideration. They have been concerned with the much more practical problem of deciding in the course of litigation whether a particular promise in particular case should be enforced . . . It seems highly probable that when the courts first used the word 'consideration' they meant no more than there was a 'reason' for the enforcement of the promise.' (2000, p.80)

Treitel rejects this argument in favour of the 'existence of a complex and multifarious body of rules known as the "doctrine of consideration" '. The proposition that consideration means the reason for the enforcement of the promise is further rejected as a 'negation of the existence of any applicable body of rules of law.'

Documents under seal

This is a document to which the maker of the document has attached their seal and which is delivered as a deed. Following the passing of the Law of Property (Miscellaneous Provisions) Act 1989, s.2 changed the formal requirements for the disposition of land. Since a contract for the disposition of an *interest in land* had to be in writing, the Act was meant to simplify the law. A document bearing the word 'deed', or some other indication that it is intended to take effect as a deed, must be signed by the individual maker of the deed. The signature must be attested by one witness if the deed is signed by the maker (there must be two witnesses if the deed is signed at his direction). It must be delivered (i.e., there must be some conduct on the part of the person executing it that he intends to be bound by it). Major construction contracts are nearly always made by deed (historically construction contracts were drafted by Chancery lawyers, so perhaps it was automatic to do so). The effect was to entitle parties to a contract under seal to bring actions up to twelve years from the date of the breach of contract. Section 1 of the Law of Property (Miscellaneous Provisions) Act 1989 has abolished the requirement that the deed be under seal.

Documents not under seal

Where a simple contract is made in writing, partly in writing or orally the limitation period is six years after the breach of contract. To be valid, some benefit must accrue to one party or some loss be suffered by the other party. English contract law is about bargains or promises. Consideration is one part of that bargain. The consideration given by one party for the other party's promise makes up the bargain.

There are two *categories* of valid consideration, as listed below.

1 *Executory* – here a promise is given in return for a promise. The contract is *executory* since it is still to be carried out or the promise performed.
2 *Executed* – here one party has carried out its obligations without reciprocal performance. This is true of the *unilateral* contract where the acceptance consists of carrying out the obligation contained in the offer, the classic example being *Carlill* v. *Carbolic Smokeball Co.* [1893] 1 QB 256. Mrs Carlill had first to purchase the smokeball and then use it. Only when the product failed did the obligation of the manufacturer to pay a substantial sum arise.

Limits to the doctrine of consideration

Past consideration is no consideration

An act done without reference to a later promise is not consideration for that promise: see *Re McArdle* [1951] Ch 669. On the death of the father the house was left to the wife for her lifetime. On her death the proceeds of the sale of the house were to be divided equally between the children. The widow lived in the house with one of her children and his wife, who spent £500 on modernising the house. After the work was done the children promised to repay the £500 out of the proceeds of the sale of the house.

On sale of the house, following the death of the widow, her executor refused pay over the sum promised. The executor denied that the promise to pay her £500 was binding since she had given no consideration for the promise. This was because the promise was made after the improvements were carried out.

Held: Since nothing was given in return for the promise, the promise was not binding.

The consideration is *not past* when services are performed at the express or implied request of someone *who later promises to pay* for those services, provided that those services are of a kind for which it would be *reasonable* to expect some payment.

In *Lampleigh* v. *Braithwaite* (1615) Hob 105, a pardon was obtained on behalf of a man who had killed another in a duel. On being granted the pardon the killer agreed to pay £100. Later he reneged on the promise, alleging that there had been no consideration for the promise.

Held: The promise was not past, since there was a category of actions that was only done in return for payment.

Another example of the category is *Re Casey's Patent, Stewart* v. *Casey* [1892] 1 Ch 104. A and B, the joint owners of certain patents, wrote to C as follows: 'in consideration of your services as practice manager in working our patents, we hereby agree to give you one-third share of the patents'. C sued for his one-third share in the patents.

Held: There was an *implied promise* to pay. Casey had done his work relying on that implied promise.

In *Pao On* v. *Lau Yui Ling* [1980] AC 614, the Privy Council identified *three conditions* to be satisfied for there to be an implied promise to pay:

(a) the promisee must have performed the original act at the *request* of the promisor;

(b) it must have been clearly understood or implied between the parties that the promisee would be rewarded for the act (this could either be by a payment or the receipt of a benefit);

(c) the promise of eventual payment or benefit must have been legally enforceable had it been promised prior to or at the time of the act.

Valuable consideration

Consideration must be valuable but it need not be adequate. Natural love and affection has no cash value and are not consideration for value given. In *Chapple & Co Ltd* v. *Nestlé Co Ltd* [1960] AC 87, Lord Somerville said: 'a contract party can stipulate for what consideration he chooses. A peppercorn does not cease to be good consideration if the recipient doesn't like pepper and throws away the pepper.' This illustrates the concept of bargain since the law of contract is not concerned with whether the parties gain equally from their exchange. A clear application of this principle is provided by *Midland Bank Trust Co.* v. *Green* [1981] AC 513. A father, to avoid an option he had granted to his son, sold his property valued at £40,000 for only £500. The House of Lords reversed the finding of the Court of Appeal that the sale at such an undervalued price could not 'amount to money or monies worth'. A court would not inquire into the *adequacy of consideration* as long as it was *real*. There was a valid contract for the sale of the property.

Consideration must move from the promisee

The Law Commission in its 1996 report on privity of contract described the maxim as ambiguous; either it meant that (i) the claimant had to provide consideration or (ii) consideration must be provided to support the promise. It adopted the second interpretation since then it was possible to reform the privity of contract rule without reforming the doctrine of consideration.

The doctrine of privity of contract

Only a person who has given consideration for a promise can enforce the promise. A contract is about private rights, duties and liabilities. It cannot for good or ill alter the legal rights of other people by imposing liability on persons who are not parties to the contract. In *Dunlop Pneumatic Tyre Co. Ltd* v. *Selfridges and Co. Ltd* [1915] AC 847, tyres were sold to a wholesale merchant who resold them to the defendant. The contract between the plaintiff and the merchant contained clauses prohibiting sales at below the list price and in default a penalty of £5 was payable per tyre sold. The plaintiff sued the defendant for £10 plus a claim for an injunction alleging they were in breach of agreement.

Held: The case had to fail. There was no contract between the parties. Although the merchant was in breach of its contract with the plaintiff, the plaintiff could not enforce a contract to which the defendant was not a party.

In *Beswick* v. *Beswick* [1968] AC 58, the nephew promised to pay the uncle a regular sum every month if he sold him the business. In addition he promised to pay his aunt a similar sum if the uncle died first. After the death of the uncle he stopped paying the aunt.

Held: There was no privity of contract between the aunt and the nephew.

The importance of *Beswick* was that the opportunity was presented to the House of Lords to *enforce a contract made for the benefit of a third party*. It could have resolved the injustice caused by the rule in circumstances where contracts were clearly made for the benefit of third parties. The rules themselves can now be modified under the provisions of the Contracts (Rights of Third Parties) Act 1999. The adoption of the Act would resolve many of the difficulties inherent in construction contracts that stem from the rule. Construction contracts due to (i) the length of the contractual chain and (ii) the multitude of parties and participants in the chain show quite clearly how the intention of the parties can be thwarted by the strictness of the rule. The reluctance of the industry to adopt the Contracts (Rights of Third Parties) Act 1999 stems in part from its resistance to change. In addition the mechanisms adopted to evade the strictness of the rule seems to work well in practice. However, the adoption of the Act would simplify contractual relations and clarify the rights of third parties. After all, the object of most construction contracts is to provide benefits for third parties.

Devices created to evade the rigours of the rule

Some of the complexities of construction contracts stem from the devices created to evade the strictness of the privity of contract rules.

1 In *Young and Marten* v. *McManus Child* [1969] 1 AC 454, the House of Lords effectively created a *chain of liability*. By making the contractor *strictly liable* for the quality of materials used, the contractor was forced to sue 'down the chain' his sub-contractors, and they in turn their suppliers and manufacturers.
2 In *Darlington Borough Council* v. *Wiltshier Northern* Ltd [1995] 3 All ER 895, the Court of Appeal held that on an assignment, the original employer was a *constructive trustee* for the third party and as such could be made to sue on their behalf.
3 The use of *collateral warranties* by the construction industry. These create contractual rights for third parties, enabling them to sue for defective work in contract.
4 *Direct warranties* given by sub-contractors to the client, thus creating a direct contractual link with the client. Unlike collateral warranties, these create *additional rights* allowing the employer to sue the sub-contractor directly instead of proceeding via the contractor.
5 Note that collateral and direct warranties are also useful tools in overcoming the hazards of contractor insolvency since they create direct links with third parties.

Part-payment of a debt

This situation usually arises when one party to a contract is being asked to accept a smaller sum in payment of the debt owed. The difficulty with doing so is that there is no consideration to support the agreement to accept less. The rule at common law can be traced back to *Pinnel's Case* (1602) 5 Co Rep 117a. It was endorsed by the House of Lords in *Foakes* v. *Beer* (1884) 9 App Cas 605. The rule states that in the absence of an *accord and satisfaction* there is nothing to stop the party accepting the smaller sum, in settlement of a debt, claiming the rest at a later date. Lord Blackburn expressed his disagreement with the rule, stating:

> 'All men of business, whether merchants or tradesmen, do every day recognise and act on the ground that prompt payment of part of their demand may be more beneficial to them than it would be to insist on their rights and enforce payment of the whole. Even where the debtor is perfectly solvent, and sure to pay at last, this is often so. Where the credit of the debtor is doubtful it must be more so.'

There are *three forms* of settlement at common law:

(a) part payment of a smaller sum before the date due for payment;
(b) part payment at a different place at the creditor's request;
(c) part payment by a third party if accepted by the creditor in full settlement of the debt.

Equity provides certain exceptions to the rule in *Foakes* v. *Beer* (1884). Where a person promises not to insist on their strict legal rights, the promise can be used as defence when sued by the promisor. See *Central London Properties Ltd* v. *High Trees Ltd* [1947] KB 130.

The exceptions are as follows:

1 The circumstances in which the promise was made had changed. It would therefore be unfair to hold the promisor to their promise (see *High Trees Ltd*).
2 The promisor gives notice within a reasonable time of an intention to insist on his strict legal rights (see *Charles Rickard Ltd* v. *Openheim* [1950] 1 KB 616).
3 The promisor acts inequitably: see *D&C Builders* v. *Rees* [1965] 3 All ER 837. The plaintiff, a small builder, was owed £482 for completed work. For several months it was pressing for payment. Finally Mrs Rees, acting for her husband and knowing the plaintiffs were in financial difficulties, offered them £300 in final settlement. If they refused they would get nothing. They agreed and subsequently sought to recover the unpaid balance. The Court of Appeal gave judgment in their favour, agreeing that Mrs Rees had *acted inequitably* in taking advantage of their financial situation. The case would now be regarded as probably an example of *economic duress*. The county court decided it on the classical grounds of the absence of consideration. The Court of Appeal held itself bound by *Foakes* v. *Beer*. An agreement to accept payment of a lesser sum was not binding since there was no consideration for the promise to accept less.

Denning LJ argued that the defendants had simply acted inequitably by taking advantage of the builder's financial circumstances.

The renegotiation of contracts

The prevalence of the system of competitive bidding within construction contract means that one of the parties may well have *underpriced* the cost of the work. The question then arises, in what circumstances it permissible for one party to initiate discussions leading to the renegotiation of the contact? Of course it is possible for both parties to agree to vary their contracts (e.g., by a deed of variation). One party may agree to pay the other more money because they genuinely accept that the other has made a mistake to their benefit. If the enforceability of the agreement is challenged then the question arises as to *whether there is consideration* to support the new agreement.

This has traditionally been the approach of English law to the enforcement of agreements to pay additional money for the performance of existing obligations. In *Stilk* v. *Myrick* (1892) 2 Camp 317, two sailors were promised extra money by the captain if they sailed the boat back to England when two other members of the crew deserted. The sailors sued for the extra money when the captain refused to pay them.

Held: They gave nothing in return for the extra money. As sailors they had signed on to meet the normal emergencies of the voyage. All they were entitled to was their usual wages.

Had the sailors done *something* over and *above* what they had been paid to do, the outcome might have been different. The facts of *Hartley* v. *Ponsonby* (1857) 7 E & B 872 were similar to *Stilk*. However, the court found as a fact that so many of the crew had deserted that the whole nature of the voyage changed. The hazardous nature of the proposed journey discharged the existing contract. The crew was thus free to enter into a new contract that was enforceable.

Williams v. *Roffey Bros & Nicholls (Contractors) Ltd* (1990) 48 BLR 69 took the case of *Stilk* a step further. A sub-contract was entered into for the carrying out of carpentry work at a price of £20,000, to be paid in interim amounts as work was completed. Some £16,000 had been paid on the completion of work on the roof, nine flats and preliminary work on the others. More than 20 per cent of the work was still *outstanding* at this stage. The sub-contractor was in financial difficulties because (i) he had bid too low a price for the work and (ii) poor supervision of the work force had compounded the problem. The contractor became aware of the problems. Conscious of the liquidated damage clause in its own contract for late completion, it made an agreement with the sub-contractor to pay an extra £10,300 based on £575 per completed flat. The sub-contractor later stopped work and brought an action for the money outstanding.

Held: The promise made by the contractor to give the sub-contractor more money to complete the work was held to be binding. The *practical benefits* gained by the employer amounted to consideration for the promise. The only exception to this rule was where there was *fraud* or *economic duress*.

Practical benefits

In *Roffey Bros* counsel for Willams submitted that the following benefits were gained in return for the promise to pay more money.

1 Ensuring that the plaintiff continued work and did not stop in breach on contract.
2 Avoiding the penalty for delay. (Note: there was no penalty in *Roffey Bros*, only liquidated damages for delay.)
3 Avoiding the trouble and expense of engaging other people to complete the work.

Purchas LJ commented on benefit 1 above, saying:

> 'It was, however, open to the plaintiff to be in deliberate breach in order to "cut his losses" commercially. In normal circumstances the suggestion that a contracting party can rely on his *own breach to establish consideration* is distinctly *unattractive* . . . I consider that the modern approach to the question of consideration would be that where there are benefits derived by each party to a contract of variation even though one party did not suffer a detriment this would not be fatal to establishing of sufficient consideration to support the agreement.'

Glidewell LJ put it differently, saying the present state of the law can be expressed in the following propositions:

(a) if A has entered into a contract with B to do work for, or to supply goods or services to, B in return for payment: (a typical arrangement in construction contracts) and
(b) at some stage before A has completely performed his obligations under the contract B has reason to doubt whether A will, or will be able to, complete his side of the bargain; and
(c) B thereupon promises A an additional payment in return for A's promise to perform his contractual obligations on time: and
(d) as a result of giving A his promise, B obtains in practice a benefit, or obviates a disbenefit; and
(e) B's promise is not given as a result of *economic duress* or *fraud* on the part of A; then
(f) the benefit to B is capable of being consideration for B's promise, so that the promise will be legally binding.

Fraud

The fraud exception is quite difficult. It is likely to be difficult for a contracting party to go down that road because of the high level of proof required. It may be possible to argue that fraudulent misrepresentation induced the payment of the additional monies: for example, the sub-contractor was never in financial difficulties and used the extra money as additional profit, or for purposes which had nothing to do with the project, such as an attempt to defraud its shareholders by hiding the extra profit made.

Economic duress

It is clear from their judgment in *Roffey Bros* that the Court of Appeal did not consider that economic duress was present. *Occidental Worldwide* v. *Skibs A/S Avati, The 'Sibeon' and the 'Sibotre'* [1976] 1 Lloyds LR 293 (a High Court decision) was the first case to recognise that duress was no longer confined to the traditional categories of persons and goods. McMeel (1996) observes that the difficulty with that decision is that it fails to adequately distinguish between 'economic duress' and 'commercial pressure'. The Privy Council in *Pao On* v. *Lau Yui Ling* [1980] AC 614 (PC) was the first appellate court to recognise economic duress. There Lord Scarman said:

'American law . . . now recognises that a contract may be avoided on the ground of economic duress. The commercial pressure alleged to constitute such duress must, however, be such that the victim had no alternative course open to him, and must have been confronted with coercive acts of the party exerting the pressure . . . American judges pay great attention to such evidential matters as the effectiveness of alternative remedies available, the fact and absence of protest, the availability of independent advice, the benefit received, the speed with which the victim sought to avoid the contract.'

Lord Scarman further stressed that the pressure must be such that the victim's consent was not a voluntary act on his part, and that economic pressure was a factor which could render a contract voidable. To do that it had to be shown that the payment made or the contract entered into, was not a voluntary act.

Lord Diplock said in *The Universal Sentinel* [1983] 1 AC 366 (HL):

'Commercial pressure, of some degree, exists whenever one party to a commercial transaction is in a stronger bargaining position than the other party. It is not, however, in my view necessary, nor would it be appropriate in the instant appeal, to enter into the general circumstances, if any, in which commercial pressure, even though it amounts to a coercion of the will of a party in a weaker bargaining position, may be treated as legitimate and, accordingly, as not giving rise to any legal redress.'

The application of *economic duress* was demonstrated by *B&S Contracts and Design Ltd* v. *Victor Green Publications Ltd* [1984] 1 CLR 94 (CA). Workers employed by B&S went on strike to demand more money during the erection of stands at Olympia for a 5-day exhibition. Victor Green agreed to pay an additional £4,500 because B&S stated they could not complete the work without additional money up front for their striking workforce. When the work was complete Victor Green sent a cheque for the work minus the £4,500. B&S sued for the £4,500.

Held: The extra £4,500 was *procured by duress*.

Evelight LJ said

'There was here, as I understand the evidence, a veiled threat although there was no specific demand, and this conclusion is very much supported, as I see it, by Mr. Barnes' reaction, which must have been very apparent to Mr. French when Mr. Barnes said, "You have me over a barrel." On 18 April what was happening was this, Mr. French was in effect saying "We are not going on unless you are prepared to pay another £4500 in addition to the contract price" and it was made clear at that stage that there was no other way for Mr. Barnes to avoid the consequence that it would endure if the exhibition could not be held from his stands than by paying the £4500 to secure the workforce.'

The limits of economic duress were considered in *CTN Cash and Carry Ltd* v. *Gallaher Ltd* [1994] 4 All ER 714. The plaintiffs owned 'cash and carry' warehouses in six towns in Lancashire. The defendants, who supplied the plaintiffs with cigarettes under separate contracts made from time to time, were the sole distributors in England of 'Silk Cut' and 'Benson & Hedges'. As a result of an error by the defendants, a consignment of cigarettes was sent to the plaintiff's warehouse in Burnley, when in fact they were ordered for delivery to Preston. In Burnley they were stolen. The defendants, wrongly believing that the risk in the cigarettes had passed to the plaintiffs, demanded the price of £17,000 saying that otherwise they would not in the future grant credit to the plaintiffs. In fact the plaintiffs were not, and never were, entitled to payment for the goods. The plaintiffs paid for the stolen goods regarding it as the lesser of two evils.

Held: the plaintiffs were not entitled to restitution of the £17,000 on the grounds of economic duress. In a commercial context a plea of 'lawful act' duress only rarely succeeds.

Steyn LJ made three important clarifications as to the circumstances where duress may arise:

1 Common law rarely recognises the doctrine of inequality of bargaining power in commercial dealings; that the defendants were in a monopoly position cannot by itself convert what is not otherwise duress into duress.
2 The defendants could have refused to enter into contracts with the plaintiffs in the future if they so wished; there was nothing unlawful in that.
3 The defendants exerted commercial pressure in order to obtain a sum that they genuinely thought was due to them: threats to withdraw credit facilities in the future were commercial self-interest.

Construction cases

Recently there have been two construction cases in which economic duress was pleaded. In one case it was successful; in the other it was not.

In *Carillion Construction Limited* v. *Felix (UK) Ltd* (2001) CILL, HHJ Dyson QC adopted the principles he set out in *DSND Subsea Ltd* v. *Petroleum Geo-services* (2000) BLR 531 as an accurate statement of the law.

'The ingredients of actionable duress are that there must be pressure, (a) whose practical effect is that there is a *compulsion* on, or lack of practical choice for the victim (b) which is *illegitimate* and (c) which is a significant cause, *inducing* the claimant to enter into *a contract*: see *Universal Tankships of Malrovi* v. *ITWF* [1983] AC 336, 400B-E, and *The Evia Luck* [1992] 2AC 152, 165G. In determining whether there has been illegitimate pressure, the court takes into account a range of factors. These include whether there has been an actual or threatened breach of contract; whether the person alleging the pressure has acted in good or bad faith; whether the victim has any realistic practical alternative but to submit to the pressure; whether the victim protested at the time; and whether he sought to affirm or sought to rely on the contract. Illegitimate pressure must be distinguished from the rough and tumble of the pressures of commercial bargaining.'

In *Carillion*, Felix was the sub-contractor for the design, manufacture and supply of cladding, and Felix approached Carillion in order to agree the final account for their works. Now the final account is normally agreed *after* the work is carried out. A formal settlement was agreed in the sum of £3.2 million. Carillion then started legal proceedings to have the settlement set aside because it had been *procured by economic duress*. Felix had threatened not to supply the cladding units that were critical to the completion of the project. This would have caused *substantial delays*, making Carillion liable for liquidated and ascertained damages (LAD) of £75,000 per week.

Held: *Carillion* agreed to the settlement only because it was determined to secure the delivery of the units. The threat to withhold deliveries was a clear breach of contact. In the absence of the threat Carillion would not have agreed to the settlement. They had *no practical choice* since there were no viable alternative suppliers.

By contrast, in *DSND Subsea Ltd* v. *Petroleum Geo-services* the same judge decided that there had been *no duress*. A riser system was to be installed in a North Sea oilfield. Installation was originally planned for early summer, the peak season for work in the North Sea. In late July 1998 it was agreed that instead of installing risers pre-installation they would be installed post-installation. The arguments which arose between the parties concerned the risk of delay caused by bad weather as well as the cost of indemnity insurance. The judge decided that even if there had been illegitimate pressure Petroleum Geo-services affirmed the contract when they could have terminated it.

Renegotiation

As stated above, parties entering into construction contracts make mistakes during the bidding process. *Roffey* recognised that it may be in the *commercial interest* of one party to offer additional payment to the other to complete the work. It stressed that such a payment had to amount to a *benefit*. Duncan-Wallace (in the context of nomination, but it is submitted the position is the same for a domestic sub-contractor), described the effect of repudiation by a sub-contractor.

Where the specialist repudiates the contract it leaves the parties with:

(a) the question of *responsibility* for the unfinished work;
(b) increased *cost* resulting from such issues as inflation;
(c) the *cost of remedying* any defective work;
(d) consequent *disruption* of the main contractor's works programme, and so on.

It is further suggested that *'benefit'* might be given a much wider meaning than adopted by the Court of Appeal in *Roffey*. Hardheaded construction managers do not agree to pay more money to contractors or sub-contractors if there is *no bene-fit* to their companies in doing so. The only issue should be *why* the payment was made in the first place. If it was in the interest of the company paying it to do so, that promise should be enforceable. As Simpson said, 'a promise which lacks any adequate motive cannot have been serious, and therefore ought not to be taken seriously'. Only where the promise was extorted by illegal commercial pressure or where the payer in reality had no choice but to pay should the law of restitution be used to set aside bargains freely entered into. Note, however, the criticism of *Roffey* made by McKendrick. Contracting parties should be encouraged to bargain their way out of their difficulties. That is only part of the story, as contracting parties should also be held to their bargain (2000, p. 94). However, this critisism ignores the practical pressures of competitive tendering.

Estoppel

There are a number of *estoppels* which arise in certain circumstances. It is treated under the doctrine of consideration because the law in certain defined circumstances will enforce 'agreements' that arise despite the absence or exchange of value. In the context of adjudication proceedings, *estoppel* has become familiar to construction professionals. The three examples discussed in this chapter are promissory estoppel, proprietary estoppel and estoppel by convention.

Promissory estoppel: summary

Promissory estoppel has been described as an enforceable agreement that is not a contract. A may be prevented from going back on a promise not to rely on his legal rights against B, subject to certain conditions:

(a) B has relied on A's promise (possibly to B's detriment);
(b) B does not need to *furnish consideration* for A's promise for it to be enforceable under this doctrine;
(c) B must have relied on the promise, which suggests an element of acceptance of the benefit of the promise;
(d) There is no requirement that this reliance has been communicated to or is known by A (*Chitty*, 1999, p.5).

Where, however, a person promises not to insist on his strict legal rights, the promise can be used as defence when sued by the promisor: see *Central London Properties Ltd* v. *High Trees Ltd* [1947] KB 130. In September 1939 the plaintiffs had let a block of flats to the defendants. In January 1940 they agreed to accept half the rent, since many of the flats were not let because of wartime conditions. In 1945 the flats were full again and the plaintiffs claimed the full rent for the last two quarters. Denning J (as he was then) gave judgment for the plaintiffs. The agreement of 1940 had ceased to operate by the middle of 1945 because of a change in circumstances. However, he added that if the plaintiffs had tried to recover the full rent from 1940 to 1945 they would have failed. The court would not have allowed them to go back on their promise.

The importance of the decision in *High Trees Ltd* is that it does not create a right of action. Promissory estoppel is essentially defensive in nature. It would be unconscionable for the promisor to go back on his promise. *Chitty* considers the doctrine as primarily dealing with the renegotiations of contracts (1999, p.250). It may operate even though the promisee merely performs a pre-existing duty and suffers no detriment in the sense that he is doing something he was not previously bound to do. It is concerned only with the variation of rights arising out of a pre-existing legal relationship between promisor and promisee. Promissory estoppel arises only out of a representation or promise that is 'clear' or 'precise'.

Promissory estoppel and consideration

In *Attorney-General of Hong Kong* v. *Humphreys Estate* [1987]1 AC 114, the Privy Council found no estoppel could be created where the negotiations were carried out 'subject to contract'. The government negotiated with Humphreys for the exchange of 83 flats in return for a Crown lease of certain properties as well as the right to develop both them and some adjoining properties. The government took possession of the flats and spent substantial amounts on refurbishing them. Humphreys in turn took possession of the site, demolished existing buildings and paid the government $103,865,609, the amount agreed by the parties reflecting the difference in value between the two properties. Humphreys withdrew from the negotiations and sued to recover its property and the amount paid over to the government. The government claimed Humphreys was estopped from withdrawing from the negotiations. Rejecting their claim the Privy Council decided that they had been unable to show that:

(a) Humphreys had created or encouraged the belief or expectation on the part of the government that they would not withdraw from the agreement in principle; *and*
(b) that the government had relied on that belief or expectation.

Their Lordships observed (pp.127–8) that it was unlikely where the agreement was 'subject to contract' that a party would be able to satisfy a court that

(a) a contract had arisen; or
(b) some estoppel had arisen which prevented the parties from refusing to proceeding with the transaction they were engaged in.

In *Waltons Stores (Interstate) Ltd* v. *Maher* (1988) 1674 CLR 387, the High Court of Australia considered the *relationship* between promissory estoppel and consideration. The parties were negotiating for lease of commercial property. Part of the negotiation included the demolition of the existing building and the erection of a new one. Solicitors for the parties began discussions in November 1983. The respondent took possession of the site and started demolition work. The appellant required that plans and specifications to suit its purposes were prepared and that work was to be completed by 15 January 1984. Contracts were *never exchanged* and when the work was 40 per cent complete the appellants cancelled the project.

At first instance the judge found that the appellant was estopped from denying the existence of a contract. Damages were awarded in lieu of specific performance of the lease. The New South Wales Court of Appeal affirmed the decision. The appellant appealed to the High Court of Australia.

At the heart of this appeal was whether promissory estoppel could found a course of action. In the process the High Court considered most of the common law authorities. It drew attention to the reluctance of the courts to allow promissory estoppel to become a vehicle for the enforcement of representations by a party that he would do things in the future. In short what they were being asked to do was to enforce a pre-contract agreement where one party had suffered a *detriment* in *reliance* upon that proposed agreement. In deciding to enforce the agreement (by a majority) Mason CJ and Wilson J said at para. 38:

> 'It was unconscionable for it, knowing that the respondents were exposing themselves to detriment by acting upon the basis of a false assumption, to adopt a course of inaction which encouraged them in the course they had adopted. To express the point in the language of promissory estoppel the appellant is estopped in all the circumstances from retreating from the implied promise to complete the contract.'

Brannan J said at para. 27 that there were differences between a contract and an equity created by estoppel.

> 'A contractual obligation is created by agreement between the parties; an equity created by estoppel may be imposed irrespective of any agreement by the parties bound. A contractual obligation must be supported by consideration; an equity created by an estoppel need not be supported by what is, strictly speaking consideration.'

The broad scope of the doctrine outlined in *Waltons Stores* does not find any support in *Chitty* (1999, p.222). Commenting on this decision it considers the actual decision hard to reconcile with English authority on the non-enforceability of gratuitous promises even when relied on. It is inconsistent with the English doctrine of promissory estoppel which does not give rise to a course of action. This observation may well be correct. In *Baird Textile Holdings Ltd* v. *Marks and Spencer Plc* [2001] EWCA Civ 274, an argument based on *Waltons Stores* failed, the Court of Appeal observing that the obligation alleged was too uncertain to found an estoppel. See also the *obiter* comments of Mance LJ that the English courts would be unlikely to follow *Waltons Stores*.

Proprietary estoppel

This estoppel is concerned with promises relating to concerning interest in land. It differs from *promissory estoppel* in that it can be used to create a cause of action. There are two categories of circumstances where it may arise. One situation that arises is where A (the owner of land) stands by and allows B to expend money on improving land in the mistaken belief by B that he is improving his own land. The second is where B relies on A's promise (to his detriment) that he will be given an interest in the land.

Estoppel by convention

This estoppel arises where the parties act on a *common assumption* that certain facts concerning their relationship are true. In *William Oakley and others* v. *Airclear Environmental limited and ors* (2002) CILL 1824, the issue arose in connection with the enforcement of an *adjudicator's award*. For further discussion of the case see Chapter 16. The county court judge found that:

> 'there was no reason in principle why an estoppel by convention should not arise where two parties proceed under a mutual assumption which [has been communicated between them] that a code or codes of dispute resolution shall be available to resolve a dispute that has arisen between them.'

Etheron J, on appeal from the county court judgment, found the judge was entitled to come to that decision on the facts of the case. The parties had all along acted under *a common assumption* that the provisions of the JCT nominated sub-contract would govern their contractual relationship. These provisions included for the resolution of any dispute by adjudication. He held further that an *estoppel by convention* could only arise where in the circumstances it would be *unconscionable* for a party to deny facts which knowingly or unknowingly he had allowed the other to act on to his detriment. On the facts of the case he found no detriment since the only detriment incurred up to the point of challenge were the costs and expense of appointing an adjudicator. These costs he found were insubstantial. There was therefore insufficient evidence to show that it was unconscionable for the respondent to resile from their common held assumptions and the appeal was accordingly upheld.

The intention to create legal relationships

It is a presumption in construction contracts that they are intended to create legal relationships. It is only in situations where a 'letter of comfort' or 'heads of agreement' are drawn up that the question of whether the resulting agreement is enforceable may arise.

Capacity

All persons of sound mind and over 18 are held in law to be capable of entering into valid contracts. Companies and other bodies are also taken as having capacity. Since we are dealing with commercial contracts we can assume this element is present. Readers are referred to the many easily available books on contract law for further information.

Summary

1 Consideration is of great importance in construction. It affects the rights of third parties and makes the renegotiation of contracts more uncertain than it need be.
2 Consideration must not be past, except where the agreement is of a commercial nature.
3 Consideration need only have value; it need not be adequate. Privity of contract arises where a party has not given consideration for the premise it is trying to enforce.
4 Construction contracts are nearly always made for the benefit of third parties. Various devices have evolved to mitigate the effect of the rule, the most important being the use of collateral warranties and direct warranties.
5 *Roffey* decided that benefits could amount to consideration where there is no fraud or duress.
6 English law uses the doctrine of consideration to enforce agreements to renegotiate contracts. The modern approach is to consider the reason for the promise. If procured by duress it can be set aside because there is no contract. Agreements which fail form part of the law of restitution.
7 Estoppel is usually dealt with under the heading of consideration. This is because there is a class of agreements can be enforced despite the absence of consideration.
8 Estoppel by convention arises where both parties have acted on the basis that certain facts were true.

5 The Role of the Architect and the Engineer

Introduction

In the traditional contract the employer appoints a construction professional – either an architect or an engineer (A/E) – to carry out the design. In addition these professionals will also be involved in making applications for planning permission, preparing the tender documents, the selection of a tenderer and (where appointed agent of the employer) carrying out the task of inspecting and approving the works on behalf of the employer. The standard forms of contract therefore regulate in *great detail* the powers of construction professionals in administering the contract. Quantity surveyors will also be appointed to prepare cost estimates and carry out the valuation of work to be certified.

The general position is that that the contractor can 'carry out its building operations as it sees fit' and that the A/E has no authority to tell the contractor *how* to do its work. However, certain powers can be expressly reserved in the contract. For example, do the drawings and contract bills require any particular item to be done in a certain way? If so, is the right to require this expressly reserved for the A/E?

The *JCT 98* in clauses 4 and 13 gives the architect powers to issue instructions and to *vary* the work. The architect, however, has no powers or authority to do the following:

1 Vary the conditions of the contract.
2 Waive any condition of the contract.
3 Vary the whole nature of the work (e.g., order a single dwelling to be turned into the construction of a block of flats).
4 Probably the architect cannot omit work contained in the contract, in order to have it done by another contractor; there is no English authority on the matter.

Limits on those powers

In *Stockport MBC* v. *O'Reilly* [1978] 1 LLR 595, the dispute was referred to arbitration. A contract to build 105 houses, garages and ancillary work under a JCT 63 standard form of contract went sour and various disputes arose. The major issues were (i) the extent to which the contract was varied and whether those variations were authorised, and (ii) the termination of the contract. The arbitrator divided his conclusions into three groups:

(a) the faults of the employer;
(b) the faults of the architect as agent of the employer;
(c) the faults of the architect otherwise than as agent of the employer.

The arbitrator's award was set aside. The trial judge said:

> 'An architect's *ultra vires actions* do not saddle the employer with liability. The architect was not the employer's agent in that respect. He has no authority to vary the contract. If he does, the parties may agree and the contract varied. But the architect cannot saddle the employer with this.'

In *Canterbury Pipe Lines Ltd* v. *The Christchurch Drainage Board* (1979) 16 BLR 76, the engineer as agent misread the contract and acted unlawfully in failing to certify progress payments. The contractor suspended the work and the Board terminated the contract and appointed another contractor to complete the works. The engineer was held to have acted unfairly and the notice given by the Board under the contract held to be invalid.

Registration

An architect has to be registered otherwise he cannot practise, which means holding out for reward to act in a professional capacity in an activity which forms a material part of his business. The title 'architect' is protected by s. 20 of the Architects Act 1977. It provides that a person cannot practise or carry out any business 'under any name, style or title containing the word "architect" unless he is a person registered under [this] Act'. Section 20 (2) (1) permits the use the designation naval architect, landscape architect and golf-course architect. The Architects Act 1977 also created the Architects Registration Board. The Board is unique in that it is the only statutory registration body in the construction industry. Amongst its responsibilities is to hold, maintain and publish the UK Register of Architects and to regulate the use of the title 'architect' and to prescribe the qualifications required for entry to the Register. The Board may prosecute persons using the title when not registered and the courts may impose fines and costs.

In sharp contrast the term 'engineer' is not legally protected. The term in the standard forms of engineering contacts refers to the person carrying out similar duties to the architect in the standard form of building contracts. There is no legal requirement requiring registration or qualifications before practising or describing oneself as an 'engineer'. The term 'engineer' as used in the standard form of engineering contracts refers usually to a chartered engineer.

Position of the architect/engineer

Usually in a construction contract, the architect will enter into a contract with the employer using the RIBA standard form of engagement. In undertaking his obligations the architect has a duty to act fairly and professionally as agent of the employer. The parties contract on an understanding that where the architect has to apply his professional skill he will act fairly and in an unbiased manner in applying the terms of the contract. This applies not only to the issue of certificates but also where a professional opinion has to be formed.

Canterbury Pipe Lines decided that the engineer, although not bound to act judicially in the ordinary sense, was bound to act fairly and impartially, the test being an objective not a subjective one: would a reasonable contractor believe the engineer's conduct was unfair?

Agency

A professional appointed to supervise the construction contract on behalf of the employer enters into a relationship of principal and agent. Furmston (2001, p.523) describes agency as 'a comprehensive word used to describe the relationship that arises where one man is appointed to act as representative of another'. As a result, the employer may be vicariously liable for the torts committed by the agent: see now the case of *Jarvis* discussed above in chapter 2).

The authority of the agent

There are two ways in which this authority arises. One is the *actual* authority of the agent. This arises out of the contract between the principal and the agent. It can arise out of an express agreement or be implied. If express, it is because he has been given authority to do certain things, or, if implied, that will depend on the conduct of the parties, and the circumstances of the case. Another is *apparent* or ostensible authority. This is the authority of the agent as it appears to others. It was described as follows in *Freeman and Lockyer* v. *Buckhurst Park Properties (Mangal) Ltd* [1964] 2 QB 549 at p.503:

> 'It is . . . a legal relationship between the principal and the contractor created by a representation, made by the principal to the contractor, intended to be and in fact acted upon by the contractor, that the agent has authority to enter on behalf of the principal into a contract of a certain kind within the scope of the "apparent" authority, so as to render the principal liable to perform obligations imposed upon him under such a contract . . . It is irrelevant whether the agent had actual authority to enter such a contract.'

Where the architect acts outside of his powers, his actions do not bind the employer: see *Stockport*. Where such a breach of the warranty of authority occurs, the architect is personally liable. Amongst the faults identified in *Stockport* were the following actions carried out or initiated by the architect as agent of the employer:

(a) failing to provide certain drawings in due time or at all
(b) providing an inaccurate site plan
(c) refusing to allow the contractor to carry out the work in the order it chose to

Acting outside of the scope of his/her authority the architect gave instructions he/she was not empowered to do by the conditions of contract. For example, giving directions to the contractor to employ its workers on particular parts of the work.

Agent and certifier (contract administrator)

When appointed under a building contract, the architect or engineer (or, in the standard forms, the contract administrator) fulfils a dual role:

1 As agent of the employer in making sure that when the work has been completed the owner will have a building properly constructed and in accordance with the contract, plans, specification, drawings and any supplementary instructions which the architect will give during the course of construction. The aim of the employer in engaging an agent is to secure completion of the work in an economical and efficient manner.
2 Where the contract contains provisions that the work is to be executed to the satisfaction or reasonable satisfaction of the architect, the architect must act independently as between the employer and the contractor.

Parties contract on the understanding that where the architect has to apply his professional skill, he will act fairly and in an unbiased manner in applying the terms of the contract. This applies not only to the issue of certificates but also where a professional opinion has to be formed: for example, what amount has to be paid to the contractor and what additions or deductions need to be made from the contractual sum? How long should an extension of time for delays caused by the employer be?

The leading case on the two different roles is *Perini Corporation* v. *Commonwealth of Australia* (1969) 12 BLR 82. Under a building contract, the Director of Works, a government employee, had powers to give the contractor an extension of time 'if he thinks the cause sufficient . . . for such time as he thinks adequate'. On application for an extension time, the contractor's request was refused because of departmental policy.

Held: As certifier, the Director of Works had duties imposed upon him by the contract. He had *discretion* to extend time or to refuse it, and while he could take departmental policy into account, he should not regard it as controlling him. A term would be implied into the contract that his employer would not interfere with the Director's duties as certifier but also ensure that he carried out them out.

Judge Thayne Forbes QC summed up the position in *Davy Offshore Ltd* v. *Emerald Field Contracting Ltd* (1992) 55 BLR 22 at 60 where he said:

'I accept that an architect or engineer appointed to a contract is obliged to act fairly in the discharge of such of his duties under the contract as require him to use his professional skill and judgement in forming an opinion or making a decision where he is, in effect, "holding the balance between his client and the contractor": see for example the speech of Lord Reed in *Sutcliffe* v. *Thackrah* [1974] AC 727 at pages 736 to 737 and *Pacific Associates* v. *Baxter* (1988) 44 BLR 33. In my judgement, it is clear that the obligation to act fairly is concerned with those duties of the architect/engineer which require him to use his professional judgement in holding the balance between the client and the contractor. Such duties are those where the architect or engineer is obliged to make a decision or form an opinion which affects the rights of the parties to the contract, e.g., valuations of the work, ascertaining direct loss and expense, granting extensions of time, etc. When making decisions pursuant to his duties under the contract, the architect or engineer is obliged to act fairly.'

There has been a great deal of *criticism* of the dual role of the architect under a building contract. Where the architect has provided the design and has then to supervise its construction, there is potential conflict between the execution of that design and the certifying role. Other professionals have argued fiercely that they could carry out the certifying role independently and that they have the requisite knowledge and understanding of contract administration to do so. The JCT has accepted these arguments and in JCT 98 has provided an option for the employer to separate the role of architect from that of contract administrator. The civil engineering contracts have not adopted this approach.

The relationship between the architect/engineer and the contractor

There is no contractual relationship between the parties and the doctrine of privity of contract applies. Since the decision in *Murphy* the position is quite clear. The contractor suing an A/E in tort for losses made as a result of its actions under the contract would be suing for economic loss. Such losses cannot be recovered using the law of tort. *Pacific Associates Inc and anor* v. *Baxter and ors* (1988) 44 BLR 33, CA, established the same position before *Murphy*.

Pacific entered into a contract with the Ruler of Dubai for the dredging of a lagoon in the Persian Gulf. The engineer supplied the borehole information on which the tender was based. The contractor found hard materials in the lagoon and made claims for extensions of time and additional expenses from the engineer. These were rejected. Under clause 67 of the contract, disputes or differences had first to be referred to and settled by the engineer. A dissatisfied contractor may then take the dispute or difference to arbitration within time limits laid down by the contract. Out of an arbitration with the Ruler of Dubai, Pacific were awarded £10 million in full and final settlement. They then sued Halcrows (an English company employed by Baxter) for an additional £45 million, being the un-recovered balance with interest of its claim against the Ruler. The contract between the Pacific and the Ruler (under the international contract called the FIDIC condition) contained an exclusion clause that read:

'Neither any member of the employer's staff nor the engineer nor any of his staff, nor the engineer's representative shall in any way personally be liable for an acts or obligations under the contract, or be answerable for any default or omission on the part of the employer in the observance or performance of any of the acts matters or things which are herein contained.'

The judge held that Halcrows owed Pacific *no duty* of care in certifying or making decisions under clause 67 of the contract. Pacific appealed.

Held, dismissing the appeal: In considering whether a duty of care existed it was relevant to look at the circumstances; and these included the contract between the Ruler and Halcrow. There had been 'no voluntary assumption of responsibility' by Halcrow relied on by Pacific which was sufficient to give rise to a liability to Pacific for economic loss in circumstances in which:

(a) Pacific had tendered against a background of a complete contractual framework which included clause 86;
(b) there was an arbitration clause.

Note that Pacific is best understood as a *policy decision*. The courts were trying to reassert the difference between contract and tort by holding that where the *contractual framework* made provision for all contractual disputes, there could be no recovery of losses using the law of tort. (For a further discussion on the position of a professional in tort see Chapter 13.) Note too the unrealistic suggestions in the judgment that the contractor could arrange a contractual relationship with the engineer.

Many commentators suggest that *Pacific* should be confined to its particular facts. It is better to regard it a part of the underlying argument which later resulted in *D & F Estates* and subsequently led to *Murphy*. Hudson states that *Pacific* clearly means that there is no duty of care owed to contractors (1995, p.178).

Trotment suggests in *Pacific Associates Inc* v. *Baxter* – Time for a re-consideration, that the case should not be followed because of *three* recent developments.

1   The decision of the House of Lords in *Beaufort*. This case decided that an architect's certificate could be challenged in either arbitration or litigation.
2   Recent extensions of the duties of the *Hedley Byrne* v. *Heller* type. This will be examined in further detail later on in this book: see Chapter 13 on professional liability.
3   The recent review of concurrent duties in contract and tort. Since *Pacific* the law on duties existing in both tort and contract have been clarified. It is clear that professionals can be sued in contract and tort.

Pacific was considered by Australian Supreme Court of Victoria in *John Holland Construction and Engineering Ltd* v. *Majorca Products and ors* (2000) 16 Const. LJ 114. The court concluded that the architects owed no duty of care to the contractor. The New Zealand Supreme Court drew the same conclusion in *RM Turton & Co. Ltd (In Liquidation)* v. *Kerslake & Partners* [2000] NZCA 115. The contractor sued the engineer to recover its additional costs arising from a faulty specification contained in the invitation to tender. By a majority it decided that the contractual framework entered into by the parties excluded any liabilities of the *Hedley Byrne* v. *Heller* type arising. Support for Trotman's position is provided by Thomas J's dissenting speech in *RM Turton* and Anthony Speaight QC in his talk to the Society of Construction Law in February 2003. See the Society's website for 'I have no direct contract with the wrongdoer – who can I sue in 2003?'

The basic unfairness in *Pacific* was not that the decision was one of policy, but that the inaccurate information provided meant that the employer ended up with a cheaper job at the contractor's expense. Had the employer known of the hard materials, it may have been prepared at tender stage, to pay a much higher price for the work. Or it may have chosen to cancel the project. But the case got caught up in the retreat from suing in tort for economic loss, so that the question of whether any duty was indeed owed to the contractor was not considered.

Duties of supervision and inspection

When employing a construction professional the following question often arises: to what extent is the professional responsible for supervision and inspection of the contractor's work under the construction contract? It is clear that the professional owes a duty of care in issuing certificates, and in giving instructions to third parties, but how far does this duty extend? Where the client *does not* employ an architect, the contractor has a duty to warn that there are defects in the plans. This is illustrated by the Canadian case of *Brunswick Construction Ltd* v. *Nowlan* (1974) 21 BLR 27.

Nowlan employed an architect to design a house and then contracted with Brunswick to build the house in accordance with that design. The design was faulty as it made insufficient provision for ventilation of the roof space and timbers. The result was serious attack of rot. No architect was engaged to supervise the construction.

Held: Brunswick was liable for the defective work. Ritchie J said:

'In my opinion a contractor of this experience should have recognised the defects in the plans which were so obvious to the architect subsequently employed by Brunswick, and knowing the reliance which was being placed upon it, I think Brunswick was under a duty to warn Nowlan of the danger inherent in executing the architect's plans, having particular regard to the absence therein of any adequate provision for ventilation.'

A further case to note is *Clark* v. *Woorf* [1965] 1 WLR 650, where Walker, a clerk of works for a local authority, designed houses in his spare time. His associate, Woorf, who was a builder, drew up the specifications for the houses. Walker introduced the Clarks to the builder, and he agreed to build a bungalow for the Clarkes carrying out Walker's design. The contract stated that multi-coloured Dorking facing bricks should be used. Because they were not available at the time, Woorf substituted them with Orkney bricks but without informing his client. They were under the impression that Walker was looking after their interests. The bricks were underbaked and of poor quality and started to flake after eight years. The Clarks consulted architects who recommend rendering over the brick faces as being the most economical solution. The builder, on being sued, argued that the case was out of time. The Clarks replied that his concealment amounted to fraud and the limitation period only started when they discovered the faulty workmanship.

Held: In the circumstances the builder knew that the Clarks relied on him to treat them in a decent and honest way. Knowing he could not get the right bricks and finding himself with poor quality bricks he still used them. His behaviour was unconscionable and in the circumstances amounted to fraud.

The architect's contractual duty: supervision

Hudson has a clear view of what the position should be. He states:

'Attention has already been drawn at a number of points in this book to a modern tendency in all countries, no doubt encouraged by contractor interest to look to the architect/engineer as the *"captain of the ship"* and the person primarily responsible if seriously defective work is discovered. Current suggestions from the contracting side of the industry in the United Kingdom are that, if better standards of workmanship are required, clients should install *a* more elaborate and expensive system of *inspection and quality control*. This is submitted in an interesting argument.' (emphasis added).

He goes on to list a number of factors why the contention that the A/E is more qualified than the contractor in construction methods, is 'unreal' and should be disregarded:

1 Courts, arbitrators, owners and contract draftsmen should recognised that contractors and sub-contractors operate under strong commercial pressures to minimise expenditure on troublesome details of good practice and workmanship.
2 There is ample opportunity inherent in construction work for doing so undetected, which is afforded by the covering-up of work.
3 There is a long period before any indication of the defects are likely to emerge.

Hudson stresses that the supervisory staff of the contractor provides 'the only effective source of control.' The 'unreal' contention that the A/E is more highly qualified than the contractor in methods of construction, should be disregarded, and so should the idea that 'in the course of the visits he is likely to make, it will be possible for him to exercise effective control so as to secure detailed compliance.' (1995, p.349). Professor Lavers (www.scl.org.uk), in a paper given to the Society of Construction Law also addressed the architect's supervisory duties. He argues that for about a 100 years, the architect's duty under the RIBA Standard Form Conditions of Engagement (as they were once called) was that of supervision. The nature of the supervision required was considered in *Jameson* v. *Simon* (1899) IF Court of Sessions 1211. The architect designed, prepared plans and supervised the construction of a house. Two years later the house developed dry rot that was traced to the cement floor of the scullery. The cause of the defect was the use of miscellaneous site rubbish including wood, instead of the specified 'dry stone' in the foundations. The owner claimed that the poor workmanship would have been discovered had there been 'even the slightest supervision' of the work and claimed that the work had been negligently supervised. In his defence, the architect claimed that he supervised the work carefully and made regular site visits. Although the amount of money involved was small due to the importance of the issues, the experts were amongst the most eminent practitioners of their day. W.W. Robertson, Architect for Scotland to Her Majesty's Board of Building Works, claimed that the duty of the architect was general superintendence. This required occasional visits to see that 'his plans were being carried out and that the contract was fairly well fulfilled.' Mr Ormiston, Dean of the Guild of Architect's Edinburgh said: 'it was his duty to pay a little more than ordinary attention, seeing that there was no clerk of works. The risk of injury arising from bad or care-less bottoming is a very obvious one in a building.' Lord Killachy giving judge-ment at first instance rejecting the view that regular visits discharged the duty of supervision:

> 'I cannot assent to the suggestions that an architect undertaking and being handsomely paid for supervision, the limit of his duty is to pay occasional visits at longer or shorter intervals to the work, and paying those visits to assume that all is right which he does not observe to be wrong.'

The Appeal Judges sought to pitch the level of supervision at a level that could be practicably achieved on site and concluded that the duty could be satisfied in the following manner:

1 The architect or his representative should see the principal parts before they are hidden from view.
2 The contractor should be required to give notice before it is done.
3 Certificates should be based upon knowledge and not assumption.
4 He is not supposed to do all the inspection personally but can delegate it.
5 The architect, before work was covered up on which another contractor was then to build on, had to ascertain by personal inspection or through an assis-tant that the work was properly done

The architect was held liable for failing in his duty in not using reasonable care in ascertaining that the bottoming was properly done.

The 1966 edition of the RIBA Conditions, in its general conditions stated that the architect was required to give periodic supervision and inspection, but that constant supervision did not form part of his normal duties. It allowed for the employment of a resident architect where the need for constant supervi-sion is agreed. In the 1970s Professor Laver suggest there was a major revision of the RIBA conditions. References to *supervision* were removed and *inspection* became the keyword defining the architect's contractual undertaking. In support he cites its 1971 Conditions (1979) Rev. which required the architect to (i) make periodic visits to the site as he considers necessary (ii) inspect the progress and quality of the work and (iii) determine if in general the work is progressing in accordance with the contract conditions. In addition the archi-tect shall not be:

1 responsible for the contractor's operational methods, techniques, sequences or procedures,
2 the safety precautions in connection with the work

3 responsible for any failure by the contractor to carry out and complete the work in accordance with the terms of the building contract between the client and the contractor: see clauses 1.33 and 1.34.

In the 1992 revision to the Architect's condition of appointment, the duty of inspection was placed within the framework of the work-stages. Work stage K deals with operations on site and specifies the Architect's obligations. Clause 1.21 'Administer the terms of the building contract during operations on site' and clause 1.22 'visit the site as appropriate to inspect generally the progress and quality of the work'. He further notes that the Concise Oxford Dictionary defines the verb 'to supervise' as to: 'oversee, superintend execution or performance of actions or work of person' while to inspect is defined as to 'look closely into; examine officially'. The thrust of his argument is not that the duty has been altered but that significant parts of the construction industry and indeed construction law continues to operate as though supervision is unquestionably part of the normal duties offered by architects. Although he acknowledges that not all architects carry out work under SFA/92, even where a judge has to imply terms into an unwritten contract, he will still look at industry practice. If judges and commentators continue to believe that supervision is part of the services offered by architects, and architects are offering only inspection, there is a mismatch of expectations. And that no real debate has taken place on the distinction between the duties. In support of his arguments he cites the case of *William Tomkinson and Sons* v. *St Micheal in the Hamlet* (1990) 6 Const. LJ 319. The architect in that case was held not liable in respect of damage done by the contractor to a church organ by opening a hole in the roof. Clause 1.33 required inspection but did not extend to supervising the contractor's work.

The appointment of a clerk of works

Supervision by the clerk of works is different from that of the architect. 'The clerk of works has nothing to do *but supervise*; that is his only function in connection with the work. It would naturally be more continuous and more thorough than the architect who has other duties to attend to.'

In *Leicester Board of Guardians* v. *Trollope* (1911) 75 JP the relationship between the architect and the clerk of works was discussed further. The clerk of works allowed the contractor to build the ground floors without the specified damp-proof course. The architects relied upon the clerk of works and the defects were not detected. The architects were held liable on two grounds:

1 'The clerk of works has to see to matters of detail . . . The architect is not expected to do so . . . the architect is responsible for seeing that his design is carried out.'
2 The architect did not check whether the work was carried out or whether it was not.

Note that in this case the clerk of works fraudulently colluded with the contractors by allowing them to lay the ground floor timbers directly on the earth without the specified precautions against damp.

Butchery on quite a massive scale was uncovered in *Gray & Others* v. *T. P. Bennet & Sons & ors* (1987) 43 BLR 63, some 19 years after construction. The John Harrison House for student nurses was a ten-storey building constructed by McClaughlin & Harvey in 1962–3. Part of the design involved the use of reinforced concrete at the gable end. These panels required concrete nibs at floor level, from the first to tenth floor. These nibs were horizontal projections of reinforced concrete across the face of the panel, the sole purpose of which was to provide support for the brickwork at approximately 2.7 m up the face of the concrete panel.

To achieve satisfactory construction it was necessary for the contractors to construct the panels accurately vertically, so that the faces of the nibs were horizontal and 2.7 m apart. The bricks had to be laid accurately so that the concrete and brickwork would fit. Inaccurate setting out would mean a poor fit where the nibs met the brickwork. In 1979, 19 years after construction, as a result of a bulge appearing in the brickwork in one of the lower panels, the brickwork was opened up.

It revealed that that the concrete nib had been hacked back very hard to fit the brickwork. All the brickwork was opened up at nib level. About 90 per cent of the nibs had been hacked back or butchered, in some cases so severely that the steel reinforcement was sticking out.

Remedial work was carried out to re-form the nibs and replace the brickwork. The hospital sued the contractors and the architects for negligence. It also claimed that the clerk of works as agent for the architects had failed to stop the builders from drilling and hammering the nibs. They had also failed to inspect the nibs before the brickwork was erected.

Held:

1 The clerk of works although recommended by the architects was an employee of the Hospital and not their agent. A man of long experience and sound competence, he was regarded as a strict clerk of works and he had other jobs to inspect as well as this one. He claimed he had never heard the sound of loud and incessant hammering with kango hammers when he had been on site.
2 The architects were held not liable for not detecting the covering-up operation. The only explanation was that there was a deliberate cover-up on his frequent visits to the site.
3 There was no evidence that the contractor, the architect and the clerk of works had deliberately conspired to conceal the damage to the nibs.
4 The contractors were held liable for the faulty workmanship. Their deliberate concealment meant that the limitation period did not start until the hospital discovered it. The bulge in the brickwork was discovered in November 1979.

Negligent and overcertifying of work for payment

The leading case is *Sutcliffe* v. *Thackrah* (1974) AC 727. An architect knew of defective work by the contractor but gave no instructions to the quantity surveyor to make deductions from the interim certificates. Expert evidence was given claiming that interim certificates were limited to approximate value, and that it was quite usual for architects to postpone final decisions on defective work to a later date in case it might be remedied. The contractors became insolvent after the client had honoured the interim certificates.

Held, by the House of Lords: The architect was in breach of his contract with his client in not giving instructions for deductions from interim certificates and certifying unreduced sums.

In *Townsend* v. *Sims* (1984) 27 BLR 26 the architect deliberately did not take into account in a late certificate certain known defects, on the ground that they were sufficiently covered by the retention still held by the client under a JCT contract.

Held, by the Court of Appeal: That whether deliberate or careless, it was professional negligence to do so, since his duty was to supervise the contract according to its terms.

In *Sutcliffe* v. *Chippendale & Edminson (a firm)* (1971) 18 BLR 149 Sutcliffe employed the defendants to design a house, prepare tender documents and undertake duties under a building contract, JCT 1963. The contract was let to David Whitbank (Builders) Ltd. Work started but progress was slow and from time to time the defendant complained to the builders. Furthermore, as the work got closer to completion it was clear that most of it was defective. Sutcliffe issued interim certificates and in particular No. 9 (£2,620) and No. 10 (£1,837).

Eventually Sutcliffe lost confidence in the builders and terminated the contract. He employed another builder who remedied the work for £7,000. He was obliged to pay the builders interim certificates 9 and 10 and the builders became insolvent soon afterwards. Sutcliffe sued Chippendale for a failure to exercise proper skill, adequate supervision and overcertifying work.

Held:

1 The circumstances at the time justified the plaintiff determining the contract.
2 An architect's duty to supervise the work required him to follow the progress of the work and to take steps to see that the works complied with the general requirements of the contract including the specification and quality.
3 *Per curiam*: an architect was duty bound to notify the quantity surveyor in advance of any work that he classified as not properly executed so as to give the quantity surveyor the opportunity of excluding it from his valuation. An architect was required, when issuing interim certificates, to exclude work which was not properly executed from the value of the work for which he recommended his employer to make payment on account, and if the work was defective and unacceptable as it then stood, it was work not properly executed until the defect had been remedied.
4 An architect, in discharging his function of issuing interim certificates, is primarily acting in protection of his employer's interest, by determining what

payment he can properly make on account and therefore he was under a duty of care to his employer in the performance of that function.

Note that this case was called *Sutcliffe* v. *Thackrah* in the House of Lords.

Negligent certification was at issue in *Turner Page Music Ltd* v. *Torres Page Design Associates Ltd* (1997) CILL 1256 HHJ Hicks QC OR. Turner engaged Torres to provide design and other professional services in connection with the redevelopment of the Shepherds Bush Empire Cinema. The contract prices agreed with the contractor were £420,000 (later re-agreed at £450,000) for the building work itself. There was a separate contract for mechanical and electrical (M&E) work. Following completion, which was a week after the planned completion date, Torres issued a final certificate for £603,000.

This made the final cost for the work, including the M&E work, £827,000. Turner by then was in financial difficulties and, after various negotiations, sold the theatre to the Break for the Border Group for £2.85 million. Turner sued, complaining of a number of breaches of duty by Torres in carrying out their professional duties in designing, costing, letting, supervising and certifying the works carried out by the contractor.

Held: What was the appropriate measure of damages where the architect negligently certifies defective work? On the facts of this case, it was the *amount of overcertification*:

> 'It must be remembered that the defendant had duties of administration as architect and valuation as quantity surveyor as well as those of certifying officer. If the contractor were in breach of their contractual duty . . . of good workmanship, the initial duty of the administrator would normally be to require that the work be done or re-done properly. Even if defective work is accepted . . . not to value or certify it at more than it is properly worth, or to withhold certification of an appropriate part pending rectification.'

Estimates given by professionals

As part of their contractual duty, professionals will give estimates of *likely costs* in two circumstances:

(a) the costs of carrying out their own professional tasks;
(b) predicting the likely costs of work to be carried out by third parties.

Webster's Dictionary defines an estimate as arriving 'at an often accurate but usually only appropriate statement of the costs'. The Chartered Institute of Builders (CIOB) Code of Estimating Practice 1983 defines it as 'the technical process of predicting the costs of construction'. Where architects (which includes engineers) give cost estimates it is accepted practice that they will normally accept the cost projections made by quantity surveyors.

It is the normal practice in the UK for professional institutions to publish, and press their members to use, approved conditions of engagement. These provide for payment under authorised scales of percentages of costs, in addition to other terms of their employment.

If there is an express agreement between the client and the professional containing specific terms as to payment, those terms will regulate the right to payment. Where there is no express agreement, the client by requesting the work will be bound to pay a reasonable sum for the work. In assessing a reasonable sum the court takes into account a number of factors, rather than merely assessing on a time basis or on a percentage of total costs.

This approach is illustrated by *Brewin* v. *Chamberlain* (1949), mentioned by Emden at p.997, where an architect sued to recover his fees.

Birkett J stated that the principles to be applied are under three heads:

(a) the difficulties encountered in preparing the design and their solution;
(b) the professional standing of the particular architect concerned; and
(c) the amount of architectural work actually done by the particular architect.

The second area where professionals provide estimates of likely costs is where these estimates predict the *likely costs* of development. They enable clients to assess the commercial risk of undertakings and also the basis on which they will seek funding.

A leading case on cost estimates is *Nye Saunders and Partners (a firm)* v. *Alan E Bristow* (1987) BLR 92. Bristow retained Saunders in 1973 to prepare and submit a planning application for the proposed renovation of his mansion. He told Saunders he had about £250,000 to spend on the work. Saunders was asked for a written approximate estimate of the likely cost of the work. He in turn consulted Parker, a quantity surveyor, and then replied with a schedule of costs amounting to £238,000. This figure did not include anything for contingencies or for the likely increase due to inflation.

Planning permission was granted in March 1974 and Saunders was appointed as architects for the development. In carrying out architectural services it became clear to Saunders that that the cost of the works would be much higher than the estimate given. At a meeting in September 1974, a new figure of £440,000 was given which also contained for the first time an allowance for inflation over the 18-month construction period. Bristow then terminated the employment of Saunders.

When Saunders sued for outstanding fees amounting to £15,582 Bristow defended their non-payment on the basis that Saunders had failed in its duty to take due care in providing a reliable approximate estimate in 1974, and in particular in failing to draw attention to the fact that inflation would increase costs beyond the £250,000 Bristow had available.

The Official Referee dismissed the claim and Saunders appealed. The Court of Appeal also dismissed the appeal and held the following:

1 The cause of the massive increase in costs was the *inflation factor*, which led to the defendant to cancel the project, about which he had not been warned.
2 Since there was in 1974 *no practice* accepted as proper by a reasonable body of architects *that no warning* as to inflation *need be given* when providing an

approximate estimate of the costs of a project, the plaintiff were in breach of their duty to the defendant.

3   The plaintiffs had not discharged their duty by consulting a quantity surveyor and providing his estimate to the client.

What this case illustrates (commentary by editor of Building Law Reports BLR) is that

'When a person is asked to give an estimate (whether architect, surveyor or engineer) he must clearly indicate the extent to which the estimate is subject to variation. All estimates are 'approximate' and most clients are unable to appreciate the variations of which a professional person is presumably aware.'

Reference should also be made to the comments of Judge Slabb in *Aubrey Jacobs* v. *Gerrard* (1981) unreported where he said:

'In my judgement the quantity surveyor's duty is to estimate the cost at the current rates to make it plain that the estimate is given on that basis and increased costs are excluded.

If then he is asked of the likely increases, he can do it with such qualifications as he wishes to add, but I do not think he is under a duty to volunteer it.'

Architects or quantity surveyors, when asked to give estimates or 'approximate' estimates should make certain that the basis is clearly stated; does the estimate include for possible increases in costs and, if so, the extent to which it does.

Note that Saunders did not recover any fees for carrying out the work.

The case of *Copthorne Hotel (Newcastle) Limited* v. *Arup Associates and Another* (1996) 12 Const LJ 402 illustrates the difficulties of proving that the estimates were faulty and therefore negligently given.

Copthorne wanted to develop a hotel at Newcastle with the benefit of grants from the Tyne and Wear Development Corporation. The original design had been done by a local architect, Mr Chipchase, and priced by a local quantity surveyor, Mr Elliot, at a total cost of £11.6 million. In order to obtain planning permission and the approval of the Development Corporation, Copthorne was obliged to approach a national firm with an established reputation, Arup, to provide a multi-disciplinary service. Copthorne and Arup met in April 1988, and afterwards they produced a number of cost plans and estimates. The construction contract was let to Rush & Tomkins, who went into receivership during construction and the work was completed by Bovis Construction in February 1991. The total costs of the hotel were £21.2 million, of which construction costs were £15.2 million, as against a cost anticipated by Copthorne at £8 million. Copthorne then sued Arup claiming, *inter alia*, that they had given inaccurate cost estimates, had failed to design within costs advised and had failed to control the cost of the project.

The judge ordered a trial of a number of preliminary issues, including the following:

1   Liability, including the existence and scope of duty, what breaches of duty occurred, if any. Liability issues fell into two groups by date: April to July 1988 estimates; and piling estimates.

2 The measure of damages for breach of contract and in tort for the breach of a duty to take care.

Held:

1 Copthorne had failed to *prove* that Arup had agreed, at their first meeting, to design the hotel to the sum of £8 million or advised they could do so. Accordingly, Copthorne had failed to provide a foundation for the allegations of breach of contract or negligence.
2 The size of the gap between the cost estimate for piling of £425,000 and the successful tender of £903,000 was not in itself sufficient to prove negligence.
3 The cost estimate in May 1989 was given after Copthorne had let the construction contract and were therefore committed to the project.
4 In the absence of any finding of liability, the other matters did not fall to be considered.

Commentary

Copthorne's argument was that at a meeting in mid-April Arup confirmed that they could design for an overall cost of £12 million plus or minus 5 per cent with construction costs of approximately £8 million. Copthorne also alleged that Arup were aware that failure to design the hotel within budget and/or any underestimate of construction costs would result in the project being aborted or becoming unprofitable.

The judge found that at the first meeting in April the parties did not have a contractual relationship. HHJ Hicks QC said at p.204:

'Whatever was said then was not, therefore, professional advice given pursuant to a contract. The most natural analysis in my view, is that if anything of the kind alleged was said it was *a representation intended to induce the plaintiff to engage Arup*. That was not expressly pleaded, and if matters rested there it might be necessary to go into interesting but difficult questions such as whether a *common law duty of care and skill arises in such a situation by reason of the close relationship* between the parties and the potential *client's reliance* on what the professional person says.' (a reference to a *Hedley Byrne* v. *Heller* type of liability for negligent misstatement).

However, the judge found that it was not necessary to consider this in any case because by the end of July the parties were in a contractual relationship under which Arup owed professional duties.

There were five persons at the first meeting. One had died, and the other four gave evidence. Unfortunately for Copthorne there was *no contemporaneous record of the meeting*.

Although a number of records and explanations was produced concerning the significance of figures quoted the judge found that 'the evidence, therefore, in no way advanced the plaintiff's case that construction costs of £8m, or indeed any specific construction was put forward or accepted at the meeting'.

Copthorne also claimed that Arup were in breach of contract and negligent in preparing the cost estimate for the piling works as part of the construction cost estimate. It allowed £425,000 but should have allowed £930,000. Reasonably skilled and competent engineers and quantity surveyors would, and Arup should, have prepared an estimate of £930,000 based on the soil investigation report, the design parameters and the scheme drawings.

Tenders were received with sums varying between £711,000 and £1,058,000 and the tender of £897,000 by Pigott Foundations was accepted. Arup also recommended that £230,000 should be allowed in the budget for various associated contingencies including £160,000 for 'piling in connection with the medieval wall'. The final sum agreed for the piling works was £975,000. Copthorne claimed a lost opportunity of obtaining an increased level of funding by grant to accommodate the increased cost of piling: 'Every one was thunderstruck by the amount of the tender prices. There was some suspicion of the operation of a ring, but no evidence of that emerged. Some consideration was given to re-tendering, but in the event that was not done' (para. 64 at p.414).

Note that the invitation to tender documentation produced by Arup was 'sent on the basis of performance of design loads at point locations. *The tenderer was to propose the design solution'.* What this meant was that it was left it to the *tenderer to produce the specification of depth and diameter of piles.*

The expert evidence on both sides was given by quantity surveyors. The expert for Copthorne based his calculations on a solution proposed by his structural engineer and argued that an estimate of £930,000 would have been arrived at by a competent quantity surveyor. The judge was unimpressed and concluded that Copthorne produced no evidence as to what depth and diameter of piles a competent adviser, combining both engineering and quantity surveying responsibilities, would have allowed for on 10 September 1988.

Neither was he, by the expert called by Arup who gave evidence that a reasonable allowance for piling at 10 September 1988 would have been £456,300. His assumptions for depth and diameter were those apparently made by Arup at the time and described by the judge as 'equally devoid of admissible evidence support by anyone qualified to speak on the subject.'

The conclusions

Copthorne's main reliance seems to have been on the gap between the cost estimate and the successful tender. At para. 68 HHJ Hicks QC said:

'The gap was indeed enormous. It astonished and appalled the parties at the time and it astonishes me. I do not see, however, that that alone can see the plaintiff home . . . Culpable underestimation is of course one obvious explanation of such disparity but it is far from the only one. The successful tender was not the lowest. The contractor may have over specified from excessive caution, or to obtain a greater profit, or to suit the drilling equipment available or for some other reason. Market conditions may have changed or been subject to some distortion outside the knowledge or foresight of a reasonably competent professional adviser.'

The measure of damages in contract and tort

In the *Banque Bruxelles case* [1996] 3 WLR 87, Lord Hoffman emphasised the importance of the distinction between warranty and negligence. 'The measure of damages in an action for breach of duty to take care to provide accurate information must be distinguished from the measure of damages for breach of a warranty that the information is accurate'. Applying this distinction in *Copthorne* would have meant that, had a warranty been proven that the hotel could be built for £11.18 million, the plaintiff would have been entitled to the excess had it cost more. However, negligent advice to the same effect does not have the same result if the plaintiff cannot show that the result of sound advice would have been different from that which ensued in any case. If they were contractually committed to the building of the hotel, the fact that the estimate was inaccurate might not have made any difference to that decision. In *Turner Page* (the facts of the case was given earlier in this chapter under negligent certifying) the same difficulties of proof surfaced. The client, Turner sued complaining of a number of breaches of duty by Torres in carrying out their professional duties in designing, costing, letting, supervising and certifying the work carried out by the contractor.

Issues and finding

1 Had Torres given a warranty as to what the works would cost? The answer to this was no, since Turner had produced no evidence that such a warranty was given.
2 Was Turner entitled to recover in the tort of negligence? The answer was no since the measure of damages in tort is not the same as that for breach of warranty. Turner had provided no evidence as whether it would still have proceeded with the work had it known the eventual outcome.
3 Was Turner entitled to recover on the basis of an implied warranty that Torres would use skill and care to ensure that the work was accomplished for the budgeted sum? Again the answer was no because for such a claim to succeed it is not sufficient to point only to the eventual cost. Turner could not succeed without pleading and proving lack of due care and skill in specific respects and the subsequent loss.
4 Could Turner recover for Torres undoubted breach in failing to produce a Bill of Quantities? They could not do so in this case. No evidence was adduced that the absence of a Bill of Quantities caused any identifiable *loss*. The judge remarked:

> 'I now turn to the absence of bills of quantities. Item 4 of phase II of the defendant's fee proposal reads [we will] Produce a full written bill of quantities for the works. It is common knowledge that no bill of quantities, full or otherwise, was ever produced. No defensible justification for the failure to do so was advanced, the reason was in fact that the defendant, which had no employed quantity surveyor, was unwilling to incur the fee which the freelance quantity surveyor whom it had held out as one of its "team" proposed to charge. There was simply no evidence that the absence of a bill

of quantities caused any identifiable part of the losses claimed. It may well have made it easier to commit, or reduced the likelihood of avoiding other breaches . . . but it caused no recoverable damage of its own.'

Summary

1 The role of the A/E can extend to design, preparation of tenders, and the selection of a contractor and acting as contract administrator.

2 The standard forms of contract give extensive powers to the A/E in the administration of the contract.

3 In carrying out its tasks the A/E has a dual role. As agent in ensuring the work is carried out according to the contract and as certifier, acting fairly in carrying out the duties that require him to use his professional skill, in holding the balance between client and contractor.

4 There is no privity of contract between the contractor and A/E.

5 The case of *Pacific* indicates that the contractor cannot sue the A/E in tort for negligently carrying out his functions under the contract.

6 Lavers argues persuasively that the role of the architect has been changed to inspection rather than supervision. Unfortunately the courts, contractors and the public still regard architects as supervising the work of contractors.

7 Hudson is highly critical of the idea that the A/E is the 'captain of the ship'. Defective work is the responsibility of the contractor and is a function of its own supervisory practices.

8 The A/E is liable in negligence to the employer if he certifies defective work for payment or overvalues work when certifying.

9 In making estimates he must warn the employer of the basis of the figure and what has not been taken into account.

10 It is very difficult to show that estimates of anticipated construction costs were made without using reasonable care and skill.

6 The Main Obligations of the Contractor

The terms of the construction contract

Parties to construction contracts use the words 'terms' and 'conditions' inter-changeably. For example, the standard forms use the word 'conditions' to describe the *obligations* and *duties* of parties to the contract. Similarly an invitation to tender may be accompanied by the 'terms and conditions' of the offereror. There are a number of ways of classifying terms. For ease of explanation this book considers construction contracts as containing three types of terms. These are express terms, implied terms and statutory ones.

Express terms

Where the parties employ professional advisers the construction contract will usually be made using a standard form of contract. Norman Rice, in 'producing a standard form' Const. LJ (CLJ, p.257) considered that 'The 1980 Standard Form is a series of compromises which none of the parties concerned is likely to regard as satisfactory but if the result is looked at objectively and as a whole, it should represent a fair balance between conflicting interest.'

Standard forms of contract will regulate in *great detail* the obligations of the parties to the contract. This particular book refers primarily to the JCT 98 and the ICE 7th. A standard form of contract such as the JCT 98 contains the express terms of the contract. These terms have sometimes been called the private code of *law* of the parties: see, for instance, *Amalgamated Building Contractors* v. *Waltham Holy Cross UDC* [1952] 2 All ER 452. Duncan-Wallace QC has frequently criticised in great detail the many deficiencies in draftsmanship of the English standard forms of contract. For examples of these see, amongst other works, *Construction Contracts*: Principles and policies in Tort and Contract, ch. 29. It is not, however, intended to rehearse those arguments here. Generally the parties are free to include anything in their contract which is not illegal or prohibited by law, but this comment should be read as now subject to the requirements of the Housing Grants Construction and Regeneration Act 1996, part II (the HGCRA 96). The statute has fundamentally altered the *allocation of risk* in construction contracts that fall into its statutory definition.

The main obligations of the parties

In *Photo Production Ltd* v. *Securicor Transport Ltd* [1980] AC 827, the House of Lords distinguished the primary and secondary obligations of the parties. A failure to perform those primary obligations (i.e., the performance obligations in the contract) is a breach of contract. Lord Diplock said: 'The secondary obligation on the part of the contract breaker . . . is to pay monetary compensation to the other party for the loss sustained by him as a consequence of the breach.' The parties are *free to allocate* those secondary obligations. This may be done by including in their contract provisions for LAD for delay, inserting an exclusion or limiting clause or making provision for loss and/or expense to be paid in certain defined circumstances.

The *primary obligations* of the *contractor* are found in clause 1.5 of JCT 98: 'the contractor shall remain *wholly responsible for carrying out and completing the Works* in all respects in accordance with the Conditions'.

The Guidance Notes to Amendment 5:1988, which introduced clause 1.5, states that the architect is not, under the standard form, made responsible for the supervision of the works which the contractor is to carry out and complete; and nothing in the conditions of the standard form makes the contractor other than responsible for the carrying out and completing the works as stated in clause 2.1.

Clause 2.1 of JCT 98 requires that:

'The contractor shall upon and subject to the Conditions *carry out and complete the works* in compliance with the Contract Documents, using materials and workmanship of the quality and standards therein specified, provided that where and to the extent that approval of the quality or standards of the workmanship is a matter for the *opinion* of the Architect such quality or standards shall be to the reasonable satisfaction of the Architect.'

Carrying out and completing the works

The contractor is obliged to carry out the works in accordance with the contract documents which includes the contract drawings, bills, articles of agreement, conditions and appendix. In addition the contractor must, where required, carry out such work to the reasonable satisfaction of the architect. The role of the architect (and engineer) was discussed in Chapter 5.

Reasonable satisfaction

The case of *Cotton* v. *Wallis* [1995] 1 WLR 1168, CA is usually cited as support for the proposition that the architect may take into account whether the price is high or low in deciding whether he is reasonably satisfied. See also *Brown* v. *Gilbert Scott* (1994) 35 Con. LR 120.

In *Cotton* v. *Wallis* a house was built using a standard RIBA form. The contract provided that the work should be carried out to the reasonable satisfaction of the architect. A number of defects were present after the completion of the work. After the defects liability period the architect instructed the contractor to put right the defects. Satisfied with the work he wrote to the owner asking him to pay the contractor's outstanding account. After the contract was completed the architect sued to recover his fees. The owner counter-claimed that the architect failed to use reasonable care and skill in supervising the construction.

By a majority the Court of Appeal held that where the work was to be carried out to the reasonable satisfaction of the architect, account could be taken of the price paid for the work. In approving the work he had not been in breach of his duties of reasonable care and skill. Evidence had been provided that the builder had 'built down to a price' and the defects were of a trifling kind. Plastering and paint work needed additional coats. The tiles in the dining room and kitchen were marked by paraffin stains dropped by the contractor's workmen. The court stressed that where the contractor had agreed to build to a high standard at a low price he would be held to his bargain.

The main obligation of the contractor under ICE 7th is contained in clause 8(1) which requires that

'The contractor shall subject to the provisions of the contract
(a) construct and complete the Works and
(b) provide all labour materials Contractor's Equipment Temporary Works transport to and from and in and about the Site and everything whether of a temporary or permanent nature required in and for such construction and completion so far as necessary for providing the same is specified in or reasonably to be inferred from the contract.'

Eggleston (2001) observes that this clause adds little to the *general obligation* of the contractor to complete the work; only the proviso that it is 'subject to other provisions of the contract' serves as a reminder that other clauses in the contract may qualify the general obligation. Examples are clause 42(2), suspension lasting more than three months and clause 64, default of the employer.

Design liability and the standard forms

Emden states that 'it is thought that provided the contractor carries out work strictly in according with the contract documents, he is not responsible if the works prove to be unsuitable for the purpose which the employer and architect had in mind' (Binder VI para 41). Where the contractor discovers a *divergence* between the statutory provisions and the works, clause 6.1.2 of JCT 98 requires him to report that to the architect. In the same manner discrepancies between the various documents of which the contractor has knowledge have to be reported as well (clause 2.3). In the light of *Plant* (2001), discussed later in this chapter, there is probably an implied duty to bring *obvious design* faults, of which the contractor has knowledge, to the attention of the architect.

The ICE 7th specifically allocates design responsibility. However, the design liability of the contractor is limited to *reasonable care and skill* (this is discussed further in ch 13). Design responsibility is allocated by clause 8(2):

'The contractor shall not be responsible for the design or specification of the Permanent Works or any part thereof (except as may be expressly provided in the Contract) or any of the Temporary Works designed by the engineer. The contractor shall exercise all reasonable care and skill or diligence in designing any part of the Permanent Works for which he is responsible.'

Where the contractor does carry out design it is important to stress that where liability is limited to reasonable care and skill, there is no obligation of fitness for purpose. Care therefore needs to be taken when allocating design liability as part of the specification. See the problems encountered in *Shanks & McEwan (Contractors) Ltd* v. *Strathclyde Regional Council* (1994) CILL 916.

Designing a standard form

The drafter of the contract starts with the existing common law and practice. The existing law can be modified or excluded altogether. Does the contract need an express term or will the law imply it into the contract? Is it wise to leave the position unclear? Since the common law results from the accidents of litigation, there will often not be any case law about certain issues: for example, can the architect give part of the work to another contractor? In Australia and Canada the courts have said the architect cannot do so. The answer in the UK is probably that the architect cannot do so. There is nothing, however, preventing the employer from putting an express clause in the contract, reserving the right to award certain work to other contractors.

Other express terms may modify the position at common law. A provision that has been much criticised is that contained in clause 2.2.1 of the JCT 98. It provides that 'nothing contained in the contract bills shall override or modify the application or interpretation of that which is contained in the articles of agreement, the conditions or the appendix'. This reverses the ordinary rule of interpretation that *written words prevail over printed* words in standard form contracts. In *English Industrial Estates Corporation* v. *George Wimpey & Co. Ltd* (1972) 7 BLR 122, Stephenson LJ suggested at p.137 that building owners using the then RIBA contract containing the identical clause recommended that the clause be either struck or amended to reflect the position at common law.

In *Beaufort Developments (NI) Ltd* v. *Gilbert-Ash* [1998] 2 All ER 778, Lord Hoffman said at p.784 that legal documents often contain superfluous words. He gave two reasons for this: (i) clumsy draftsmanship and (ii) the lawyer's desire to cover every conceivable point. Of the *JCT* standard form of contract he said the following:

1 Since it was periodically renegotiated, amended, and added to over many years, it was unreasonable to expect there to be no redundancies or loose ends.
2 Careful examination of the contract as a whole is required to discover its 'true meaning'.
3 Regard should of course be taken of earlier judicial authority and practice on the construction of similar contracts.
4 The evolution of the standard forms of contract reflected the interaction between the draftsmen and the court. The efforts of the draftsmen cannot be properly understood without reference to the meaning that the judges gave to language used by their predecessors.

Uff (1993) suggested that the draftsmen should produce a document setting out the stated intentions of the drafting body.

The nature of implied terms

Where the parties have used a standard form of contract or have drafted their own, the express terms may not contain the whole of the agreement. Another common situation that often arises is that although a contract is found to exist, the terms of that contract need to be implied (see, for example, the cases discussed in Chapter 3). This work classifies implied terms as being of three types: those implied *in fact*, in *law* and terms *imposed by statute*. Some of the laws concerning implied terms in construction contracts have been codified by the Supply of Goods and Services Act 1982. Part II states:

> 'Section 4.2(A)
> For the purposes of this section and s. 5 below, goods are of satisfactory quality if they meet the standards that a reasonable person would regard as satisfactory, taking into account the price (if relevant) and all other the other relevant circumstances.
>
> Section 4 (5)
> In that case [a transfer in goods in the course of business s 4(4)] above there is (subject to subsection 6 below) an implied condition that the goods supplied under the contract are reasonably fit for that purpose, whether or not that is the purpose for which such goods are normally supplied.
>
> Section 4(6)
> Subsection (5) does not apply where the circumstances show that the transferee does not rely, or that it is unreasonable for him to rely, on the skill and judgement of the transferor (or credit broker).'

However, these clauses do not advance the common law position in any significant way. The Act also contains uncertainties (e.g., warranties are not accurately distinguished from conditions). The result is that it has not been greatly used and is often ignored in favour of the common law principles. There have recently been some cases on the Act but they have not added much to the existing position at common law. See for instance *QV Limited* v. *Frederick F. Smith and Others* [1998] CILL 1403. In that case, s.13 – the implied term about *care and skill* – was applied to a design prepared by the project manager, which proved to be defective when built. The section states that in: 'a contract for the supply of a service where the supplier is acting in the course of a business, there is an implied term that the supplier will carry it out with reasonable care and skill'. The judge applied the test set out in *Bolam* (see Chapter 13). In not following the guidelines produced by the manufacturer, the designer was found to have been professionally negligent. The contractor, who was sued under the implied terms of the Supply of Goods and Services Act 1982 for providing defective work, and the supplier of the materials, who was sued under the Sale of Goods Act 1979, were both held *not liable*.

In *BP Refinery (Westport)* v. *Shire of Hastings* [1978] 52 ALJR 20 PC, Lord Hastings summarised at p.26 the conditions to be satisfied for a term to be implied:

(a) it **must** be reasonable and equitable;
(b) it **must** be necessary to give business efficacy to the contract, so that no term will be implied if the contract is effective without it;
(c) it **must** be so obvious that 'it goes without saying';
(d) it **must** be capable of clear expression;
(e) it does not contradict any express term of the contract.

(a) Terms implied in fact

The question to be asked is whether the parties would have *agreed* the term? Any ruling made by the court only applies to *this* particular contract. This is because the decision is based on the *presumed intention* of these particular parties (see, for instance, *Greaves & Co. (Contractors) Ltd* v. *Baynham Meikle & Partners* (1975) 4 BLR 56). Both parties were found to be in agreement that the engineers had promised to produce a result. Giving judgment Lord Denning said that: 'the evidence showed that both parties were of one mind on the matter. Their common intention was that the engineer should design a warehouse fit for the purpose for which it was required. That common intention gives rise to a term to be implied in fact.' Sometimes the term contended for may be what is termed 'the *usual* terms'. These are readily implied provided that they are not inconsistent with the express terms. An example of such a term is a contract where the contractor builds dwelling houses for sale. It will be implied that they will be fit for human habitation.

In other cases the argument is that the contract lacks *business efficacy*. The court is asked to imply a term to make the contract work. Without it the contract may be inefficacious, futile and absurd. An example where such an argument failed is provided by *Trollope & Colls* v. *North-West Metropolitan Regional Hospital Board* (1973) 9 BLR 60, [1973] 1 WLR 601, [1973] 2 All ER 260.

A hospital extension was to be built in three phases. Phase 3 was to start 6 months after the completion of Phase 1. Phase 3 had a fixed completion date. Phase 1 was completed late but the contractor had been given an extension of time. The contractor argued that there was an *obvious* term in the contract that the completion date for Phase 3 should be similarly extended. The House of Lords *held* that that the express terms were clear and workable. The parties must have known the risks in their arrangement. No further term would therefore be implied in their contract. Lord Pearson said at p.609:

'Function [of the court] is to interpret and apply the contract which the parties have made for themselves. If the express terms are clear and free from ambiguity there is no choice between different possible meanings. The clear terms must be applied even if the court thinks some other term would have been more suitable. Non expressed terms can be implied if and only if the parties intended that term to form part of their contract. It is not enough for the court to find that such a term would have adopted by the parties as reasonable men if it had been suggested to them: it must have been a term that went without saying, a term necessary to give business efficacy to the contract, a term which, though tacit, formed a part of the contract which the parties made for themselves.'

(b) Terms implied in law

The essence of a term implied in law is that the parties would *never* have agreed to it. Its application is not the result of any agreement by the parties but imposed by the general law. Some examples are:

- the duty of reasonable care and skill care owed in contract.
- the duty of competence of an employee.
- a landlord remains responsible for the common parts of building that it has retained.

(c) Statutory terms

The following are examples of statutory material discussed in this book.

- The Sale of Goods Act 1979 as amended
- The Supply of Goods and Services Act 1982
- The Unfair Contract Terms Act 1977
- The Housing Grants Construction and Regeneration Act 1996 and the Scheme for Construction Contracts (England and Wales) Regulations 1998 SI No. 649.
- The Defective Premises Act 1984
- The Contracts (Rights of Third Parties) Act 1999
- The Limitation Act 1980 and the Latent Damage Act 1986.

Some of these cannot be contracted out of whilst others can be excluded. The parties can decide that the Contracts (Rights of Third Parties) Act 1999 will not apply to their contract. They can alter the limitation periods by agreement but cannot contract out of the Housing Grants Construction and Regeneration Act 1996. They are, however, free to modify its provisions.

The construction process

Numerous issues may arise for resolution during the construction of the work. This section deals with those issues that arise as a consequence of construction activities and the standards to be applied to the work being carried out.

Possession

It is an implied term that the employer will give possession of the site in a reasonable time to allow the contractor to complete by the contractual date. Most contracts provide expressly for possession (see Chapter 7). The contractor usually has a licence to occupy the site. This licence normally ends on completion of the work. Under the Occupiers Liability Act 1957, the occupier owes a duty of care to his *lawful visitors*. The contractor also owes certain duties at common law to its employees. These include the provision of adequate supervision, materials, training, and a safe system of work. Visitors to the site without invitation are owed duties under the Occupiers Liability Act 1984.

The contractor as occupier must have some degree of control and supervision of the site or premises. *Wheat* v. *Lacon & Co.* [1966] AC 552 established that it is possible for there to be more than one occupier of the site. Where refurbishment work is carried out in an occupied building, the contractor and the employer will both be 'occupiers' of the premises. The duty of the occupier is to protect *lawful visitors* from *dangers caused by the physical defects* in the condition of the premises. 'Premises' has been widely interpreted and includes land, buildings, fixed and moveable structures such as ships, pontoons and even ladders. In *Bunker* v. *Charles Brand* [1969] 2 QB 480, a cutting machine used by the contractor for the tunnelling of the Victoria underground line was held to constitute premises for the purposes of the act.

The occupier owes a *common duty of care* to his lawful visitors. Section 1 (2) of the Occupiers Liability Act 1957 provides that these include invitees, licensees, those with contractual rights to enter and persons given that right by the operation of law. The occupier has to be especially aware of children. Their presence on the site will usually involve the application of the Occupiers Liability Act 1984. Prior to the passing of the act the *doctrine of allurement* was used to turn child trespassers into lawful visitors since otherwise they had no rights to compensation on being injured by the defects in the premises. The doctrine was applied recently in *Jolley* v. *Sutton London Borough Council* [2000] 1 WLR 1083. The council was held liable under Occupiers Liability Act 1957 for injuries sustained by a teenage boy. An abandoned boat left lying in the grounds of a block of flats collapsed on to the boy while he was attempting to repair and repaint it. On construction sites the concept is not so important in that the protection provided by the Occupiers Liability Act 1984 is quite strict with regard to children. The Act was passed to provide a clearer application of the decision of the House of Lords in *British Railways Board* v. *Herrington* [1972] AC 877. There it was held that the occupier owed the trespasser a 'common duty of humanity'. Section 1(a) of the act states that it applies to persons other than visitors. Section 1 (3) provides that the common duty of care is only owed if:

(a) the occupier is aware of the danger or has reasonable grounds to believe it exists;
(b) the occupier knows or has reasonable grounds to believe that the (trespasser) is in the vicinity of the danger concerned, or may come into the vicinity;
(c) the risk is one from which, in the circumstances of the case, the occupier may reasonably be expected to offer the (trespasser) some protection.

The Act effectively imposes a duty on the contractor to fence the construction site and ensure that children are unable to enter it. In that respect the duty is as strict as that provided by the 1957 Act with regard to child trespassers.

With regard to adult trespassers, the House of Lords in *Tomlinson (FC)* v. *Congleton Borough Council and ors* [2003] UKHL 47 reviewed the scope of the duty. Adults are now held responsible for their own actions in circumstances where it is clear that a known danger exists.

The contract at common law

Where the parties do not make express provisions for matters such as design, quality, workmanship or payment terms the common law will imply such terms.

Design

In the traditional form of building contract the employer and his professional designers assume responsibility for the design of the building and the contractor agrees to build what has been designed. In such a contract the contractor does not undertake liability for design. See *Cable (1956) Ltd* v. *Hutcherson Brothers Pty Ltd* (1969) 43 ALJR 321. The High Court of Australia held that the obligation of the contractor was to carry out the work agreed in the drawings. Under a *JCT* standard form of building contact (JCT 80/98) the contractor has no express liability for design. Under the ICE forms the contractor has design liability for temporary works. Where the standard form of building contract does impose a design liability the difficulties are illustrated by *John Mowlem & Co* v. *British Insulated Callenders Pension Trusts Ltd* (1970) 3 Con LR 63. The question for the court was whether the contractor had any design responsibility. Mowlem contracted to build a warehouse and office block for British Insulated Callenders Pension Trusts Ltd (BICP) under a JCT 80 contract. Jampel were the consulting structural engineers and carried out the design of the reinforced floor slabs in the basement area.

The bills of quantities read:

'watertight construction –
Retaining walls forming external walls to the building and basement slabs are to be constructed so that they are impervious to water and damp penetration, and the contractor is responsible for maintaining these in this condition.'

It adds:

'the Engineer does not require a mix richer than 1–2–4 as specified previously. The position and the arrangements of the construction joints shall be agreed between the engineer and the contractor, and how the construction joints shall be constructed, and the watertight work shall be in concrete lengths not exceeding 25 feet, and adjoining sections shall only be cast after a time interval of ten days.'

The basement floor developed cracks and water penetration, necessitating expenditure on remedial work and resulting in loss and damage from delay in completion and occupation.

At trial two issues had to be determined:

1 Was the defect the result of inadequate design and/or faulty workmanship?
2 If the lack of watertightness was due to a design failure, was Mowlem in any case liable because of the performance specification in the bills of quantities?

Held: On the evidence and in the circumstances of the failure, Jampel, in designing the basement slab, failed to exercise reasonable care and skill and the failure was due to inadequate design. *Design was no part of the contractor's obligation* and so far as the performance specification attempted to place such liability on them, clause 2.2.1. of JCT 80 made this ineffective. Judge William Slabb QC observed in his judgment:

> 'I should require the clearest possible contractual conditions before I should feel driven to find a contractor liable for a fault in design, design being a matter which the structural engineer alone is qualified to carry out and which he is paid to undertake and over which the contractor has no control.
>
> I agree that the construction that [Jampel] contends for, places the contractor in an impossible position. He cannot (a) alter a faulty design without being in breach of contract, for the fault in design is not, in my view, a discrepancy or divergence between the contract drawings and/or the bill of quantities, and yet (b) if he complies with the design he would still be in breach.
>
> I decline to hold that the specification in the bill of quantities makes the contractor liable for the mistakes of the engineer.'

Note that in practice the contractor's obligation as to materials and workmanship may include a measure of liability for what can be called *'second order design'* (i.e., detailed decisions as to the suitability of materials and methods of construction). The contractor may expressly assume liability for design especially under *'package deal'*, *'turnkey'* or *'design and build'* contracts.

Materials and workmanship

In any building contract the contractor impliedly agrees:

(a) to do the work in a good and workmanlike manner;
(b) to supply good and proper materials; and
(c) (where the contract is to build a house) that the house will be reasonably fit for human habitation.

Progress and completion

Normally the contractor undertakes to complete the work in its entirety. If the contract is an *entire* one, the contractor who does not substantially complete will not be entitled to payment. The practical effect of this rule is often modified by the provision for payment against certificates. The risk for a contractor who stopped work without good reason is shown by *Lubenham Fidelities* v. *South Pembrokeshire District Council and Wigley Fox Parts* (1986) 33 BLR 39, where the contractor suspended work because the certificate issued by the architect was incorrect. The court held that the only right under JCT 63 (now JCT 98) was to have the certificate corrected in the next interim certificate and, failing that, to go to arbitration. By suspending work the contractor was in breach of contract. *No right to suspend work for non-payment existed in English law*. In the light of HGCRA 96, part II, *both* of these *propositions are wrong* as regards contracts falling within the legislation.

1 There is now a *statutory right to interim payments* and if the contract does not contain such a provision, the Scheme for Construction Contracts (a statutory instrument) applies.
2 The contractor has a *statutory right to suspend work* for non-payment.

Examples of no implication

Below are listed some examples of no implication:

1 In a lump sum contract there is no implication that the contractor will be paid a reasonable sum. The contractor will be paid for the contract entered into.
2 There is no implication that VAT will be paid.
3 If an architect is appointed there is no implication that the architect will issue interim certificates of payment. Certificates will only be issued if the contract says so. See now the HGCRA 96.
4 There is no warranty given that the work of the architect or employer will be fit for its intended purpose.

The level of competence of the contractor

The contractor has a duty of reasonable care and skill in carrying out the work. The level of skill required is that of an *ordinary competent contractor*. Part of this competence is whether the contractor has a duty to warn the employer that (i) the design is defective and (ii) that the material specified is not suitable for its purpose.

The contractor's duty to warn

Where the employer carries out the design, the contractor's only duty is to build what has been designed. There is no requirement to warn the employer that the design is faulty. However, in *Lindenberg* v. *Joe Canning and others* (1992) 62 BLR 147 it was held that a reasonably competent contractor would have warned the employer that the plan was defective.

Conversion work was carried out in the basement of a large block of flats in Hyde Park Gate, West London. The work included the demolition of a wall including a chimneybreast. The drawings were prepared by a building surveyor employed by the employer, and they showed that the wall was non-load bearing. Relying on this information, the contractor began demolition work without any temporary propping for the ceiling. The wall proved to be load bearing and substantial deflection occurred in the upper floors. The developer settled a claim by the owner of damages amounting to £390,000.

In this case the walls to be demolished were 9 inches thick. HHJ Newey QC concluded that the builder should have realised that 9-inch walls cost more than $4\frac{1}{2}$ inch walls and were not erected simply to separate stores. He should have also been aware of the steel joists in the ceiling and proceeded with extreme caution. Acting with less care than expected from an ordinary competent builder, he was in breach of contract.

The builder's defence was that he had carried out the work in accordance with instructions given on the drawings was rejected. The court held the following:

1 The walls were load bearing and props were needed before their removal.
2 It was an implied term of the contract that the builder would exercise the reasonable care expected of an ordinary competent contractor.

3 The plan was inadequate and the builder at fault for taking no precautions whatsoever.
4 The building surveyor was liable for 75 per cent of the damage and the builder for the remaining 25 per cent.

This decision can be explained on the grounds that the contract contained no express terms dealing with the matter or excluding it. *Lindenberg* was, however, approved by the Court of Appeal in *Plant Construction* v. *Clive Adams and JHM Construction Ltd* [2000] BLR 137. Plant contracted with Ford Motor Company to construct pits for engine mounts at their research and engineering centre. JHM were sub-contracted to dig the pits. One pit required the removal of part of a concrete base of a stanchion that supported the roof of the centre. This required: (i) underpinning of the stanchion and (ii) provision of temporary support for the stanchion and the roof. JHM were responsible for the design of the temporary support. Ford's in-house engineer vetoed their suggested design. The engineer issued instructions dictating the manner of support to be adopted. The method of support – acrow props – proved to be inadequate and the roof collapsed.

The appeal raised the following issues:

1 Did JHM have a *duty to warn* that the system proposed by Ford's engineer was inadequate?
2 What *steps* should they have taken in the face of the instructions to proceed with an inadequate system?

Held: JHM had a duty to warn Plant that the propping was inadequate, as part of its implied contractual duty of reasonable care and skill. To discharge this duty they should have *protested more vigorously*. Giving judgement it was reiterated that:

> 'Any analysis of implied terms in a building contract must start with a proper account of its express terms. Subject to the express terms, there will normally be an implied term that the contractor will perform his contract with the skill and care of a competent ordinary contractor in the circumstances of the actual contractor.'

The position is also clear where the client does *not employ an architect*. The contractor has a duty to warn that there are defects in the plans. This is illustrated by the Canadian case of *Brunswick Construction Ltd v. Nowlan* (1974) 21 BLR 27 (discussed in Chapter 5).

The supply of materials

The leading case of *Young & Marten* v. *McManus Child* (1969) 9 BLR 77 concerned the liability of the contractor for the supply and installation of defective material *specified* by the employer. The owners of dwelling houses bought from the developers McManus Child Ltd successfully sued them for the cost of re-tiling their roofs. Construction of the houses was undertaken by Richard Saunders Ltd who was treated by the court as the agent of the developers. Young & Marten were employed as nominated roofing sub-contractors for the supply and laying of roofing tiles called 'Somerset 13', an expensive and prestigious make of tile. Young & Marten in turn sub-contracted the supply and fixing to Acme Roofing Co. (London) Ltd. They obtained the tiles from John Browne & Co. (Bridgewater) Ltd., the sole makers of the product. Twelve months after installation, the tiles started to disintegrate because of a latent defect. Browne accepted liability and supplied replacement tiles, which were fixed by Acme. These attempts to resolve the problem failed. The owners re-roofed their houses and recovered their costs from McManus Child. They in turn sued Young & Marten on an *implied warranty* to recover their costs. The Official Referee refused to imply such a term into the contract. The Court of Appeal reversed the judgment and Young & Marten appealed to the House of Lords.

The House decided that in a contract for *work and materials* two terms would be implied:

(a) material (Somerset tiles in this case) should be fit for its purpose provided there was *reliance* on the skill and judgement of the contractor;
(b) the material should be free from latent defects.

Two important results

This was a *policy decision*: the courts were protecting the rights of the consumer, according to the theme of government and public policy at the time. It *creates a chain of liability*, enabling the consumer, where there was a chain of contracts, to sue the party with whom it had a contract, leaving *that* party with the problem of seeking redress from the supplier or manufacturer.

An important consequence of *Young & Marten* v. *McManus Child* [1969] is that there is no need to distinguish between sales of goods contracts and contracts for *work and material*. The Sale of Goods Act 1979 implies obligations of merchantability (now satisfactory quality) and fitness for purpose into contracts for the sale of goods. Similar obligations were implied into contracts for work and material. In *Young & Marten* v. *McManus Child* Lord Pearce observed at p. 91:

> 'It is frequent for builders to fit baths, sanitary equipment, central heating and the like, encouraging their clients to choose from the wholesalers' display rooms the bath or sanitary fittings they prefer. It would, I think, surprise the average householder if it were suggested that by simply exercising a choice he lost all right of recourse in respect of the quality of fittings against the builder who normally has better knowledge of these matters. Of course if the builder warned him against a particular fitting or manufacturer and he persisted in his choice, he would obviously do so at his own risk: and the builder can always make it clear that he is not prepared to take responsibility for a particular fitting or material.'

It is important to realise also that *Young & Marten* v. *McManus Child* effectively made the contractor *strictly liable* for the quality of the materials supplied. There are, however, circumstances in which the implied warranties do not apply because the chain has been broken, as explained below.

1 Where the limitation period has expired. *Young & Marten* is itself, an example of this since by the time the case was decided in the House of Lords, the limitation period had expired and the sub-contractor was unable to sue the next party in the chain.
2 Bankruptcy of the next person in the chain will leave a claimant without an effective remedy.
3 Where the contract contains an exclusion clause: see *Gloucester County Council v. Richardson* [1969] 1 AC 480 HL. Concrete units provided by a nominated supplier proved to be defective. The quotation provided to the architect contained a limitation of liability for defectively supplied material. The contractor was instructed to accept the nomination.

Held: It would be unreasonable to impose liability on the contractor because (i) the terms were agreed by the employer and supplier before the nomination and (ii) the contractor had no right of reasonable objection to the nomination of the supplier.

The relationship between the implied terms and the specification

The relationship between these implied terms and the express terms of the contract are illustrated by *Rotherham MBC* v. *Frank Haslam Milan & Co Ltd and MJ Gleeson (Northern) Ltd* (1996) 78 BLR 1. The issue for the court was a simple one: what is the relationship between the *implied warranties* of *Young & Marten* v. *McManus Child* and the *express terms* of the contract dealing with the same issues?

In 1979 Rotherham began to develop a site in the city centre. The work included advance site preparation and the construction of a new five-storey office building. Phase 1 was awarded to Haslam to carry out advance site preparation. This included the excavation and removal of old foundations and cellars, the installation of bored cast in place piles, the construction of external retaining walls and the supply and placing of imported fill material to the level of approximately the underside of the ground floor slab construction. The contract was let under JCT 63 and the terms of the relevant clauses are almost identical to JCT 98. Gleeson was employed in Phase 2 of the work in late 1979 to construct the remaining foundations of the superstructure. The terms of the contract were also JCT 63. Gleeson's work reached practical completion on 7 September 1982 but no final certificate was ever issued (for the significance of the final certificate see below).

Steel slag was used as fill material for the cellars. Samples were provided to the architect for testing and approval. Subsequently it was discovered that the heaving of the ground floor had been caused by the expansion of the steel slag. This led to the cracking of the reinforced concrete slabs of the five-storey building.

Rotherham sued both contractors alleging that they were in breach of an implied term; the grounds were (i) that the steel slag should be of merchantable quality *and* (ii) the steel slag should be fit for its purpose. At first instance Rotherham succeeded and the contractors appealed against that finding.

The express terms

Clause 1(1) of JCT 63 required the contractor to: 'carry out and complete the Works shown upon the Contract Drawings and described by or referred to in the Contract Bills . . . in compliance therewith, using materials . . . of the quality therein specified'.

Clause 6(1) of JCT 63 stated: 'all material, goods and workmanship shall so far as procurable be of the respective kind and standards described in the Contract Bills'.

Bill B of the Contract Bills read:

'B Granular hardcore shall be well graded or uncrushed gravel, stone, rock fill, crushed concrete or slag or natural sand or a combination of any of these. It shall not contain organic material, material susceptible to spontaneous combustion, material in a frozen condition, clays or more than 0.2 per cent of sulphate ions as determined by BS 1377.'

Experts for both sides agreed the following.

1 The heaving of the ground floor was caused by the expansion of the underlying imported fill materials.
2 It was well known at the time by the Building Research Establishment and also by specialist slag production companies that steel slag was not a safe or satisfactory product for use in confined conditions because of its *expansive properties*. None of the parties involved in the project was aware of this. The steel slag had also been visually inspected and approved by the architects.

Held:

(a) A *term of fitness* for purpose will not be implied without good reason. In this case the steel slag was *within the wording* of the employer's specification; the employer had the means of knowing that steel slag *was not inert*. Therefore it was *unreasonable and unjust* to impose a term of fitness for purpose on the contractor.

Fitness for purpose in a contract for work and materials can be compared with a contract for sale. The crucial questions here were whether the employer *trusted to the judgement of the contractor* and not to his own? Here the purpose of the fill was made known to the contractor. The freedom to choose was not to allow the supplier to use skill and judgement *but because the architect believed that no further stipulations were necessary*. Leggatt LJ said at p.17:

'In my judgement Rotherham as employers expressly and by implication made known to the appellants the purpose for which the hardcore was required. It was to constitute fill underlying the concrete slab at the base of the new building, and that purpose was clearly disclosed by the contract. The question then is not merely whether the particular purpose for which the fill was required was made known to Rotherham as it undisputedly was, but whether the circumstances show that Rotherham did not rely on the appellants' skill and judgement.'

At p.21 of his judgment he added that if an officious bystander had attracted the attention of the contracting parties and asked them if it was left to the contractor to determine the suitability of the fill materials, both parties would have replied 'Of course not!'

Rioch and Ward LJJ both agreed with that conclusion, Rioch LJ accepting without hesitation the evidence of Mr Ward that the definition of hardcore in the contract bills included steel slag 'because the design team thought they know but did not actually know the true nature of steel slag'.

(b) An implied warranty of *merchantability* can survive although the implied term as to fitness for purpose is excluded. *There was no express term that excluded such a warranty*. The fact that provision was made for testing and that the contractor was to pay for the tests if the material was defective showed the contractor bore the risk of the material not being of good quality. However, the warranty of merchantability was satisfied if the material was fit for some of the purposes *within the description* under which it was sold *without a reduction* in price. Since the steel slag was perfectly good for use as hardcore in road building or other situations where it was not confined and could have been sold as such without reduction price it was *merchantable*.

Ward LJ put it clearly at p.41:

'As I understand it, merchantability is a multi-faceted term comprising aspects of description, purpose, price etc. Here the description was of "'granular" hardcore, the purpose was as infill around the foundations but it was only one of several purposes for which steel slag was commonly used, and the price was presumably a market price whatever the particular purpose.'

The experts agreed that steel slag had a number of applications in the construction industry. It could be used in unconfined areas and in the manufacture of bituminous macadam. Ward LJ added that, as such, it was commercially saleable and therefore of merchantable quality, a conclusion with which the other judges agreed.

Express terms dealing with defective performance

The JCT 98 contract provides for three periods in which defects may occur and provides methods of dealing with those defects. These are (i) defective work occurring during the construction period (ii) defective work surfacing during the defects liability period and (iii) defects discovered after the issue of a final certificate. All three periods are governed by different clauses and have different outcomes.

Defects occurring during construction

Clause 8 applies here. The clause is headed 'Work, materials and goods'. Clause 8.1.1 requires that 'all materials and goods shall, as far as procurable, be of the kinds and standards described in the Contract Bills . . provided that materials and goods shall be to the reasonable satisfaction of the Architect where and to what extent this is required in accordance with clause 2.1'. This clause states that where the 'approval of the quality of materials and/or of standards is a matter for the opinion of the Architect such quality and standards shall be to the reasonable satisfaction of the architect.'

Clause 8.1.2 deals with the standards of workmanship to be carried out. It requires that 'all workmanship shall be of standards described in the Contract Bills . . . [including the requirement of reasonable satisfaction of Clause 2.1 where the workmanship is not specified]'.

The effect of this provision was analysed in *Crown Estate Commissioners* v. *John Mowlem* (1995) 70 BLR 1, by Stuart-Smith LJ (at p. 15) where he identified three cases for dealing with defects in quality of the work:

1   Case A – the criteria stipulated in the contract documents, (for example, to a British standard specification).
2   Case B – standards and quality not stated in the contract documents, in which case there is an implied term that materials will be of a reasonable quality and fit for their purpose and workmanship will be to a reasonable standard.
3   Case C – the standards and quality is expressed to be to the architect's satisfaction.

Cotton LJ p. 483 supported this summary in *London Borough of Barking & Dagenham* v. *Terrapin Construction Ltd* (2000) BLR 479. Note that the contractor's duty is likely to be limited by the words 'as far as procurable'. It is unclear what is meant by the phrase. Can 'procurable' extend to the provision of sub-standard material and goods? Or does it merely mean that the contractor has to inform the architect of their non-availability? If the contractor suggests suitable alternatives, which in turn prove to be defective, the employer may claim that he relied on the skill and judgement of the contractor not on his architect.

What is the nature of the implied term to carry out the work in a 'proper and workmanlike manner' imposed on the contractor? Clause 8.1.3 provides that all work shall be carried out in 'a proper and workmanlike manner'. This clause was originally inserted by Amendment 9 (1990). It expresses, what was previously thought to be implied and is a direct response to the outcome of *Greater Nottingham Co-operative Society* v. *Cementation Piling and Foundation* (1988) 41 BLR 43. A contract under the then JCT 63 standard form of building contract for the extension and refurbishment of the client's premises in Skegness was let to Shepherd Construction as main contractor. Cementation was nominated as the sub-contractors for the design and installation of the piling works. They entered into a direct warranty with the client in which they promised to use reasonable care and skill:

(a)   in the design of the sub-contract work; and
(b)   in the selection of goods and materials selected by them.

Piling started in July 1979. Cementation had won the contract by proposing the use of 450 mm diameter piles to be driven to design depth using a continuous flight auger technique. This contract was their first successful attempt to break into the small size piling market and they had undercut Shepherd's preferred sub-contractor by some 50 per cent.

Piling started in July 1979. Immediately adjacent to the site was a restaurant called 'The Windsor'. As a result of the piling operations structural damage was caused to The Windsor. Its owners brought an action against the Greater Nottingham Co-operative Society (Co-op), which was settled for £124,000. Cementation agreed that they were liable to indemnify the Co-op for this sum. Work was suspended in order to deal with the problem. This was caused by the presence of silt on the site. Silt layers and their extent are notoriously difficult to identify during a site investigation. In only one site investigation in a long career has the writer ever encountered silt layers *in situ*. Piling restarted on 17 October using 130 × 300 mm diameter piles. Work was completed on 20 November.

The Co-op sued Cementation in both contract and tort. It won its case before the Official Referee and was awarded the following damages:

(a) additional costs paid to Shepherd for the cost of the revised piling scheme – £69,000;
(b) additional costs paid to Shepherd for loss and/or expense for the 33 weeks' delay – £79,000;
(c) consequential economic loss due to delayed completion – £283,000.

Cementation appealed against the judgment and the Court of Appeal held in its favour. In the course of doing so, it addressed the arguments made to it based on contract and in tort.

The contract   It found that the warranty given by Cementation dealt only with *how* the work was to be *designed* and not with *how* it was to be *carried out*. All parties agreed that Cementation had not been negligent in the design. The Court pointed out that had the warranty stated that the *work* was to be *carried out with due care*, then Cementation would indeed have been liable.

Tort   The basic argument of the Co-op was that this case was identical to the facts of *Junior Books* v. *Veitchi Co. Ltd* [1982] 3 WLR 477. Here the relationship was even closer because *there was a contract*: the warranty. (In *Junior Books* the House of Lords said that the relationship between the employer and the nominated sub-contractor was so close it was almost like a contract.)

The Court of Appeal rejected this argument although *Junior Books* was binding on it; it declined to follow that decision. It therefore accepted the argument of Cementation that this case was different precisely because *there was a contract* in place. It held that although Cementation had carried out the piling operation negligently, the parties had had an opportunity when the contract was made to allocate responsibility. Since it had not done so, no greater liability could exist in tort than there was in contract between the parties. In any case there was no liability in tort for economic loss unconnected with any physical damage to property. There had been no physical damage to the property of the Co-op.

The response of the Co-op to the argument that it had not expressly made provision for what happened was to argue for an *implied term*. In any building contract the contractor impliedly agrees (i) to do the work in a good and work-manlike manner and (ii) to supply good and proper materials. There was no dispute about (ii) since it was covered by the warranty. With regard to (i) the court was reluctant to imply this into the warranty agreement. The Court declined to imply a term into the warranty that the work should have been carried out in a *proper* and *workmanlike manner.* Between the contractor and the nominated sub-contractor it would have had no difficulty in doing so.

Judge Smous QC said at p.50.

> 'I mention in passing that I do not feel any need to imply in this employer/nominated sub-contractor form of agreement as a matter of business efficacy a term to the effect that the nominated sub-contractor warrants his ability to carry out the work with reasonable care and without damage to the surrounding buildings. Such an implication does not have to be read into the somewhat artificial arrangements between the property owner and the nominated sub-contractor, although it is implicit between. the contractor and the nominated sub-contractor.'

The response of the JCT

It responded to this decision by issuing amendment 9, which amended clause 8. Clause 8.1.3 was inserted to make express what was usually implied: all work shall be carried out in a *proper and workmanlike manner.* In addition it added clause 8.5, which deals with a *failure to comply* with 8.1.3. Where there is any failure to comply with clause 8.1.3 in regard to the carrying out of the work in a proper and workmanlike manner 'the Architect . . . after consultation with the contractor . . . *shall issue* such instructions whether requiring a variation or not or otherwise as are reasonably necessary as a consequence thereof . . . no addition to the contract sum and no extension of time shall be given'. Note that for such a clause to have any value it needs to be inserted in the sub-contract as well. The JCT in its notes accompanying the amendment states that instruction under clause 8.5 may be of two kinds.

1 Where there has been a failure to carry out the work in a proper and work-manlike manner and some change has to be made to the works by way of a variation (in the case itself for example, the pile sizes had to be changed by means of a variation).
2 The work is in accordance with the contract but not carried out in a proper and workmanlike manner (e.g., it is unsafe or a potential hazard to third parties or to the employer's existing structures). Here the instructions would be to carry out the work in a proper and workmanlike manner in the future. (In this case this would have been damage to The Windsor restaurant adjoining the site. The onus would therefore have been on the sub-contractor to change their methods of working).

The supply and use of materials   Clause 8.2.1 requires that the contractor 'shall . . . provide . . . vouchers to prove that materials and goods comply' with clause 8.1.

Where the quality of materials, goods and workmanship has been left to the reasonable satisfaction of the architect, any dissatisfaction with unsatisfactory work, has to be expressed within a reasonable time (clause 8.2.2). Under Clause 8.3 the Architect has power to 'issue instructions to open up for inspection any work covered up or to arrange and carry out any tests of any materials or workmanship . . . the costs . . .(together with the cost of making good . . .) shall be added to the Contract Sum unless the materials, goods or work are not in accordance with this contract.' Note now amendment 2 issued January 2000 and the extension of provision for adjudication to deal with opening up or testing. See also the default procedure on nomination of the adjudicator.

Where the work does not conform to the contract the architect may respond in one of two ways using powers granted by clause 8.4:

(a) issue instructions clause 8.4.1 for the removal from site of any such work, materials or goods; *or*
(b) using the powers under clause 8.4.2 and after consultation with the contractor and with the agreement of the employer, 'allow any such work, materials or goods to remain . . . and an appropriate deduction shall be made'.

Where the architect orders the removal of defective work, the contractual duty of the contractor is still to complete the work. No further instruction is needed. It is not sufficient however to merely order a correction: see *Holland Hannen & Cubitts (Northern) Ltd* v. *Welsh Health Technical Services Organisation* (1981) 18 BLR 80.

Where defective work is allowed to remain on site, the architect can deduct an appropriate sum from the contract sum. The deductions should neither penalise the contractor nor equate the costs of making good the defective work, in accordance with the contract. Note the strictness of the requirement to remove work not complying with the contract from the site. The employer may determine the contractor's employment under the contract after following the notice procedure required by clause 27.2.1.3. where the contractor 'neglects or refuses to comply with a written notice to remove any work, materials or goods'.

Opening-up   Clause 8.4.4 enables the architect to issue instructions requiring the opening for inspection and testing of workmanship and materials. The architect has to have regard to the Code of Practice produced in 1988 to ensure the fair and reasonable operation of the provisions. The requirements of the code are set out below.

1 The parties to agree the amount and type of opening up or testing where possible.
2 Criteria:
   (a) Is non-compliance unique and unlikely to occur in similar elements?
   (b) Alternatively, what is the extent of any similar non-compliance in work already constructed or still to be constructed?
   (c) Is the failure that of workmanship or materials? Should rigorous testing of similar elements be made? Can it be repaired or is there an inherent weakness requiring selective testing?

(d) Is the non-compliance significant with regard to the nature of the work carried out?
(e) Is the safety of the building affected, that of adjoining buildings, the public at large or any of the statutory requirements?
(f) A number of questions have to be asked of the contractor with regard to its operations. What is the level and standard of supervision and control of the works by the contractor taking into account (i) its relevant records, (ii) what other codes of practice require (iii) whether the contractor failed to carry out other tests (iv) why it failed to comply (v) what technical advice it had and (vi) what were the current recognised testing procedures?

The ordering of tests does not entitle the contractor to the costs of inspection or any loss and/or expense. Only where, after opening up, the workmanship and material is found to be satisfactory, is the contractor entitled to an extension of time and the right to make a claim for loss and/or expense.

Defects occurring after practical completion of the contract

The JCT contracts deal with defects that are found after the completion of the works by first scheduling a completion date called the date of practical completion. This is done by the use of clause 17.1 which states that 'when in the opinion of the Architect practical completion of the works is achieved . . . he shall forthwith issue a certificate to that effect and practical completion shall be deemed for all purposes of this contract to have taken place on the day named in such certificate'.

Comment

The following points can be made:

1 There is no definition of practical completion given in JCT 98 or any of the previous versions of the contract.
2 The issue of the certificate is at the discretion of the architect. Once the opinion has been formed it must be issued immediately.
3 The use of the word 'deemed' means that the date on the certificate is taken as the date of practical completion regardless of whether it took place earlier or later.

The meaning of practical completion   In the absence of any clear-cut definition in the standard forms, the courts have had to grapple with the phrase. The House of Lords have considered the issue in two cases. In *Jarvis & Sons* v. *Westminster Corporation* [1971] 1 WLR 637, Viscount Dilhorne at p. 646 drew two conclusions about the meaning of the phrase:

'One would normally say that a task is practically complete when it is almost, but not entirely finished, "Practical completion" suggests that that is not the intended meaning and what is meant is the completion of all the construction work that has to be done.'

He then added:

'first, that the issue of the of the final certificate determines the date of completion, which may, of course be before or after the date specified for that in the contract and, secondly, that the defects liability period is provided in order to enable defects not apparent at that date of practical completion to be remedied. If they had been apparent no such certificate would have been issued.

It follows that a practical completion certificate can be issued when owing to latent defects the works do not fulfill the contract requirements and that under the contract Works can be completed despite the presence of such defects. Completion under the contract is not postponed until defects which become apparent only after the work had been finished had been remedied.'

The phrase was considered further in *P & M Kaye* v. *Hosier & Dickson* [1972] 146 at 165, where Lord Diplock said:

'By issuing his certificate of practical completion he signifies his satisfaction with the works at the end of the construction period; but this is subject to any latent defects which become apparent to him during the defects liability period.'

At first instance, the cases of *Nevill (Sunblest) Ltd.* v. *William Press & Son Ltd.* (1981) 20 BLR 78 and *Emson Eastern* v. *EME Developments* (1991) 55 BLR 114 also considered the meaning of the phrase. In *Big Island Contracting* v. *Skink* (1990) 52 BLR 110 (Hong Kong CA), a clause provided for the payment of 25 per cent of the price on practical completion. It was held that practical completion could not be distinguished from substantial completion.

On a practical level the phrase can mean two things: (i) the owner could take over the building and use it for its intended purpose, or (ii) the building is reasonably safe for use and not unreasonably inconvenient. Legally, the position is not so clear cut. Keating on Building Contracts (sixth edition p. 673) in discussing the phrase 'when the work is practically complete' in clause 15 of JCT 63 says that no clear answer can be given to the meaning of the phrase. The words have been changed in JCT 80 (now 98) to 'practical completion of the work' but its meaning is unclear. In the seventh edition it mentions four authorities at p.673, para. 18–172, but states the practical completion is easier to recognise than to define. It cautions an architect when certifying, to obtain written assurance of the any incomplete work and an express undertaking to complete them.

In *Nevill (Sunblest)*, HHJ Newey QC OR had to decide what the words 'practically complete' in clause 15 of JCT 63 meant. By a JCT 63 contract dated 7 December 1973, Press agreed to carry out works consisting of site clearance, piling, foundation and drainage work prior to the building of a bakery in Walthamstow. This was Phase 1 of the works. The work was carried out between September 1973 and April 1974 when new contractors (Trenthams) started work on Phase two (the building of the bakery). A certificate of practical completion dated 1 May 1974 was issued to Press. A certificate of making good defects dated 21 May 1975 and a final certificate were also issued. In November 1975, Nevill's architect discovered that the drains were defective and that there were defects in the hard standing. Press returned to the site and repaired their defective work. Trenthams, however, were delayed by four weeks and Nevill had to pay them for that delay and additional work arising from the defective work. Nevill also incurred additional architect's fees and losses because the bakery was late in opening. They sued Press to recover their additional expenditure of £100,000. In their defence Press argued their case in a number of ways:

1 Although their work had been defective, Nevill's remedies were limited to clause 15 (now clause 17 of JCT 98).
2 They did not have to achieve perfection by the completion date because the contract recognised that 'defects, shrinkages and other faults' might occur during the defects liability period.
3 Defects were therefore not breaches of contract and they did not have to pay damages. They had also a right to re-enter the site and make good any defects.
4 If the defects were breaches of contract, clause 15 (now 17 of JCT 98) limited their effects.
   (a) Press was obliged to return to site and make good the defects.
   (b) Nevill were obliged to allow them to do so.

Held:

1 The plaintiffs' remedies were not limited to the remedies specified in Clause 15 (now 17 in JCT 98) since the defects were breaches of contract.
2 Clause 15 merely created a simple way of dealing with part of a situation created by breaches of contract and were not to be read as depriving the injured party of his other rights.
3 The plaintiffs could therefore claim damages for breach of contract to include consequential loss:

'in my judgement, since clause 21(1), and the appendix requires William Press to complete the works by 18 February, 1974 or within such extension of time as was granted by the architect, that was the date by which they had to be completed. I think the word "practically" in clause 15(1) gave the architect a discretion to certify that William Press had fulfilled its obligations under clause 21(1), where very minor de minimis work had not been carried out, but that if there were any patent defects in what William Press had done the architect could not have given a certificate of practical completion.'

This case restates the earlier decisions in *Jarvis* and *P & M Kaye* but provides clearer guidance of what the term means in practice. Giving judgement in *Emson Eastern* v. *EME Developments* (1991) 55 BLR 114, HHJ Newey QC OR reached the following conclusions having reviewed the authorities mentioned above:

'The JCT contract is a standard form of contract very widely used in the construction industry in this country and abroad: Its wording has been unanimously agreed by a committee consisting of representatives of all sections of the industry. I think that the matrix of facts against which it should be construed is what happens on building sites generally, rather those, which prevailed when the plaintiffs and defendants entered into their contract. In any event, however, the present case seems to be a typical one. I think the most important background fact which I should keep in mind is that building construction is not like the manufacture of goods in a factory. The size of the project, site conditions, the use of many materials and the employment of various kinds of operatives make it virtually impossible to achieve the same degree of perfection that a manufacturer can. It must be a rare new building in which every screw and every brush of paint is absolutely correct.'

The issue of a certificate of practical completion has a number of effects. The following events are triggered by the issue of such a certificate:

(a) The contractor's liability to pay LAD comes to an end.
(b) The employer has no right to issue further instructions to the contractor.
(c) The contractor's responsibility to insure the building ceases.
(d) Half the retention money is repaid.
(e) The defects liability period starts.
(f) For all practical purposes the limitation period starts.

The defects liability period   Clause 17.2 describes what is to happen after this period, which is usually six months or a year. 'Any defects, shrinkages or other faults which shall appear within the Defects Liability Period and which are due to materials or workmanship not in accordance with this contract . . . shall be specified in a Schedule of Defects . . . not later than 14 days after the expiry of the Defects Liability Period . . . and shall be made good within a reasonable time.' It may well happen that serious defects requiring immediate rectification may surface during this period. Clause 17.3 therefore gives the architect powers to deal with this situation. It states that 'notwithstanding clause 17.2 the Architect may whenever he considers it necessary issue instructions requiring any defect, shrinkages or other faults which shall appear within the defects liability Period . . . shall be made good within a reasonable time'.

The certificate of making good defects   Once a schedule of defects (commonly called a 'snagging list') has been issued and the required repairs made, clause 17.4 applies. This states that 'when in the opinion of the Architect any defects, shrinkages etc.' identified under clauses 17.2 and 17.3 'have been made good', he shall issue a certificate to that effect. This is the Certificate of completion of Making Good Defects. The balance of the retention money is due after the issue of such a certificate. As described, this is a quite detailed and comprehensive code for rectifying defective work. What then is the position where no notification of defects is given? Does this exclude a right to claim damages and if it does not, is a damages claim limited by the failure to give such a notice? The Court of Appeal considered the matter in *Pearce & High Ltd* v. *John P. Baxter and Mrs Baxter* (1999) CILL 1488. The contract conditions were the JCT Agreement for Minor Works and used for the exterior and interior alterations made to a domestic dwelling. Clause 2.5 required 'defects and other defaults' appearing within six months of the issue of the final certificate to be made good by the contractor. No notice of the presence of defects was ever issued. In an action the contractor claimed monies outstanding from certified sums and the defendants raised the claim of defective work in their counter-claim. It was held that the notice was a condition precedent to the contractor's liability to repair the defects. This did not prevent a claim for damages being made. However, the damages recoverable were limited by the failure to give notice to the cost to the contractor of carrying out the work.

The final certificate   Clause 30.8 requires the architect to issue the Final certificate not later than 2 months after whichever the following occurs last:

(a) the end of the Defects Liability Period;
(b) the date of issue of the certificate of completion of making Good defects;
(c) the date on which the Architect sent a copy to the Contractor of direct loss and/or expense or other adjustments to the contract sum.

The problem with final certificates

In *Colbart* v. *Kumar* (1992) BLR 59 it was successfully argued that the issue of an architect's final certificate (under a JCT intermediate form of contract) gave the contractor a great deal of immunity for defective workmanship and materials. If approval of the work was during construction, 'inherently' a matter for the opinion of the architect, then giving that opinion gave the contractor immunity from being sued after the certificate was issued. In *Darlington Council* v. *Wiltshier Northern* (1993) 37 Con LR 29 at 47 HHJ Newey QC referring to 'conclusive evidence' type of clauses said:

'The desire of both contractors and employers in the construction industry is to bring about early conclusion to disputes between themselves. Contractors, having finished the works and having put right any defects which have arisen during the defects liability periods, wish to put the contracts behind them and be free from future claims. Employers are content that should be the position, so long as there are no deliberately concealed or otherwise hidden defects.'

The present wording of the clause results from amendment 15:1995, in response to the decision of the Court of Appeal in *Crown Estate Commissioners* v. *John Mowlem* (1995) 70 BLR 1. A £9 million retail and commercial development was built on the site of the old Kensington Palace barracks. It consisted of shops, offices, a penthouse flat and restaurant. The contract conditions was the JCT 80 with quantities. Clause 30.9.1 provided that a Final Certificate had effect as 'conclusive evidence' only to the extent that where 'the quality of materials or standards of workmanship are to be to the reasonable satisfaction of the architect the same are to such satisfaction.'

A number of problems occurred during and after construction. Complaints were made by one of the tenants, WEA Records about defective glazing. Problems also occurred with slates and courtyard paving. Crown Estates sued Mowlem for £2.5 million. The judge gave judgment for Crown Estates on the basis that the Final Certificate was only conclusive on matters left to the reasonable satisfaction of the Architect. Mowlem appealed: the Court of Appeal held that the certificate was conclusive evidence that all work had been properly carried out. This ruling caused great surprise in the industry. The JCT responded by issuing Amendment no 15, which amended the wording of clause 30.9.1. This now reads as follows:

'the Final Certificate shall have effect in any proceedings under or arising out of or in connection with this contract . . . as described in the following subclauses. Clause 30.9.1.1. describes the effect that the certificate is to have. It provides 'conclusive evidence that where and to the extent that any of the particular qualities of any materials or goods or any particular standard of an item of workmanship was described expressly in the Contract Drawings or the Contract Bills . . . (etc.) to be for the approval of the Architect, the particular quality or standard was to the reasonable satisfaction of the Architect, but such Certificate is not conclusive evidence that such or any other materials goods or workmanship comply or complies with any other requirement or term of this contract.'

(In essence the clause now states that the final certificate only covers work specified to be left to the reasonable satisfaction of the architect).

Many commentators argue that the amendment does not override *Crown Estates*. They argue that the ruling does not apply to latent defects. The concept of a defect that is latent was described by Steel J in *Baxall Securities Ltd and Narbain SOC Ltd* v. *Sheard Walshaw & parts* [2002] EWCA Civ 9 as not a difficult one. It simply meant 'a concealed flaw', an 'actual defect in the workmanship and design, not the danger it represents'. Wharton, Construction Law (1995) Vol. 6, states that in his opinion 'nothing in the final certificate, or indeed the certificate for making good defects, can operate to exclude liability for latent defects'. This is simply because a supervisory officer in signing off items contained in a 'snagging list' can only expected it to apply to patent defects, items he could see, not hidden flaws which could surface later. Anthony Lavers in (Final Certificate revisited) www.scl www.scl. org agrees with this view.

The Court of Appeal considered the question posed by the commentators in *London Borough of Barking & Dagenham* v. *Terrapin Construction Ltd* [2000] BLR 479. The parties entered into a design and build contract to carry out new and refurbishment work at a school in Dagenham. The contract was made on 28 April 1992, using the JCT standard form of building contract with contractor 's design 1981 edition. Work started in December 1992, and practical completion was achieved on 6 November 1993. The final account and final statement was subsequently agreed. Clause 30.8.1 provided that: 'The Final Account and Final Statement when they are agreed . . . shall have the effect in any proceedings arising out of this contract . . . as conclusive evidence that where it is stated in the Employer's requirements that the quality of the materials or the standard of workmanship are to the reasonable satisfaction of the Employer the same are to such satisfaction.'

The employer issued proceedings against the contractor (amongst other things) for breach of contract and negligence in respect of alleged defects in design, workmanship and materials. On a trial of a preliminary issue, HHJ Newman QC held that: 'All the claims by the employer are subject to the final certificate by reason of the operation of clause 30.8.1 of the JCT 1981 conditions.' The employers appealed against this judgment and the Court of Appeal held that:

1 Clause 30.8.1 of the JCT 1981 form was to be construed in the same way as the Court of Appeal construed it in *Crown Estate*.
2 Clause 30.8.1 was not to be construed as relating to or including design defects; it did not apply to the contractor's breach of its obligation to complete the design of the works.
3 Clause 30.8.1 made no distinction between latent and patent defects; the agreement of the final account and final statement provided conclusive evidence of the employer's satisfaction as to the quality and standards of all materials.

Professor Ian Duncan Wallace QC said of the decision in *Crown Estates* in 'Not what the RIBA/JCT meant' (1995) Const LJ 184 at 188:

'the unreasonable and anti-consumer aspects of the "wider construction", covering all defects whether patent or latent . . . must be obvious to any but the most blindly partisan of observers in the construction industry . . . The inevitably rapid covering up of many defects or omissions of workmanship, and the often insuperable difficulties of detecting latent defects in many types of goods and materials must, if an unqualified immunity is to be obtained on the Final Certificate, play straight into the hands of those contractors and subcontractors prepared to cut corners on quality of workmanship and materials as against their more responsible competitors . . . By any responsible standards, the "wider construction" of the post–1976 wording represents a policy disaster for all private and public owners reasonable interests in a vitally important area of construction, by removing at a stroke the pre–1976 exceptions to immunity of latent defects in materials or workmanship, and by benefiting not merely contractors maximizing their profits by policies of under-compliance . . . but also . . . suppliers of sub-standard work or materials with similar policies.'

The Court of Appeal felt itself unable to accede to these arguments put by counsel for the employers. On the proper construction of clause 30.8.1 covered both patent and latent defects by the issue of the Final Account and Final Statement. It added that the employer could have expressly excluded latent defects had it chosen to do so.

In amendment 18 to the JCT 80, further confusion was added to the debate whether the Final certificate was to be treated as conclusive evidence of the quality of workmanship and materials. A new clause 41 was added giving the court the power to open up, review and revise certificates. This was added to side-step the restrictions introduced by the decision in *Northern Regional Health Authority* v. *Derek Crouch Construction Co Ltd* [1984] 2 All ER 175. That case decided that only an arbitrator could open up and review certificates given by the architect. In

Beaufort Developments (NI) Ltd v. *Gilbert-Ash NI Ltd and anor* [1998] 2 All ER 788, decided after the issue of the amendment, the House of Lords overruled the *Derek Crouch* decision.

Cockram (2000) suggested that anyone using amendment no 18 should add an overriding 'for the avoidance of doubt' provision to their contract. It should state that the final certificate is to be treated as conclusive, except:

1   in arbitration, litigation or adjudication started before the final certificate is issued;
2   in arbitration, litigation or adjudication started not more than 28 days after the certificate is issued;
3   in any arbitration or litigation taken for the purpose of overturning an adjudicator's decision and commenced within 28 days of the decision (regardless of the final date of the certificate).

Defects and the ICE 7th edition

Clause 36 (1) requires that 'All materials and workmanship shall be of the respective kind described in the contract and in accordance with the engineer's instructions and shall be subject to testing from time to time as the engineer shall direct'. The phrase 'described in the contract' will include the drawings, bills of quantities and the specification. The engineer has no general power to issue instructions with regard to the quality of materials and workmanship (Keating, p.1121).

Clause 36 (3) and (4) deals with the costs of carrying out tests. Under (3) the costs of tests is borne by the contractor where it was included in the tender. Where it was not (4) provides that the employer will pay the costs unless the tests result from contractor's default or failure to perform his obligations. A case in which the employer paid the costs of tests is *Hall & Tawse Construction* v. *Strathclyde* (1990) SLT (51 BLR 88).

Clause 38(1) states that 'no work shall be covered up or put out of view without the consent of the Engineer'. There are two circumstances provided for in the clause. Where work is to be covered up the contractor has to give 'due notice' to the engineer that the work is ready for examination. The other is where the engineer is to measure the work before it is covered up.

Clause 39(1) 'The Engineer shall during the progress of the work have the power to instruct in writing the

(a)   removal from Site within such time or times specified in the instruction of any material which in the opinion of the engineer is not in accordance with the Contract;
(b)   substitution with materials in accordance with the Contracts and
(c)   removal and proper replacement (notwithstanding any previous tests thereof or any interim payment therefor) of any work which in respect of
     i.   material or workmanship or
     ii.   design by the contractor of which he is responsible
     is in the opinion of the Engineer not in accordance with the Contract.'

Keating states at p.1124 that an express power is needed for the removal, substitution and re-execution of non-conforming work. However, it warns that the power should be exercised sparingly since a carelessly drafted instruction could trigger an application for a variation. Clause 39(2) deals with default by the contractor in obeying the instruction. 'The employer shall be entitled to employ and pay other persons to carry out the same . . . all costs consequent . . . shall be recovered.' Note also the power given by clause 65 (1) (h) to terminate the contractor's employment for failure to remove defective goods or materials from the site. See also the position where the contractor refuses to allow such persons access to the site. Under a JCT contract the employer was able to obtain the grant of an injunction: *Bath and North East Somerset DC.*

Certificate of substantial completion

Civil engineering contracts avoid the uncertainties created by the completion regime under the JCT form of contracts. The ICE 7th, in clause 48(1), provides for the contractor to give notice to the engineer when he considers the work to be *substantially complete* or that it has passed any test prescribed by the contract. As in JCT 98 no definition is given of substantial completion. Unlike the JCT 98, the engineer cannot issue a certificate of substantial completion until given notice by the contractor. The notice must also give an undertaking to complete any outstanding work.

Clause 48 (2) requires the engineer to respond to the notice within 21 days by either (i) issuing the certificate or (ii) giving instructions in writing specifying work which in the opinion of the engineer is required to be done by the contractor.

The meaning of substantial completion

At common law, where there has been a substantial completion of the contract, the contractor is entitled to payment less a deduction for the costs of repairing the defects. In deciding whether the work is substantially complete, account must be taken of (i) the nature of the defects and (ii) the difference between the cost of rectifying them and the contract price. In dealing with practical completion under the JCT 98, the approach of the courts has been to decide that it occurs when most of the work has been done and the employer can use the building safely. By contrast substantial completion under the *ICE* is a more flexible instrument for two reasons: one is that the completed work has to be acceptable to the engineer; a second reason is that the contractor is required to give an undertaking to carry out incomplete work.

Defects correction certificate

Under clause 61(1) the engineer issues a defect correction certificate when all the outstanding work under clause 48 and all repairs under clauses 49 and 51 have been done. Clause 61(2) states that the issue of the certificate does not relieve the parties of any liability towards each other arising out of their performance of their obligations under the contract.

Note also the power of the engineer to order remedial or other work or repairs urgently needed during the defects correction period (clause 62). Where the contractor is unwilling or unable to carry out the work upon given notice, the employer may employ others to do the work required.

Status of the certificate

It terminates the licence the contractor has to enter the site. The engineer has no power after its issue to give further instructions to the contractor. It also establishes the last date for the submission of the contractor's final account under clause 60(4).

Status of the last certificate

This certificate does not have the *evidential value* of the final certificate under the JCT contract. Unfilled obligations are dealt with by clause 61(2) which states that the issue of the Defects Correction Certificate 'shall not be taken as relieving either the Employer or the Contractor from any liability the one towards the other arising out of or in any way connected with performance of their respective obligations under the contract'. The certificate has administrative value only, but affects the working of other clauses. The contractor's right or duty to carry out further works is limited by clause 49. There is nothing to stop the parties agreeing that the contractor should complete its work. In the absence of such agreement the employer can carry out the work and sue for damages.

Summary

1 The main obligation of the contractor is to carry out the work in accordance with the contract documents.
2 The standard form of contract regulates in great detail the obligations of the parties under the contract.
3 Terms may be implied into the contract by statute, by law and in fact.
4 The duty of the contractor is to carry out the work with the skill and care of an ordinary competent contractor.
5 Duties by the contractor to warn the employer that the design is faulty have been implied into contracts.
6 Materials supplied by the contractor are subject to two warranties. These will be implied where it is reasonable to do so. By making the contractor strictly liable, the contractor is forced to sue down the contractual chain to obtain redress.
7 *Rotherham* illustrates that the element of reliance on the skill and care of the contractor is vital for the imposing of liability. The more precise the specification is, the easier it is to show that the material had only one purpose.
8 Practical completion under the JCT contracts is not defined but seems to happen when the employer is able to use the building safely.
9 The final certificate, when issued, creates an evidential bar preventing the employer from suing the contractor for defects despite the limitation period having not yet expired.

10 The employer has 28 days from the issue of the certificate to initiate proceedings either in arbitration or ligation.
11 The final certificate covers both patent and latent defects.
12 The provisions of the ICE escape many of the uncertainties of the JCT final certificate provision.

7 The Main Obligations of the Employer

The primary obligation of the Employer is to pay for the work carried out by the contractor. All standard forms of contract now incorporate the provisions of the Housing Grants Construction and Regeneration Act 1996 (part II). All contracts covered by the statute are required to make provision for interim payments where the contract period exceeds 42 days. An example of the incorporation of such a provision is clause 30.1.1.1 of JCT 98, which requires that:

> 'The Architect shall from time to time as provided in clause 30 issue interim Certificates stating the amount due to the Contractor from the Employer specifying to what the amount relates and the basis on which that amount was calculated; and the final date for payment pursuant to an interim Certificate shall be 14 days from the date of issue of each interim Certificate.'

In addition to payment, the employer has further obligations arising out of the nature of construction contracts. In order to complete the contract, the contractor requires the co-operation of the employer. This duty itself can be divided into two contrary aspects in the sense that they are both positive and negative. The positive aspect of the duty requires that where the contractor is to do a piece of work which requires the employer's co-operation, that it will be forthcoming. See *Luxor (Eastbourne) Ltd* v. *Cooper* [1941] 1 All ER 33 where such a term was implied into the contract. Furthermore, the employer will do all things necessary on his part to bring about completion of the contract.

The obligation under the negative aspect of the duty is not to hinder or wrongfully interfere with the performance of the contract. Such a term is implied into the predecessor of the JCT 98. Implied into the contract is a term that the employer will neither hinder nor prevent the contractor carrying out its obligations. This extends to the duties of the architect or engineer, employed to administer the contract. The employer must do all things necessary to enable the contractor to carry out the work. In *London Borough of Merton* v. *Stanley Leach* (1985) 32 BLR 51, the completion of a contract to build 287 dwellings was delayed. The contractor claimed that the delay was entirely due to the architect's lack of diligence and care, as well as a lack of co-operation by the employer's architect. The contractor argued that there was an implied term in the then JCT 63 that the employer would co-operate in ensuring completion on time. The delay, it argued, was due to a lack of co-operation by Merton's architect.

Held:

1 Merton would not hinder or prevent the contractor from carrying out its obligations in accordance with the terms of the contract and from executing the work in a regular and orderly manner; and

2 Merton would take all steps necessary to enable the contractor to discharge its obligations and execute the work in a regular and orderly manner.

These implied terms might be limited by the express terms of the contact. *Leach* was considered in *Scottish Power Plc* v. *Kvaerner Construction (Regions) Ltd* [1998] www.scotcourts.gov.uk. The appendix to a subcontract required the sub-contractor to 'complete all the site works within 24 weeks. All to suit the requirements of the main contractor and in accordance with the main contract programme current at the date of execution of the sub-contract'. A further clause required the subcontractor to start on site when instructed by the main contractor and proceed diligently in conforming to the requirements of the main contractor. No guarantee of continuous work was given and in addition, the subcontractor was required to carry out the work in such stages and sequences as required by the main contractor. On completion, the subcontractor brought a claim for disruption and delay caused to the performance of their works by the main contractor. Giving his opinion, Lord Macfadyen said it was necessary to guard against too readily implying terms where the parties had expressly dealt with the matter. He then went on to find that the express terms did not *wholly exclude* the implied ones. Whilst the main contractor had express powers under the appendix and clause F, they were not wholly free to obstruct and disrupt in any way the regular and proper performance of the subcontract. In addition, the main contractors owed a duty to regulate the timing and continuity of the subcontract work and to take all steps necessary to enable the subcontractors to discharge their obligations.

Possession

It is an implied term that the employer will give possession of the site in a reasonable time to allow the contractor to complete by the contractual date. In *Freeman* v. *Hensler* (1900) 64 JP 260 CA. Collins LJ said: 'There is an implied undertaking by the building owner, who has contracted to have buildings placed by the contractor on his land, that he will hand over the land for the purpose of allowing the [contractor] to do that which he bound himself to do.' In *London Borough of Hounslow* v. *Twickenham Garden Developments Ltd.* [1970] 7 BLR 81, Magarry J, held that the contractor was entitled to such possession, occupation or use as is necessary to enable him to perform the contract.

The degree of possession

The contractor is entitled to sufficient degree of possession to permit execution of the work unimpeded by others: *Queen in Right of Canada* v. *Walter Cabott Construction Ltd* (1975) 21 BLR 42 Canadian Federal Court of Appeal. The contractor contracted with the Crown to build a hatchery building. It was one of six contracts let for the whole project. Work on two of those other contracts interfered with its work because they encroached on the site. The contractor claimed damages for breach of an implied term relating to possession. It was held that the failure to provide sufficient possession was a breach of the contract. It was fundamental to a construction contract that sufficient working space should be provided and that other contractors working on the site would not impede its activities.

Express terms – JCT 98

Clause 23.1 states that 'on Date of Possession possession of the site shall be given to the contractor ... who shall ... proceed regularly and diligently ... to complete on or before the completion date'. The employer in the appendix to the contract specifies the actual dates. The words, possession of the site, were held to mean possession of the whole site and that piecemeal possession by the employer was a breach of contract: see *Whittal Builders* v. *Chester-le-Street District Council* (1987) 40 BLR 82.

Regularly and diligently

The meaning of the phrase was considered in *West Faulkner Associates* v. *London Borough of Newham* (1994) 71 BLR 1. It was held that the phrase meant orderly planned progress towards the completion date. A very wide construction of the phrase was adopted. Simon Brown LJ said at p.14:

'Taken together, the obligation upon the contractor is essentially to proceed continuously, industriously and efficiently with appropriate physical resources towards completion substantially in accordance with the contractual requirements such as time, sequence and quality of works.'

Deferring possession

Under the provisions of clause 23.1.2 the 'Employer may defer the giving of *possession* . . . up to 6 weeks.' Clause 25.4.13 makes this delay a relevant event entitling the contractor to an extension of time to complete the work, which together with clause 26.1 entitles contractor the to compensation for any loss and/or expense suffered as a result. The employer requires such a clause in case there are delays, which may mean that the site is not ready for occupation. For example, demolition work may not be complete, permits have not been obtained, design work is incomplete or finances are not in place, and so on.

Powers to postpone and suspend the work

To postpone is to put off to a later date or defer to a later date, and to suspend is to debar or to put off, usually for a short time. Clause 23.2 further gives power to the architect to issue instructions with regard to the postponement of any work under the contract. This power is limited by clause 28.2.2, which provides that suspension of the work for reasons specified in the clause provides grounds under which the contractor can determine its employment under the contract. The length of time allowed for this is subject to the appendix. If no period is stated in it, the period of delay is limited to a month.

Further powers to suspend arise out of the operation of clause 28A. It allows either party to determine the contractor's employment after the expiry of the period of suspension permitted. This period differs depending on the provision under which it is applied:

1 Under clause 28A.1.1.1 to 1.1.3 where, unless specified in the appendix, the period permitted is three months. Events that trigger it are *force majeure*, damage resulting from specified perils and civil commotion.
2 Under clause 28A.1.1.4 to 1.1.6 where the period is one month unless a different period is specified in the appendix. Triggers are architects instructions under clause 2.3, 13.2 or 23.2 'due to negligence or default by local authority or statutory undertakers carrying out statutory obligations', hostilities (whether war is declared or not) or terrorist activities.

Suspension under the ICE 7th

The ICE 7th contains wider powers to order the suspension of the work. A general power is given to the engineer by clause 40 to suspend the work 'whenever the engineer considers it necessary'. This may be for a number of reasons:

(a) design changes by the employer
(b) problems with possession and access – this is more likely in civil engineering contracts than in building work
(c) unforeseen conditions encountered.

Civil engineering contracts specify the time for completion rather than a specified date as in the JCT 98. The ICE 7th states in clause 41 that the work commencement date shall be either:

• the date specified in the appendix, or where no date is fixed
• a date between 14 and 28 days of the award of the contract, or
• such date as agreed between the parties.

Failure to give possession of the site

Most writers have considered that failure to give possession of the site was a repudiatory breach of contract: see Emden (Binder II Ch. 9 para. 31) views such a failure as a fundamental breach of contract. In this aspect it adopts the reasoning in *Walter Cabott*. However, it points out that this approach resembles the long discredited doctrine of fundamental breach of contract. In *The Wardens and Commonality of the Mystery of Mercers of the City of London* v. *The New Hampshire Insurance Company* (1992) 60 BLR 25 Parker LJ stated that the failure to give the contractor possession of the site was not a repudiatory breach of contract. Delay under a building contract was not unusual and the contract made provision for delay caused by 'relevant' events. Although the relevant events did not include delay after the six-week period, in giving possession of the site, the possibility of delay was inherent in a building contract. The contract in that case being JCT 80, clause 23 which made extensive provision for delay in providing possession of the site.

What is the site?

Neither the JCT 98 nor ICE 7th defines the word site. The extent of the site is a question of fact to be determined in the light of all circumstances. In *Walter Cabott* the Federal Court of Appeal adopted the conclusion of the trial judge that:

> 'Subject to express exclusions, the "site for the works" must in the case of a completely new structure, comprise not only the ground actually to be occupied by the completed structure but so much of it around it as is within control of the owner and is reasonably necessary for the carrying of the work efficiently and accordance with generally accepted construction practice.'

The extent of the site is likely to be more restricted for building work than it is for works of civil engineering construction. Whether the site extends to borrow pits for fill material for example is unclear. Quite often fill materials have to be imported. The borrow pits or quarry containing material complying with the specification may be some distance from the construction site.

Site conditions

Generally the employer owes no duty to disclose to the contractor the condition of the site. Hudson (1995, p.571) states that 'the owner owes no duty to the contractor to do work to render the site easier to work upon, or to conduct surveys or sink boreholes or make other investigations.' A view echoed by Keating (2001, p.1,082) who considers that the words 'the contractor shall be deemed to have inspected and examined the site . . . and satisfied himself' as re-affirming the general principle under English law that the contractor takes the risks in the ground and in the sub-soil. There are a number of ways in which the contractor may challenge the strictness of the rule. In addition the standard forms of contracts transfer a substantial part of this risk to the employer.

Challenging the rule

There are a number of the ways in which the contractor may challenge the general rule described above. This can be by arguing for (i) an implied term (ii) a warranty (iii) a duty of care and (iv) claiming misrepresentation.

An implied term

The courts have been reluctant to imply terms. Two old cases demonstrate the difficulty of establishing entitlement to compensation, where the ground conditions proved to be unfavourable. These are *Thorn* v. *Corporation of London* (1876) 1 App Cas 120 and *Bottoms* v. *Lord Mayor of York* (1812) 2 Hudson's Building Contracts (4th edition) p.208. In *Bottoms* the contractor abandoned the work on discovering the ground conditions were unsuitable and required extra work. The engineer refused to issue a variation. It was held that the contractor was not entitled to abandon the work on discovering the ground conditions. The decision must be qualified in that (i) neither party to the contract carried out a soil investigation and (ii) the employer had warned the contractor that he was bound to lose money due to the expected soil condition.

In *Thorn*, the Corporation employed the contractor to demolish an existing bridge and construct a new one using plans and specification prepared by the Corporation's engineer. In constructing the bridge the contractor was to use the existing caissons to construct new ones. This proved impossible and the contractor sued to recover his extra costs for completing the works. The House of Lords rejected his claim that the contract contained an implied warranty that the bridge could be built inexpensively in accordance with the plans and specification. *Thorn* and its ambit was considered by the Supreme Court of New Zealand in *RM Turton & Co Ltd (In Liquidation)* v. *Kerslake & Partners* [2000] NZCA 115. A clause headed 'sufficiency of tender' read: 'the contractor shall have satisfied himself before tendering as to the correctness and sufficiency of the tender for the works to cover all his obligations under the contract and all matters and things necessary for the proper completion and maintenance of the works'. The contractor commenced an action against the engineer who had prepared the specification for mechanical and engineering work. The specification proved to be incorrect and the contractor incurred additional costs in completing the work in compliance with the contract. It sued the engineer in the tort of negligence, for the costs of carrying out the remedial work. By a majority, the Court of Appeal decided that the rule in *Thorn* applied, that no warranty was given by plans and specifications in the tender, and that the 'sufficiency' clause would not 'offend against the general principle'. As far as the general claim against the engineer was concerned, it held that there was no justification for imposing duty of care on the engineer. It rejected any application of the principle of *Hedley Byrne* v. *Heller* to the contractual matrix surrounding the relationship of the parties to the project. This was the traditional construction contract in which the duties of the parties were fixed by their respective contracts. Note the powerful and wide ranging speech delivered by Thomas J in *RM Turton* in which he dissented from the majority. This is considered further in Chapter 13 below.

A case where the contractor successfully argued for a warranty/implied term is *Bacal Construction (Midlands) Ltd* v. *Northampton Development Corporation* (1975) 8 BLR 88. The contractor, as part of its tender, submitted sub-structure designs and detailed priced bills of quantities for foundation work. The design was based on the soil conditions found in boreholes made adjacent to the site. The tender documents stated that the site conditions were a mixture of Northamptonshire sand and upper Lias clay. Tofu was instead discovered on the site during the work. As a result, a redesign was necessary as well as the carrying out of additional work. The contractor successfully claimed that there had been a breach of an implied term or warranty by the employer that the soil conditions would accord with the hypothesis upon which they had based their design.

A duty of care

There is no English authority on whether the employer owes a duty of care in tort to the contractor to supply accurate or sufficient information. This preposition was not entirely dismissed by the Court of Appeal in *Jarvis* (discussed further above in Ch.2). There is however, authority from other common law jurisdictions that such a duty may exist but the position is by no means clear-cut.

Australia

The High Court of Australia has declined the invitation to say whether a duty of care existed and in what circumstances. In *Morrison-Knusden* v. *Commonwealth of Australia* (1972) 13 BLR 114, it accepted the difficulty of contractors in obtaining site information for themselves during the tender period. Barwick CJ commented at p.121:

'The basic information in the site information documents appears to have been the result of much highly technical effort by a department of the defendant. It was information which the plaintiff had neither the time nor the opportunity to obtain themselves. It might even be doubted whether they could be expected to obtain it by their own efforts as a potential or actual bidder.'

The court declined to hold that the words in the invitation to tender that 'It is clearly understood by tenderer that the Commonwealth will not be responsible for any interpretation or conclusion drawn by the tenderer in regard to the site conditions based on the information contained herein' amounted to a disclaimer as to the accuracy of the information provided. In *Dillingham Construction Pty Ltd* v. *Downs* (1972) 13 BLR 97 however, the Supreme Court of New South Wales accepted that the principle in *Hedley Byrne* v. *Heller* could apply between contracting parties. Hardie J observed that the fact that the parties were in a pre-contract relationship at the time the duty was said to have been created, did not by itself exclude the application of the principle. The court held that there was a no clear misrepresentation by the employer in that it did not disclose to the contractor, records of underground mine workings stored in its archives.

Canada

Morrison-Knusden v. *State of Alaska* [1974] ALJR 265, one of the tenderers disclosed information about the site to the employer's engineer. The court declined to hold that a duty to disclose the information to all tenderers existed. Such a duty was however, held to exist in *Brown & Huston* v. *York* (1985) 5 CLR 240. The employer was held liable for the failure of the engineer in omitting water levels from the soil information provided to the contractor. The engineer himself was sued in *Edgeworth Construction Ltd* v. *N D Lee & Associates* (1994) 107 DLR (4th) 169, the Supreme Court of Canada. The winning tenderer successfully sued the engineer for losses suffered as a result of faulty specifications and drawings. This was a pre-trial motion on pleaded facts, and it was held that liability arose under the principle in *Hedley Byrne*. *Edgeworth* itself was rejected by the majority in *RM Turton* on the ground that no chain of contracts was present. Hudson too is highly critical of the decision in *Edgeworth*. Part of his argument is that the contractual background in a construction contract is incompatible with duties arising under *Hedley Byrne* (1995 p.166).

Misrepresentation

A misrepresentation is an untrue statement that induces a contract. The statement must have acted as an incentive for the other person to enter into the contract.

The statement must be one of fact and not of opinion. However, half-truths may sometimes amount to a misrepresentation.

The common law position

Up to the decision in *Hedley Byrne* the common law only recognised fraudulent and *innocent* misrepresentation. *Hedley Byrne* introduced a third category, namely negligent misrepresentation. Fraudulent Misrepresentation is where the statement was known to be false or the defendant did not care whether it was true or not. If the claimant entered into a contract to his detriment, he could recover damages under the *tort of deceit*. If the claimant had entered into a contract he could rescind it. Whereas if the defendant makes a statement that he believes is true, even though he should have known it was untrue, this is an innocent misrepresentation. The defendant was unable to sue for deceit for damages. Equity intervened and allowed the defendant to rescind the contract, that is to say, set aside the whole of the transaction as if the parties had never entered into a contract.

The Misrepresentation Act 1967

The Act provides a third mechanism for challenging the accuracy of the site information provided by the employer at tender stage. Section 2 (i) dealing with damages for negligent misrepresentation, provides a defence that the defendant had reasonable grounds for the belief. Section 2 (ii) deals with innocent misrepresentation. The section does not provide for damages, only for rescission of the contract. It does however give the court discretion. It can in certain circumstances award damages and keep the contract alive. It is easier to sue under the Misrepresentation Act 1967 than at common law. This is because the Act reverses the burden of proof. Where a misrepresentation has induced a contract it is up to the defendant to prove it had reasonable grounds for making the statement.

Fraudulent misrepresentation and disclaimers

Tender invitations for piling work are frequently issued with disclaimers. They absolve the employer from any liability for the accuracy of the site information. In *Pearson* v. *Dublin Corporation* (1907) AC 351, the size of the sea wall was inaccurately marked on the drawings. The engineer failed to check the accuracy of the dimensions. It was held that a clause in the contract requiring the contractor 'to satisfy himself' as to the accuracy of the information only applied to honest mistakes made by the engineer. The failure to check the information amounted to fraudulent misrepresentation. Lord Loreburn L.C. observed at p. 353: 'Now it seems clear that one can escape liability for his own fraudulent statements by inserting in a contract a clause that the other party shall not rely upon them.'

Innocent misrepresentation

The leading case under the Act is *Howard Marine Dredging Co Ltd* v. *Ogden & Sons (Excavations) Ltd* (1977) 9 BLR 34. During negotiations for the hire of two barges, the payload of the barges was incorrectly stated to be much higher that it was. Bridges LJ stressed that where 'in the course of negotiations leading to a contract, the statute imposes an absolute obligation not to state facts which the representor cannot prove he had reasonable grounds to believe.'

Modification of the general rule

This general rule that the contractor takes the risk in the site conditions, is modified in the standard forms in three ways. First, the tender itself may be subject to a 'sufficiency clause.' Second, the preparation of the bills of quantities may incorporate standard methods of measurement. Third, the contract may contain provisions dealing with responsibility for unforeseen ground conditions.

Sufficiency of tender

The ICE 7th in clause 11 requires the contractor 'to have inspected the site and its surroundings and information available in connection therewith and to have satisfied himself . . .' of the correctness and sufficiency of his rates and prices. This restates a general principle of English law that it is for the contractor to satisfy himself of the circumstances of the work to be undertaken. Clause 12 however, relieves the contractor from physical conditions or artificial obstructions, which in his opinion could not have been foreseen by an experienced contractor. Hudson expresses deep reservation about the ambiguous nature of the wording of such a clause. It considers that the lack of precision in the wording and the highly technical nature of the subject matter, explain the scarcity of case law on its interpretation. The wide discretion conferred on engineering arbitrators mean the courts are unlikely to interfere with their awards (1995 p.1001).

Standard methods of measurement

Both the JCT and the ICE contracts incorporate the standard methods of measurement. JCT 98 in clause 2.2.2.1 states that the contract bills are to have been prepared in accordance with the standard method of measurement. ICE clause 57 states that the bills of quantities are deemed to have been prepared and that measurement shall be made according to the procedures set out in the 'Civil Engineering Standard Method of Measurement third edition 1991.' The effect of this provision was that where the contract incorporates a method of measurement, the contractor may be entitled to extra payment for unexpected ground conditions. *C. Bryant & Sons Ltd* v. *Birmingham Hospital Saturday Fund* [1938] 1 All ER 503 illustrates the application of the provision. The JCT 63 conditions in clause 11 provided that the quality and quantity set out in the bills of quantities had been prepared in accordance with the then standard method of measurement. Rock, which the method of measurement required to be given separately, was discovered. The contractor was held entitled to be paid separately for excavating the rock, plus an allowance for a fair profit.

Effect of a programme

Whether a programme provided by the contractor is a contractual document will depend on the wording of the contract. JCT 98 in clause 5.3.1.2 requires the contractor to provide the Contract Administrator with 'two copies his master programme for execution of the works.' The provision of such a programme does not bind the contractor to perform in accordance with it. Often misunderstood, the purpose of such a requirement is for the benefit of the employer. It enables the employer's agents, to make arrangements for the supply of information and working drawings, the co-ordination of the work of other contractors, consultants and suppliers. Hudson (p.1128) refers to the use made by contractors of overoptimistic programmes to advance unmeritorious claims.

In *Glenlion Construction* v. *Guinness Trust* (1987) 111 Con LR 126, the contract bills required the contractor to provide a progress chart. This had to be in the form of a bar chart in the approved form and to be kept up to date. A residential development at Bromley in Kent was let under JCT 63. The completion date was 114 weeks after the granting of possession. The bill of quantities required a programme chart to be produced one week after possession. The contractor programmed to complete within 101 weeks. At issue was whether the requirement was a term of the contract. Clause 6.3 (1) (now JCT 98 clause 2.2.1) provides 'nothing . . . can override, modify or affect the printed conditions.'

The Contractor's programme showed a completion date before the contractual completion date. It argued for an *implied term* that the employer should perform the contract so as to enable the contractor to carry out the work, in accordance with the programme by the programme completion date. It was held that the supplemented printed conditions did not override the printed conditions. The progress clause was not a contract provision and the contractor was entitled to finish early whether a programme was produced or not. There was *no requirement* (i.e., no implied term) that the employer had to do everything necessary to enable the contractor to complete by the earlier date programmed. Note now that JCT 98 in clause 5.4 requires the employer to provide the contractor with an 'Information Release Schedule' which states what information the Architect will release and when.

Like the JCT 98 the ICE 7th, in clause 14 also requires the contractor to provide a programme. Such a programme has to be approved by the engineer. However clause 14(9) makes clear that the consent of the engineer to the programme does not relieve the contractor from his contractual duty or responsibilities under the contract. There is, of course, no reason why the employer should not put a clause in the contract, making the provision of a programme, a contractual document.

Instructions

In order to complete by the contractual date, the contractor requires instructions, nominations, information, plans and details. If it is required, the employer must supply it at a proper time to enable him to complete the work. Where the matter is not regulated by an express term, the employer's obligation is to supply it within a reasonable time. What then is a reasonable time? This depends on the express terms of the contract and all the circumstances. What is reasonable does not solely depend on the convenience and financial interest of the contractor. Diplock J (as he was then) said in *Neodox Ltd* v. *Borough of Swinton & Pendlebury* (1958) 5 BLR 34, though it was in their [contractors] interests to have every detail cut and dried on the day the contract was signed, the contract did not contemplate that situation.

It also depended on the point of view of the employer, the architect, their staff and other employees including other contractors. The test is when does the contractor need it?

Other obligations

1 The employer must obtain planning permission and building regulation consent in sufficient time for the contractor to proceed without delay.
2 If an architect is to supervise the work, one must be appointed. If the architect dies another must be appointed.

Payment

Payment is usually made by instalments as the work proceeds, normally against interim certificates issued by the architect or where appointed, the contract administrator. Failure to pay a certificate when due does not entitle the contractor to repudiate the contract or amount to a repudiation. The contractor's remedy is to sue on the debt created by the contract. Note however, that the common law rules have been modified by the HGCRA 96. Every construction contracts falling within the legislation must contain:

1 A payment schedule.
2 A procedure for notifying contractor of the amount to be paid, when payment is to be made and how the amount has to be calculated.
3 Failure to pay the amount due gives the contractor a right to suspend work till paid.
4 The parties have a right to immediate adjudication

See above, for the implementation of this provision into clause 30 of JCT 98. Clause 30.1.1.1 provides for payment against certificates. Additional rights are contained in clause 28.1, which gives the contractor power to determine its employment under the contract. Clause 60 regulates payment under the ICE 7th and incorporates similar provisions.

The position at common law

In an entire contract, performance is a condition precedent to payment. The JCT/ICE contracts are not entire ones as far as payment is concerned (since payment is by instalments as the work is carried out). However, the contract is an entire one as far as the retention money is concerned. This means the contractor must complete the work before it is entitled to payment of the retention monies deducted from interim certificates of payment. The rules about entire contracts exist largely in relation to small, informal contracts. A householder employs a jobbing builder for a particular item of work. Experience has shown that they do not always finish the work. The only effective sanction then is withhold payment until the builder returns to complete the work outstanding. It would however, be a mistake to imagine that only small informal contracts attract the rule. A recent example of a large commercial contract where the rule was applied is *Discain Project Services Ltd* v. *Opecprime Developments Ltd.* [2001] EWHC Technology 450.

The claimants entered into contract with the defendants for the fabrication and installation of part of the steel balconies and the building of a deck for a block of flats being refurbished. The value of the contract was initially near £600,000. A dispute arose as to (i) the terms of the contract between them and (ii) exactly what work the claimant had originally agreed to undertake. The defendants terminated the contract and the claimant brought an action for wrongful termination of the contract and for the unpaid balance. The defendant advanced a counterclaim alleging defective and incomplete work. HHJ Richard Seymour QC found that a contract came into existence after an oral acceptance by the defendant of an offer contained in a letter contract written by the claimant. The letter summarised the agreement reached by the parties at an earlier meeting. The resulting oral contract was later varied, with the claimant agreeing to provide further work and material to the balconies. No contract was found to exist for the work carried out for the fabrication and erection of the deck.

The termination of the contract

The judge found that there was no express term in relation to construction of the balconies as to when the work of fabrication and erection was to be done.

> 'Consequently, as a matter of law the obligation was to perform the contract within a time that was reasonable in all the circumstances in which the contract work was done – see *Hick* v. *Raymond and Reid* [1893] AC 22. If the contract was not performed in a reasonable time, it was, as a matter of law, open to Opecprime to serve a notice making time for completion within a time that was reasonable as at the date of the giving of the notice of the essence of the contract – see *Charles Rickard Ltd* v. *Openheim* [1950] I KB 616. If the contract was not performed within the time fixed by such notice, then Opecprime could have treated Discain as having repudiated the contract.'

By failing to give any notice the termination of the contract was made for no good reason. Since the claimant had not made claim for damages for wrongful termination, the defendant could not claim damages for the work that was outstanding at the time of determination of the contract.

The defective work

The judge held that implied into the contract, as a matter of law, were terms that the balconies when complete would be of (i) merchantable quality, (ii) fit for its intended use and (iii) erected with reasonable care and skill. The defendant argued that as no balcony had been complete because of defective work, (the parties agreed there were some defective and incomplete work) no payment was due. To this argument the judge concluded:

> 'What I have found provided for was payment on installation of the steel works for every four balconies. That did not require that the installation had to be complete in every detail before there was a right to payment. Under the

doctrine of substantial performance, illustrated by the well-known case of *Hoenig* v. *Isaac* [1952] 2 All ER 176, if a party has substantially performed his obligations under the contract, he is entitled to payment, although exposed to a claim for damages in relation to those aspects in which his performance of his obligations is less than complete. In my judgement that doctrine is applicable to the circumstances of the present case.'

He held that in the light of those findings, the claimant was in breach of contract. The damages awarded against it were the reasonable costs of their outstanding claim, less the cost to them of undertaking the remedial work. See below for a further explanation.

The claim in restitution

The judge found that no contract had been made for the fabrication and erection of the deck. As a result the defendant was entitled to be paid a reasonable sum on a restitutionary basis. The defendant had no claim for damages for breach of contract. He then added:

'However, it is convenient from a practical point of view, to calculate what sum was due to Discain by calculating what would be a reasonable price for the Deck, if it were not defective in the respects admitted on behalf of Discain, and then to deduct the reasonable cost of remedying the admitted defects.'

Where there is an entire contract, the rule in *Cutter* v. *Powell* (1795) 6 TLR 320 applies. In *Cutter* a sailor died before completing the return journey. It was held that his widow could not recover his wages, as he had not completed the contract. The justness of this result is often defended on the basis that he was to be paid substantially more on his lump-sum contract, than had he been paid a normal wage for a sailor at that time. In other words he took the *risk of non-completion* and *non-payment* in return for a higher rate of pay. As a result of his early death, his services were never completed and thus no obligation to pay ever arose.

Hoenig v. *Isaacs* and the doctrine of *substantial completion* have diluted the harshness of this doctrine. Where there has been a substantial completion of the contract, the contractor is entitled to payment less a deduction for the costs of repairing the defects. In deciding whether the work is substantially complete, account must be taken of (i) the nature of the defects and (ii) the proportion between the cost of rectifying them and the contract price. The test is not that the defects are trifling compared to the contract price. It is clear however, that where the defects are major, the rule in *Cutter* still applies: see *Bolton* v. *Mahadeva* [1971] All ER 1322. A central heating system installed by the contractor failed to heat the house adequately. Cairns LJ said in the Court of Appeal:

'If a central heating when installed is such that it does not heat the house adequately and is such, further, that fumes are given out, so as to make the rooms uncomfortable, and if the putting right of those defects is not something that can be done by some slight amendment of the system, then I think that the contract is not substantially performed.'

Such a result might nowadays be subject to a claim in restitution, where the employer receives the benefit of the work. Guidance as to the circumstances in which an entire contract arises, and when payment is due, was given in the following case. In *Holland Hannan & Cubitts (Northern) Ltd.* v. *Welsh Health Technical Services Organisation* (1981) 18 BLR 80, HHJ Newey J OR QC said: 'It seems to me the law is clear and can be shortly be stated as follows:

1   An entire contract is one in which what is described as "complete performance" by one party is condition precedent to the liability of the other party: *Cutter* v. *Powell* (1795), and *Munro* v. *Butt* (1858).

2   Whether a contract is an entire one is a matter of construction; it depends upon what the parties have agreed. A lump sum contract is not necessarily an entire contract. A contract providing for interim payments, for example as work proceeds, but for the retention money to be held until completion is usually entire as to the retention moneys, but not necessarily to interim payments: Denning LJ in *Hoenig* v. *Isaac* (1952).

3   The test of complete performance for all purposes in an entire contract is in fact "substantial performance": *H. Dakin & Co Ltd* v. *Lee* (1916) and *Hoenig* v. *Isaac* (1952).

4   What is substantial is not to be determined on a comparison of cost of work done and work omitted or done badly: *Kiely & Sons Ltd* v. *Medcraft* (1965) *and Bolton* v. *Mahadeva* (1972).

5   If a party abandons performance of the contract, he cannot recover payment for the work he has completed: *Sumper* v. *Hedges* (1898).

6   If a party has done something different from that which he has contracted to perform, then, however valuable his work, he cannot claim to have performed substantially: *Foreman & Co Proprietary Ltd* v. *The ship 'Liddesdale'* (1900).

7   If a party is prevented from performing his contract by default of the other party, he is excused from performance and may recover damages: dicta by Blackburn J in *Appleby* v. *Myers* (1866); *Mackay* v. *Dick* (1881).

8   Parties may agree that, in return for one party performing certain obligations, the other party will pay him on a quantum meriut.

9   A contract for payment of a quantum meriut may be made in the same way as any other type of contract, including conduct.

10 A contract for a quantum meriut will not be readily inferred from the actions of a land owner in using something which has become physically attached to his land: *Munro* v. *Butt* (1858).

11 There may be circumstances in which, even though a special contract has not been performed, there may arise a new or substituted contract: it is a matter of evidence: *Whitaker* v. *Dunn* (1887).'

Interim payments in informal contracts

In such contracts, the parties normally make no arrangement for payment by instalments. Hudson claims that there is a class of contracts for work and services in which, from the nature of the work, it is necessary to imply a term for reasonable payments on account of the work, at reasonable intervals (at p. 493). *Appleby* v. *Myers* (1866), states that 'generally, and in the absence of something to show a contrary intention, the bricklayer . . . is to be paid for work and materials he has done and provided, although the work is not complete'. Similarly in *The Targeste* (1903) it was said that 'a man who contracts for a long and costly piece of work, does not contract unless he says so, that he will do all the work, standing out pocket until he is paid at the end'. Note also *Charon (Finchley)* v. *Singer Sewing Machine Ltd* (1988) 112 SJ 536, where vandals broke into a shop being converted, the day before it was due to be completed. Acting under a provision in the specification, the employer's agents ordered the contractors to make good the damage at the earliest possible moment. Nield J held that as it was an entire contract, all the work needed to complete before the employer was liable to pay the price. The contractor was therefore bound to reinstate the damage and could not recover any extra payment.

The contract sum

This is defined by the JCT 98 in its articles of agreement as the sum the employer will pay the contractor exclusive of VAT in accordance with Article 2. A clear sum is specified at the outset, but it may be altered as work proceeds, as provided for in the conditions. Clause 3 provides that adjustments to the contract sum shall be taken into account in interim certificates. This can be a result of variations under clause 13 or the payment of loss and/or expense claims under clause 26. Hudson summarises the position under the English standard forms of contract (1995 at p.989) as follows:

1 After 1973 the ICE conditions expressly permit an increase in the price where it is unreasonable or inapplicable because of an increase in the as-built quantities. In theory, this should also trigger a decrease in the price.
2 Contracts pre the 1980 RIBA/JCT contract permitted expressly a 'fair valuation' under clause 11 (4) (b) and a possible 'direct loss and or expense' claim under 11(6).
3 The post 1980 and 1998 JCT contracts contain 'an almost embarrassing choice of financial claims'. These include:
    (a) 'fair allowance' or 'significant changes of quantities' claim
    (b) allowances for adjustment of preliminary items
    (c) similar allowances if there is substantial change in the conditions under which the work is executed
    (d) a possible claim for disturbance of regular progress
    (e) a final fair valuation under clause 13.5.6.

The commercial purpose of these provisions should always be borne in mind because they alter the allocation of risk. If the contractor is being reimbursed for these events, the tendered price should be lower because the contractor does not have to include a price for every risk it foresees.

Market fluctuations

These provisions reimburse the contractor for input price changes over which it has no control at all. JCT clause 37 brings the fluctuation clause into operation. Clause 38 covers increase in contribution, levies and tax fluctuations; clause 39 deals with labour and material cost and tax fluctuations; and clause 40 concerns the use of price adjustment formulae. Once again, these clauses should reduce the tender price because they relieve the contractor from making allowances for events which may not occur in the life of the project.

Retention money

This is a practice widely used in construction contracts. Where interim payments are made, it is usual for a small percentage, typically 5 per cent of the value of certificate, to be deducted. A fund is thus created to rectify defects or to induce the contractor to return and rectify any defects arising after completion. Under JCT 98 clauses 30.2 to 30.4 create the retention fund.

The status and treatment of retention funds

Since construction contracts are usually carried out over quite lengthy periods of time, the retention monies are often kept for periods of two years or more after completion. If the employer becomes insolvent, the contractor would want to claim priority over the general creditors. One way of ensuring this is to create a trust of the retention monies. JCT 98 provides for this in clause 30.5.1: 'the Employer's interest in the Retention money is fiduciary as trustee for the Contractor and any Nominated Sub-Contractor (but without the obligation to invest)'. A very important limitation, meaning that the employer has no duty to enlarge the fund.

The retention money becomes a trust fund of which the employer is the trustee. The employer cannot use the money in its own business and has a duty to safeguard the fund in the interests of the beneficiaries: Bedlam LJ in *Wates Construction* v. *Franthom Property* (1991) 53 BLR 23 at 37, CA. However, there is no duty to invest and account as trustee for the fund. A trustee of property normally has an express duty to increase the value of the fund. Any liquidator appointed where the employer goes bankrupt, or trustee in bankruptcy, must hand over the retention money in full. Where no fund has been established, no mandatory injunction will be granted to create such a fund if the employer is insolvent. The contractor becomes merely an unsecured creditor.

Practical problems

Since Clause 30.5.1 on its own is not sufficient for such a claim to be made as the money must be 'appropriated and set aside.' This requires it to be in a separate trust fund and usually in a separate bank account. Under clause 30.5.3 application can be made for a court order requiring the employer to keep the money separate. The employer can keep the interest earned in the separate account. The position is the same even if there is no specific provision in the contract. *Raynack* v.

Lampeter Meat Co (1979) 12 BLR 30. Note however the special facts, the retention was 50 per cent and the defects liability period five years long. The application for a court order must be made before liquidation proceedings have started. Under clause 30.1.1.2 rights of set-off can be exercised against the retention fund. If a certificate of delay has been issued and the liquidated and ascertained damages are in excess of the fund, the Employer cannot be made to pay the retention into a separate fund.

Set-off

The right of the employer to refuse payment on the basis that the employer has some counter-claim, which would extinguish/reduce, the amount that is owed. Again, the position of set-off has been clarified by the HGCRA 96, and will be further examined in Chapter 10.

Summary

1 The main obligation of the employer is to pay for the work completed by the contractor.
2 Positive and negative terms are implied into the contract. The employer must do all things necessary and not hinder or prevent the contractor in completing the work by the completion date.
3 The employer has to provide the site by the contractual date and where no date is fixed within a reasonable time to allow completion by the date set.
4 The contractor takes the risk in the ground conditions. This risk is modified by the (i) standard forms of measurement and (ii) clauses relieving the contractor from physical impossibility.
5 An entire contract is one where the obligation to pay arises only when the work is complete. This rule is modified in the standard forms by provisions for interim payment.
6 In such a contract the materials provided shall be of merchantable quality and fit for its intended use and the work carried out with reasonable care and skill.
7 Defective work carried out under an entire contract is subject to the doctrine of substantial completion. The general rule is that payment may be made less a deduction for the costs of carrying out the defective work.
8 Where no time for completion is specified, the employer after a time which is reasonable in the circumstances, may issue a notice to make time of the essence for the completion of the work.
9 Retention monies are often withheld for long periods of time after completion. The risk of employer insolvency is dealt with by creating trust funds. To be effective the money has to be kept in a separate account.
10 The problem of set-off is now dealt with by the HGCRA 96.

8 Time and Provisions for Delay

Time for completion

Where parties to the contract want to ensure its completion or performance by a specified date, it is usual to do so expressly. This is due to the normal rule of English law that time itself, is not of the essence in a contract. Where the parties have not done so, or the time for performance has passed, a party may well argue for an implied term that the contract should be performed within a reasonable time. It is then a question of construction for the court, bearing in mind the subject matter and its importance to the contract as a whole, whether time is indeed of the essence. In a variety of contracts, the courts have held that the requirement that time is of the essence, should be applied to the respective oblig-ations of both parties (examples are between (i) buyer and seller (ii) ship owner and charterer, (iii) purchaser and vendor, and (iv) contractor and owner). Time is normally of the essence where:

(a) The parties expressly provide in their contract for it to be so: see *Peak Construction* discussed below.
(b) Where the innocent party gives notice to the delaying party that unless performance is made in a reasonable time, it will regard the contract as at an end; see *Charles Richards*.
(c) The surrounding circumstances make it vital that the date fixed should be met.

In contracts for the sale of goods, exercising of options to purchase and in contracts for the sale of land, the courts have had no difficulty in making time of the essence. This is because in these types of contracts, market prices can fluctuate quickly. The time of performance may be crucial since a lapse of time may make the bargain entered into worthless by then. Within the context of a construction contract, the matter is much more complex. For a start, time being of the essence in such a contract will in practice, provide the employer with little benefit. What will be gained by insisting on completion by a given date when the project as a whole is in delay, especially when the likelihood of delay becomes apparent to both parties long before the completion date is likely to be reached? This awareness is complicated by other factors listed below:

1 The contractor will have expended large sums of money in carrying out the work.
2 The act of fixing the work to the land passes ownership to the owner, (even if the contractor is never paid).

3 Provisions for an extension of the time for completion together with provisions for Liquidated and Ascertained Damages (LAD) are incompatible with any intention that time should be of the essence.

Express or implied terms as to progress

The courts have been quicker to imply terms into the contract that the contractor must proceed with reasonable diligence and expedition. Most sophisticated modern contracts, including the standard forms, have express terms that allow the owner to terminate the contract for failure to complete the work with due diligence or reasonable expedition (per Hudson at p.1115). At common law, the Employer must give possession of the site in a reasonable time to allow the contractor to complete by the contractual date see *Freeman* v. *Hensler* (1900).

Express term

In the JCT 98, clause 23.1 requires that 'on date of Possession possession of the site shall be given to the Contractor who shall thereupon begin the Works and regularly and diligently proceed with the same and shall complete the same on or before the completion date'. The meaning of the phrase regularly and diligently was considered in *West Faulkner Associates (A firm)* v. *The London Borough of Newham* (1993) 61 BLR 81. The employer's contract administrators brought an action against the Local Authority (the employer) for outstanding fees and damages for the wrongful determination of their contract. The employer in turn alleged that they had failed to act with reasonable care and skill in administering the building contract in not advising them that there were grounds for determining the contractor's employment. The employer entered into a £3 million contract with the contractor to refurbish 132 flats and associated houses. The contract was to be completed in 61 weeks. After 40 weeks only 6 flats were completed. The architects insisted there were no grounds for determining the contract since the contractors were carrying out the work regularly, if not diligently. The quantity surveyor, on being consulted, said that it was 'inappropriate to determine'. The employer consulted lawyers who reported that the case was a borderline one and that without the support of the professional team, it was unwise to determine the contract. The employer resolved the matter by making an agreement with the contractor to terminate the contract and employed another to complete the work. In construing the words regularly and diligently, HHJ Newey QC OR said at first instance that the obligation of the contractor was to do the work in such a way as to complete in time. Factors to be taken into account included the following:

1 The contractor had to plan the work
2 Manage its work force.
3 Provide sufficient and proper materials.
4 Employ competent tradesmen.

The Court of Appeal held that the phrase meant orderly planned progress towards the completion date. The contractor must go about work in such a way as to achieve its contractual obligations. The work must be done in sequence; there must be a work plan and orderly progress.

The ICE 7th uses the phrase 'failing to proceed with the Works with due diligence' as a ground for determining the contractor's employment: see clause 63 (1)(b)(iv). In the Australian case of *Hooker Construction Pty Ltd* v. *Chris's Engineering Contracting Company* [1970] ALR 821, Blackburn J said of a provision that allowed the employer to determine the contractor's employment for failing 'to proceed with the works with reasonable diligence':

'I think that a sensible commercial construction of the phase is that the actual extent of the work completed is of some significance. Diligence in this context means, it seems to me, not only of the personal industriousness of the defendant himself but his efficiency and that of all who work for him.'

In accordance with clause 23.1.2 the 'Employer may defer the giving of possession . . . up to six weeks.' The purpose of such a provision is to provide against the eventuality that the employer may not be able to give possession. For example if demolition work is being carried out beforehand by another contractor and the work may not be completed in time. Clause 25.4.13 deems this a relevant event entitling the contractor to an extension of time and Clause 26.1 entitles the contractor to a claim for loss and/or expense caused by the delay.

Controlling time

Standard forms of contract in the construction industry make provision for the payments of Liquidated and Ascertained damages arising from delays in completion, caused by the contractor (LAD). The provisions of clause 24.1 of JCT 98 provide for 'Damages for non-completion.' It states that:

'If the Contractor fails to complete the Works by the Completion Date then the Architect shall issue a certificate to that effect. In the event of a new Completion Date being fixed after the issue of such a certificate such fixing shall cancel that certificate and the Architect shall issue such further certificate under clause 24.1 as may be necessary.'

It then makes complex provisions for the deduction of such sums. A number of factors complicate the matter further:

1 Under clause 24.2.1 the architect has to issue a certificate under Clause 24.1 stating that the contractor has failed to complete by the contractual date.
2 The Employer has to inform the contractor *in writing* before the date of the Final Certificate that he may require payment, or may withhold or deduct, LAD.
3 The Employer may, not later than 5 days before the final date for payment of the debt due under the Final certificate, do one of two things;
   (a) require in writing that the contractor pay LAD at the rate specified in the appendix and recover as the sum a debt or

(b) give a notice pursuant to clause 30.1.1.4 or clause 30.8.3 to the Contractor that he will deduct from monies due to the contractor, LAD at the rate stated.

Since the contract makes provision for the granting of extensions of time, the completion date can be extended. Where this happens, clause 24.2.2 provides that monies deducted as LAD shall be repaid to the contractor. Where further extensions of time are granted, and the contractor fails to complete by the new date, the original notice of intention to deduct remains effective: see clause 24.2.3. No new notice to deduct has to be issued. The HGCRA Act 96 requires a withholding certificate to be issued before LAD can be deducted. Clause 24 makes provision for this notice to be given five days before payment is due. In the *Construction Centre Group Ltd* v. *The Highlands Centre* [2002], OH, Court of Sessions, the defenders argued that they should be allowed to set-off their claim to LAD against the adjudicator's award. Lord Macfaden held that s.111 of the HGCRA 96 did not permit this to be raised as a case for resisting summary judgment of the adjudicator's award.

The main objective of LAD

The function of LAD is to act as an inducement to 'due performance of particular contractual obligations, or to regulate beforehand in an agreed and certain manner the rights of the parties'. Failure to do so leaves the parties the less predictable remedies available for breach of a contract. The classic example of such an inducement is a provision for the payments of LAD for late completion of a construction contact. Note the practice in some types of Facilities Management Contracts to use LAD as an incentive to perform.

The advantages of LAD provisions

Fixing the damages payable for delay beforehand has many advantages. It removes the uncertainty, cost and risk of suing at common law for damages caused by the breach of contract. In order to demonstrate this difficulty, damages for breach of contract is discussed at this stage. A claim for damages has two different aspects. First, the remoteness of the damage which is being claimed must be assessed and second, separated from the monetary assessment of that compensation. English law uses the tests formulated in *Hadley* v. *Baxendale* to decide whether a particular loss in the circumstances of the case is too remote to be recovered.

There are two limbs to the *Rule* in *Hadley* v. *Baxendale* [1854][1843-60] All ER 461:

'Where two parties have made a contract, which one of them has broken, the damages the other ought to receive in respect of such breach of such a contract should be such as may
(a) fairly and reasonably be considered as either arising naturally, i.e., from the usual course of things, from the breach of contract itself, or

(b) such as may reasonably be supposed to have been in contemplation of both parties, at the time they made the contract, as the probable result of breach of it.' (per Alderson B).

Balfour Beatty Construction (Scotland) Ltd v. *Scottish Power plc* (1994) 71 BLR illustrates the application of the two limbs of *Hadley* v. *Baxendale* in a construction setting. The main contractors were employed to construct the roadway and associated structures of the Sighthill section of the Edinburgh City by-pass. A concrete batching plant was built near the site and an agreement made with Scottish Power for the temporary supply of electricity. Part of the work was the construction of a concrete aqueduct to carry the Union canal over the road. During the contraction the batching plant broke down due to the rupturing of the fuses provided by Scottish Power in their supply system. Watertight construction of the aqueduct required a continuous pour. Once power was restored, attempts were made to continue the construction by cutting back the old concrete and adding fresh concrete. Despite these efforts the contractors were unable to meet the specification for a watertight aqueduct. They were ordered by the engineer to demolish the structure and start again. In an action against Scottish Power they claimed damages for breach of contract. In appeal to the House of Lords against damages of £229,102.53 plus interest, the House of Lords confirmed that:

1 Damages should either arise naturally from the breach or have been in the contemplation of the parties at the time they made the contract.
2 What one party knew about the other's business was a question of fact.
3 The demolition and reconstruction of the aqueduct was not in the contemplation of Scottish Power (since they did not know, and were unlikely to know, that a continuous pour was required make it water-tight).

Note that damages were awarded under the first limb of *Hadley* v. *Baxendale* for the damages that arose naturally when the fuses failed. An example of this was the costs of cutting back unsuccessfully the concrete in an abortive attempt to restart the work. The fact that the employer has suffered *no loss* does not mean that LAD is not payable. In *Clydebank Engineering and Shipbuilding Co* v. *Castaneda* [1905] AC 6, the appellant argued that the LAD clause was invalid because the Spanish Government had suffered no loss. In fact they actually had gained by the late delivery. Had the ships been delivered in time they would have been sunk by the United Sates along with the rest of their fleet. The House of Lords dismissed their arguments. The absence of loss was irrelevant in a LAD clause. Note that LAD provisions seldom reflect the actual loss arising from the breach of contract: see for instance the *Bath* v. *Mowlem* case. The Council was able to demonstrate that their actual losses would be far greater that the LAD of £12,000 per week (at para. 9).

LAD and penalties

On construction sites, the phrases LAD and penalty clauses are used interchangeably. The comment that penalties are illegal is usually met with surprise if not incomprehension. So what then is a penalty? The leading case is *Dunlop Tyre* v. *New Garage* [1915] AC 60 where it was said:

> 'The distinction between penalties and liquidated and ascertained damages depends on the intention of the parties to be gathered from the whole of the contract. If the intention is to secure performance of the contract by the imposition of a penalty or a fine, then the sum is a penalty; but if, on the other hand, the intention is to assess the damages for breaches of contract, it is liquidated and ascertained damages.'

Dunlop laid down a number of tests to be applied:

1 The description of the clause as one for LAD or a penalty is not conclusive. What has to be examined is the effect of the clause. In other words, even if it is called a penalty in the contract, but it is in fact a genuine pre-estimate of damages does not mean it is a penalty clause.
2 Is the sum extravagant compared to the cost of the likely breach?
3 Does the sum apply to all breaches whether they are minor or major ones?

In the *Oresundvarvet Aktiebolag* v. *Marcos Diamantis Lemos* (the *'Angelic Star'*) [1988] 1 LLR 122, Gibson LJ stated that the introduction of penalties was not a rule of illegality. It was a rule of public policy by which the courts refuse to sanction legal proceedings for the recovery of a penalty. This comment does not however, explain why the distinction exists. It appears that the distinction is purely a historical one as indicated by Lord Woolf in *Phillips Hong Kong Ltd* v. *The Attorney General of Hong Kong* (1993) 61 BLR 41: Const LJ 202. At p.55 he said that there was a great deal of disagreement about the origins of the rule. Such relief was granted in contracts where the sum agreed to be paid as penalty for non-completion of a contract could be estimated. This principle was particularly applied where it concerned the non-payment of a bond. Equity and the common law had long maintained 'a supervisory jurisdiction, . . . to relieve against provisions which are so unconscionable or oppressive that their nature is penal rather than compensatory.' The court in such a case could refuse to enforce what it considered a penalty. The case of *Stanor Electric* v. *Mansell* (1988) CILL 399, is an example of the application of the rule today. The contractual sum for LAD was £5000 per week for late completion of electrical work on two houses. One house was completed on time and the employer charged the contractor the sum of £5000 (£2500 a week) for the late completion of the second house. No provision for sectional completion having been agreed, the court held the sum charged to be a penalty. To qualify as LAD the contractor must know the figure payable from the start.

Two issues arose in *J. F. Finnigan Ltd* v. *Community Housing Association Ltd* (1993) Const LJ 311:

1 Community Housing Association (CHA) did not follow the requirements of clause 24 of the JCT 80 when deducting LAD and was therefore not entitled to deduct the £47,500 claimed.

2 In any event, the sum of £2,500 per week claimed as LAD amounted to a penalty.

The court came to the following conclusion:

1 Notification that the employer may 'require in writing' a sum to be paid or deducted as LAD was a condition precedent to the employer's entitlement to deduct. A note sent with the payment cheque did not comply with the requirement of the clause.
2 On the evidence given by the parties the figure of £2,500 was a genuine attempt to estimate the loss which CHA was likely to suffer if the contractor did not complete in time.

Note that the Court of Appeal subsequently decided that the notice need only be sufficient for a literate contractor to understand and was not a penalty. In addition, this situation could not happen now because of the HGCRA 96: see above in relation to the express terms in JCT 98 clause 24.

Phillips Hong Kong illustrate that provisions made by the parties to fix damages in situations where it is difficult to compute the likely loss, will not attract the penalty rule. The Government of Hong Kong employed the company in the construction of a highway. The contract works included the design, testing, delivery, installation and commissioning of a processor based supervisory system for the approach roads and twin tube tunnels. These were to be constructed under Smuggler's Ridge and New Needle Hill Mountain in the New Territories as part of the Shing Mun section of the project. The Government entered into seven separate contracts including the one with Phillips. By dealing directly with the contractors, the Government was seeking to exercise greater control over the whole project than would be possible with a single contract. The total value of the seven contracts awarded was HK(649 million and Phillips own contract was valued at HK(51 million. Each contract contained flow charts setting out the programme each other contractor had to meet. Each of the contractors therefore should have been aware of the activities which the other contractors would be engaged in at each stage of the work and the possible consequences of delay on the part of one contractor on the other contractors. LAD was to be paid at two rates:

1 A daily rate for being late on your own contract and
2 A daily rate if the overall project date was not met.

Phillips challenged the deduction of LAD on the basis that the contract clauses amounted to a penalty and that they were under no obligation to pay it. Held on appeal to the Privy Council:

1 The correct test for a penalty was whether the sum was a genuine pre-estimate of the likely loss. It was not the correct test to consider whether there were possible circumstances where a lesser loss would be suffered.

2 The fact that the Government might obtain on two grounds did not make it penalty.

Note the observation of the Privy Council that '. . . parties . . . should be able to know with a reasonable degree of certainty the extent of their liability and the risks which they run as a result of entering into contracts. This is particularly true in the case of building contracts and engineering contracts'. It is also particularly so in situations where it is difficult for the employer to quantify its actual losses. Provisions for LAD remove this uncertainty and it makes commercial sense for both parties to agree the actual losses recoverable beforehand. For the relationship between penalties and extension of time clauses see the case of *City Inn* below.

Calculating the LAD

In making the pre-estimate of likely loss, the figures may include the following: The employer's own supervisory staff costs. Additional costs:

- rent/rates on present premises
- rent/rates for alternative premises
- charges for equipment
- movement of equipment
- movement of staff (including travel costs)
- site charges (employer responsibility)
- extra payments to directly employed staff
- insurance
- additional administrative costs
- anticipated profit (if not covered above)
- interest
- inflation

The calculation of LAD in *Finnegan* above is a good example of how a genuine pre-estimate of damages can be made:

$$\frac{\text{Estimated total scheme cost} \times \text{HC Lending rate} \times 85 \text{ per cent}}{52}$$

Estimated total costs equal the latest estimate of the costs of construction figures plus acquisition costs, professional fees, legal fees plus any other costs incurred e.g., tender costs. This total amounted to £895,000. To this was added the Housing Corporation's lending rate of 10.75 per cent multiplied by 85 per cent (this is the likely amount of work complete when LAD becomes payable). The division by 52 produced a sum £1,730 per week, rounded up to £1,750 to allow for inflation. Ten per cent was added to this figure for CHA's expenses for late completion, rounded up to £200. Loss of rents on 18 units at £530 was also added giving a total of £2,480, rounded up to £2,500. Note the reason for the multiplication by 85 per cent is that by the time LAD is levied, that amount of work has usually been completed. The contractor argued that the rounding-up figure of £65 in the total meant that the figure produced was not a pre-estimate of likely costs but in fact a penalty.

The Judge concluded:

1 The sum of £2,500 was a genuine attempt to estimate in advance the loss arising if the contract works were not completed on time.
2 The figure was neither extravagant nor unconscionable.
3 Where parties rely on funding by third parties it is not unreasonable if they insist that the contract should contain a clause protecting their rights.

Preserving the right to liquidated damages

Where an act of prevention for which the employer is wholly or partially responsible, prevents the contractor from completing on time, the employer cannot recover LAD unless the contract provides otherwise. Phillomere LJ in *Peak Construction (Liverpool) Ltd* v. *McKinney Foundation Ltd* (1970) 1 BLR 111 commented that:

> 'A clause providing for liquidated damages are closely linked with a clause which provided for an extension of time. The reason for that is that when the parties agree that if there is delay the contractor is liable, they envisage that that delay shall be the fault of the contractor and, of course the agreement is designed to save the employer from having to prove the actual damage he has suffered . . . if delay is the fault of the employer . . . the problem can be cured if allowance can be made for that part of the delay caused by the employer; and it is for this purpose that recourse is to be had to a clause dealing with extension of time.'

In *Peak* the contractor agreed to construct multi-storey flats using the Corporation's own standard form of building contract. The completion period was 24 months and the Corporation's Director of Housing was appointed architect in charge. Provisions in the contract stated that 'time was to be of essence under the contract' and liquidated damages shall be '25s per dwelling for each and or part of a week for delay'. The architect was empowered to grant extensions of time for a number of reasons including 'extras and additions, *force majeure* and other unavoidable circumstances'. McKinney was the nominated piling sub-contractor and major defects were found in the piling after they had completed their work. After a site meeting, all parties agreed to submit the problem to consulting engineers and to accept their solution. A month later, the Housing Committee resiled from that position and only agreed to accept the engineer's report 'without prejudice'. The Corporation did not appoint the engineer for several months and several delays added to that period. McKinney was not able to recommence work until 58 weeks after the work stopped.

Held:

1   Since part of the delay was attributable to the Corporation's own delay, there would be a new hearing in order to assess how much of that period of delay was due to the defendants.
2   So far as the employers were concerned, they were not entitled to recover liquidated damages against the plaintiffs under the terms of the main contract. The Corporation could not recover liquidated damages for delay for which it was partly to blame

Some problems with liquidated damages

Construction professionals often misinterpret the right to recover LAD provisions when there is no obvious loss to the employer. In *BFI Group of Companies* v. *DFB Integration Systems Ltd* (1997) CILL 348, the contractor disputed the deduction of LAD. In an agreement under the JCT minor works contract, alterations were required to a transport depot. A delay of six weeks occurred in completing two out of six loading bays. The contractor raised the argument since the employer had to fit-out the building after completion; the contractors had caused no loss of revenue. HHJ Davies QC OR held that it was quite irrelevant to consider whether there had in fact been any loss. *Temloc Ltd* v. *Erril Properties* (1987) 39 BLR 30 illustrates some of the complexities in this area. The construction of a large shopping development at Plymouth was awarded under the JCT 1980. The contract sum was £840,000. Clause 24 as left in the printed contract as executed but the relevant figure in the appendix was completed as '£ NIL'. The contractor overran by some five weeks and the prospective tenant brought claims for delay against the developer. The developer in turn claimed against the contractor.

Held by the Court of Appeal:

> 'The effect of "£ NIL" was not that clause 24 was infective or to be disregarded, but that it was left to apply in a negative way. The true interpretation of the contract was that the parties had agreed that no LAD was payable for late completion. Since the LAD clause had exhausted the developer's remedies for late completion, there was no claim available for general damages on proof of actual loss.'

Extension of time and relevant events

The extension of time clause in JCT 98 is clause 25. The contract provides a comprehensive list of matters that give rise to a relevant event, which may entitle the contractor to an extension of time. Such a clause protects the contractor from blame for what can be called 'acts of prevention' by the employer since they prevent the contractor from completing the work on time. Its primary purpose is however, to protect the employer's right to LAD. The contractor is also under a duty to use his 'best endeavours' to avoid delay no matter how caused (clause 23.3.4.1). Under the provisions of clauses 25.4.1 to 25.4.18, the following relevant events entitle the contractor to make an application for an extension of time to complete the contract. (In order not to repeat the words clause 25.4, the references to the individual clauses are their last numbers only).

1 *'Force majeure'*. The meaning of the phrase in English law is rather unclear; it can be defined as an 'act of God'. It may have a wider meaning in the context of the standard forms and includes man-made events or interventions, which may be beyond the control of the party relying on it

2 'Exceptionally adverse weather conditions'. In *Walter Lawrence* v. *Commercial Union Properties* (1984) 4 Con LR 37 within the context of JCT 63, the phrase 'exceptionally inclement weather conditions' was considered. What had to be taken into account was not whether the extent of time lost was exceptional, but whether the weather was so exceptionally inclement so as to give rise to delay caused. Adverse includes extremes of heat and dryness. Weather records are required to prove that the weather was exceptional compared to the norm.

3 'Loss or damage occasioned by any one or more of the Specified Perils'. Clause 1.3 defines these as: 'Fire, lighting, explosion, storm, tempest, flood, bursting or overflowing of water tanks, apparatus or pipes, earthquake. Aircraft and other aerial devices or articles dropped therefrom, riot and civil commotion, but excluding excepted risks'. These are defined in the same clause and include nuclear exposition and pressure waves caused by aircraft or other aerial devices travelling at sonic or supersonic speeds

4 'Civil commotion' etc. This includes strikes, riot and war, including local conditions affecting trades involved in the contract.

5 'Compliance with architect's instructions'. Under the provisions of the contract, the architect is given wide-ranging powers to issue instructions to the contractor. The following instructions may entitle the contractor to an extension of time:
   (a) Clause 2.3 – where there is a difference or discrepancy between two documents, the contractor has to bring it to the attention of the architect.
   (b) Clause 2.4.1 – difference or discrepancy between Statements in Performance Specified work and later instructions.
   (c) Clause 13.2 – the architect may issue instructions requiring a variation. Where instructions are given amounting to a variation the contractor will be given an extension of time, etc.
   (d) Clause 13.3 – instruction with regard to provisional sums etc.
   (e) Clause 13.A.4.1 – where the employer does not accept a clause 13A quotation, the work may be done as variation under 13.3.
   (f) Clause 23.2 – instruction to suspend work for up to a month. See its effect in clause 28.2, a ground for terminating the contractor's employment under the contract.
   (g) Clause 8.3 – where the opening up for inspection of work covered up for inspection or testing show the work complies with the contract.
   (h) Clause 34 – antiquities. Clause 34 requires the contractor on discovering or during the excavating of the site 'all fossils, antiquities and other objects of interest or value' to use its best endeavours not to disturb such objects. It must also take any step if required to preserve the objects in the same position as they were found. It shall permit third parties to examine, excavate or remove such objects when instructed by the architect.

Clearly the contractor has to stop working immediately on discovering Roman remains.

(i)   Clause 35 – the nomination by the employer of sub-contractor for any specialist work.

(j)   Clause 36 – the nomination by the employer of a supplier of either equipment or specified materials.

6   A failure by the architect to comply with the requirements of the information release schedule (where provided under clause 5.4.1) or to comply with clause 5.4.2 in providing instructions, drawings and details.

7   'Delay on the part of a nominated sub-contractor or nominated supplier'. It is irrelevant what the cause is (whether due to sloth, the remedying of defects, bad luck, etc.): see *Jarvis Ltd* v. *Westminster Corporation* [1970] 1 WLR 637. The contractor must however, take all practicable steps to avoid or reduce any delay.

8   Delay in the execution or supply of work and goods by the employer and or persons employed by him.

9   Delay due to statutory powers coming into force during the contract.

10   Failure to secure essential labour and secure goods and material which could not be reasonably foreseen.

11   Work carried out by statutory undertakers in pursuance of their statutory obligations. The words were considered in *Henry Boot* v. *Central Lancashire Development Corporation* (1980) 15 BLR 1. The court held that statutory undertakers do their work under contract with the employer. Their work therefore falls under clause 25.4.8. Where statutory undertakers under their statutory powers, hinder the contractor, the work falls within clause 25.4.11. Most major developments will make specific provisions for the work of statutory undertakers. Where employed by the employer clause 24.5.8 applies. Where employed by the contractor, they are at the contractor's risk unless nominated. Then, any delay caused by them comes within clause 25.4.7.

12   A failure to give in due time ingress to or egress from the site. This clause applies where the contractor's right to exit or enter the site from and to land in the possession of the employer, is affected.

13   Where clause 23.1.2 is stated in the appendix to apply. This entitles the employer to delay giving possession of the site for up to 6 weeks from the date of possession.

14   The execution of work for which an approximate quantity is included in the contract bills, which is not a reasonably accurate forecast of the quantity of work required.

15   Delay caused to any performance-specified work by statutory requirements coming into force, which necessitates some modification or alteration in the work.

16   The use or threat of terrorism and/or the activity of the relevant authorities in dealing with it.

17   The Construction (design and management) Regulations 1994 (CDM) require the employer to appoint a planning supervisor. Compliance or non-compliance by the employer with the requirements of the CDM regulations, which

result in delay, is covered here. Clause 6A.1 states that the employer shall ensure the planning supervisor carries out his duties under the regulations.

18 'Delay arising from the suspension by the contractor of his obligations under the contract with the employer pursuant to clause 30.1.4' – a failure to pay certificates on the dates specified. This clause was inserted by amendment no. 18 which incorporated the HGCRA 96 into the JCT 80 (now JCT 98). No such provision was required by the Act.

Types of delay

HHJ Fay QC OR summarised the broad scheme of clauses 23 and 24 (now 24, 25 and 26) in *Henry Boot* v. *Central Lancashire Development Corporation* (1980) 15 BLR 1:

'The broad scheme of these provisions is plain. There are cases where the loss must be shared, and there are cases where it should be wholly borne by the employer. There are also those cases which do not fall into either of these conditions and which are the fault of the contractor, where the loss of both parties are wholly borne by the contractor. But in those cases where the fault is not that of the contractor the scheme clearly is that in certain cases the loss is to be shared: the loss lies where it falls.'

The scheme

1 Where delay is caused by the contractor, no extension of time is granted and there is no right to extra money.
2 Neutral events. The contractor gets an extension of time but no extra money. No LAD is payable.
3 Delay caused by the employer or its agent. The contractor gets an extension of time and extra money. No LAD is payable for the period of delay.

Granting extensions of time

The contractor is required by clause 25.2.1.1 to give written notice 'when it becomes reasonably apparent that the progress of the work is being or likely to be delayed'. The clause has number of requirements:

(a) notice is to be given forthwith i.e., immediately
(b) the written notice must include the cause or causes of delay
(c) the notice must identify any event which in his opinion is a relevant event.

Clause 25.2.2 further requires the contractor to provide with the notice or as soon as practical after giving of the notice:

(a) particulars of the effect of the relevant event
(b) an estimate of the extent of the delay.

Clause 25.3.1 requires the architect 'upon receiving the notice, particulars and estimate' to give in writing extension of time if in his opinion the event[s] is a relevant event and it is likely to delay the completion of the work. The fixing of the new completion date has to be fair and reasonable and take into account:

(a) which relevant event has been taken into account
(b) the addition or omission of work required under a clause 13.2 variation or the omission of work under 13.3 or any performance specified work and
(c) fix such a date not later than 12 weeks from the receipt of the notice or within 12 weeks of practical completion.

Extensions of tine and penalty clauses

In *City Inn Ltd* v. *Shepherd Construction* [2003] May 22, (unreported), the parties entered into a contract for the construction of a hotel in Bristol. The conditions of contract was the Scottish version of JCT 80 but with special conditions including a clause 13.8 and provision for adjudication. LAD was fixed at £30,000 pro-rata, which both parties agreed later was an accurate pre-estimate of likely damage for delay. The contractor applied for an extension of time and was granted a four-week extension. The dispute as to whether the initial completion date was ever changed was referred to an adjudicator who awarded the contractor an additional five weeks. The employers brought an action claiming that the provisions of clause 13.8 had not been followed and sought repayments of sums granted for loss and expense, and seeking LAD for late completion. The contractor defended the action, alleging amongst other things, that the provisions amounted to a penalty and was unenforceable. At first instance Lord Macfadyen had to construe the provisions of clause 13.8 before considering whether in effect it amounted to a penalty. Before dealing with his conclusions, it has to be pointed out that the clause is remarkably like clause 13A 'Variation instruction – contractor's quotation' under JCT 98. The crucial difference between the two being that if the contractor did not follow the requirements of the clause, the contractor lost its right to claim an extension of time.

Clause 13.8.1 required that the contractor, on receiving any instruction, that in the opinion of the contractor constitutes an instruction that requires an adjustment to the contract sum and the completion date, shall within 10 working days, in writing, submit to the architect the following initial estimates:

1 the adjustment to contract sum or completion date
2 additional resources needed (together with a method statement)
3 the length of extension of time required under clause 25 and the new completion date
4 the amount of any direct loss and/or expense

Within 5 working days of receipt, the architect may do one of two things: (i) instruct the contractor to comply with the instruction (in that case the clause 13.5 valuation rules, and clauses 25 and 26 apply or (ii) instruct the contractor not to comply and re-imburse the contractor the reasonable costs for preparation. The architect may also give instructions to the contractor to dispense with the requirements of clauses 13.8.1 and in complying with those instruction, clauses 13.5, 25 and 26 apply (in other words, it becomes an ordinary variation). In the absence of a notice of dispensation, where the contractor fails to comply with 13.8.1, the contractor is not entitled to any extension of time (and by implication no loss and/or expense) and further, may incur liability to pay LAD.

The main argument the contractor made was that the effect of the clause was that of a penalty. Lord Macfadyen having reviewed the case law came to the conclusion that, applying the tests laid down in *Dunlop*, the clause was penal. In electing not to comply with the clause, the contractor also deprived the architect of an opportunity to review the instruction in the light of its consequences. It was therefore of material value to the employer in proving an opportunity to rethink the instruction. He concluded that LAD was payable because of late completion and not a penalty for failing to take the steps required by the clause. The contractor appealed to the Court of Session. Giving its opinion, the Court rejected the contractor's motion. Clause 13.8 was a condition precedent to the right of the contractor to claim an extension of time. 'In short, Clause 13.8 provides the contractor with an additional right, in the specific case to which it applies, that would not be available to him in the case of an instruction issued under the general provisions of clause 4. But clause 13.8 does not oblige the contractor to invoke its protection. It merely provides the contractor with an option to take a certain action if he seeks the protection of an extension of time in which the clause applies.'

Concurrent delays

Where there are two causes of delay, one caused by the employer and the other by the contractor, if it is impossible to separate delays, the contractor gets the benefit. See *Walter Lawrence* v. *Commercial Union Properties* (1984) 4 Con LR 37. The contractor was in delay, due to exceptionally inclement weather. The architect took the view that it was the contractor's fault that the effect of weather was greater than it would otherwise have been. Held that this was an erroneous view. The contractor was entitled to extension of time because of the inclement weather whenever it occurred.

Direct loss and/or expense

Clause 26 of JCT 98 allows the contractor to claim 'loss and/or expense' caused by matters materially affecting the regular progress of the works.

Clause 26.1 states that 'if the contractor makes written application to the Architect stating that he has incurred or is likely to incur direct loss and/or expense (of which the contractor may give his quantification) in the execution of this Contract for which he would not be re-imbursed by payment under any other provision of this contract . . . and if, and as soon as the Architect is of the opinion that the direct loss and/or expense has been incurred . . . by matters referred to in clause 26.2 . . . then the Architect shall ascertain or shall instruct the Quantity Surveyor to ascertain, the amount of such loss and/or expense.'

Extension of time and money claims

Reimbursement of direct loss and/or expense and the grant of an extension of time are quite separate and distinct matters. The purpose of provisions for money claims and those for extension of time are quite different. The grant of an extension of time merely entitles the contractor to relief from paying Liquidated and ascertained damages from the date for completion stated in the contract. There is no automatic entitlement to compensation because the supervising officer or architect has determined an extension of time. Some events giving rise to an extension of time are neutral events and the responsibility of neither party. Some contracts (IFC 98 for example) emphasise the lack of connection between the two by widely separating the provisions. JCT 98 places the two provisions sequentially.

The meaning of direct loss and/or expense

The words direct loss and/or expense was considered in *Wraight Ltd* v. *PH &T (Holdings) Ltd* (1968) 13 BLR 26. It was held to mean that the sums recoverable are equivalent to damages at common law, see *Hadley* v. *Baxendale*.

The Court of Appeal in *F.C. Minter* v. *WHTSO* (1980) 13 BLR 1 CA considered the words in relation to the JCT 1963 contract. It held that direct loss and/or expense is loss that arises naturally, and in the ordinary course of things, as stated in the first limb of *Hadley* v. *Baxendale*. It approved the definition of 'direct damage' in *Saint Line Ltd* v. *Richardson* [1940] KB 99 at 103. 'That which flows naturally from the breach without any other intervening cause and independently of any special circumstances whereas indirect damage does not so flow'. Any claim put by a contractor is subject to the question as to whether it falls within the first limb of *Hadley* v. *Baxendale*. Keating (2001) considers that the use of other formula by contractors does not displace or detract from this principle. Examples of claims loss and/or expense are head office overheads, loss of profit, interest on borrowed capital (finance required for particular operations will either be borrowed and thus incur interest or may be available from won sources thus foregoing interest), idle plant or machinery, etc.

Judicial interpretation

F.C. Minter v. *WHTSO* (1980) 13 BLR 1, finance charges were claimed as direct loss and/or expense. The finance charges arose out of the claimants being kept from their money. The defendant argued that the claim was for interest and was not direct loss. In holding that the JCT terms contained an implied term to pay interest as a constituent part of direct loss and/or expense, Stephenson LJ added:

> 'In the context of this building contract and the accepted "cash flow" procedure and practice, I have no doubt that the two kinds of interest claimed here are direct loss and/ or expense unless there is something in the contractual machinery for paying direct loss and/ or expense which excludes this loss and expense of interest by the claimants. It was not obviously indirect, like the loss of a generous productivity bonus paid by the contractor to his men. It was . . .

interest reasonably paid on capital required to finance variations or work disrupted by a lack of necessary instructions and as such are direct loss and/or expense in which the claimants had been involved.'

Rees & Kirby Ltd v. *Swansea Corporation* (1985) 5 Con LR 34, decided that an application for payment under clause 26 (the then JCT 63 clause 11(6 and 24 (11) (a)), must make it clear that some loss or expense was incurred by reason of the contractor being out of pocket. Such notice must be given within a reasonable time of the expenses being incurred. It was proper to take into account the fact that banks calculate interest with periodic rests. See also *Tate & Lyle Food & Distribution Ltd* v. *Greater London Council* [1982] 1 WLR 149, where it was decided that an interest claim for direct loss and /or expense should be calculated at a rate equivalent to the rate of borrowing. Any special difficulties the contractor had in borrowing money should be disregarded.

Matters which entitle the contractor to be reimbursed

Clause 26 of JCT 98 contains the 'matters' which entitle the contractor to make a claim for direct loss and/or expense due to these materially affecting the regular progress of the work. Clause 26.2 states that the following are the 'matters' referred to in clause 26.1. They are found in clauses 26.2.1.1 to 26.2.1.10 (only the last number being used in the clauses below):

1 Failure to meet the dates of the Information Release Schedule (clause 5.4.1 or drawings, which explain or amplify the Contract Drawings, included in the Information release Schedule (clause 5.4.2.1.).
2 Opening up for inspection and testing work materials and goods (which then turn out to comply with the contract).
3 'Any discrepancy or divergence' between documents such as drawings, numbered documents or contract bills.
4 Delay due to other persons employed by the employer to execute work under clause 29 – works by other persons employed by the employer (clause 26.2.4.1) and failure to supply materials or goods which the employer has agreed to supply (clause 26.2.4.2).
5 Instruction under clause 23.3 by the architect to postpone any work to be carried out under the contract.
6 A failure to give the contractor ingress to or egress from the site, from land in possession of the employer in due time (provided the contractor has given any notice required by the conditions).
7 Architect's instructions issued: under 13.3 or 13A.4.1 or 13.3 (variations).
8 The execution of work of an approximate quantity included in the contract bills, which proves to be inaccurate.
9 The compliance or non-compliance with clause 6A by the employer. Under the clause the employer has to ensure the planning supervisor carries out its duties under the CDM regulations. Where the contractor is not the principal contractor ensure that the principal contractor carries out those duties.

10   Suspension of work by the contractor for non-payment under clause 30.9.1.4 provided the suspension was not frivolous or vexatious.

Variation when the contractor is in delay

When the employer causes a delay, and the contract contains no extension of time clause, the employer loses the right to claim LAD. This is because the completion date falls away. Time then becomes at large. The contractor has only to complete in a reasonable time. To prevent time being at large the contract must contain an extension of time clause. Until *Chestermount* it was unclear whether time could ever become at large under a JCT 80 contract. The general consensus was that it probably could not become at large. *Chestermount* confirmed this view. The comprehensive nature of the drafting of Clauses 24, 25 and 26 means that any situations not expressly covered by it are covered by implication. Note that this case is discussed in greater depth in chapter 9. In *Chestermount* the contractor argued that an act of prevention i.e., the ordering of a variation instruction after the completion date (when the contractor was already late in competing the contract) was to destroy the completion date/LAD regime and put time at large. Coleman J:

> 'in my judgement *the construction that the contractor* contends involves legal and commercial results which are inconsistent with other express provisions and with the contractual risk distribution regime applicable to pre-completion date relevant events that, in the absence of express words compelling that construction, it *cannot be right.*' (emphasis added).

Provisions excluding indirect and consequential losses

Some contracts contain provisions excluding indirect and consequential losses. This is most commonly found in computer contracts or contracts for the supply and installation of goods and materials. In *British Sugar plc* v. *NEI Power Projects Ltd* (1997) 87 BLR 87 the clause provided that:

> 'The Seller will be liable for any loss, damage, cost or expense incurred by the Purchaser arising from the supply by the Seller of any such faulty goods or materials or any goods or materials not being suitably fit the purposes for which they are required save that the Seller's liability for any consequential loss is limited to the value of the contract.'

The importance of the clause for the parties is shown in sharp relief when the facts of the case are considered. The contract between the parties was for the design, supply, delivery, testing and commissioning of electrical equipment at a final price of £106,585. The buyer claimed that the equipment was poorly designed and badly installed and caused the power supply to break down. Damages of over £5 million were claimed for the increase in production cost and the losses of profits from the breakdowns. The clause mentioned above had been agreed after lengthy negotiations.

Recent cases have made the meaning of the phrase much clearer. In *British*

Sugar and also *Deepak Fertilisers & Petroleum Corporation* v. *Davy McKee (London) & Anor* [1999] LLR 387 the Court of Appeal had once again to consider the formulation adopted in *Croudace Construction Ltd* v. *Cawoods Concrete Products Ltd* (1978) 8 BLR 20. This was that the phrase 'consequential loss' did not exclude losses arising naturally from the breach of contract. In each case the Court of Appeal concluded that 'consequential loss' and 'indirect and consequential loss' refer to damages falling into the second limb of *Hadley* v. *Baxendale*. In *British Sugar* it also did two other things:

1 It adopted the view that once a court has in a similar context, authoritatively construed the phrase, the 'reasonable businessman' must intend the phase to bear that meaning.
2 It confirmed that the phrase is concerned with damages that are too remote unless they are within the actual contemplation of the parties at the time they made the contract.

The result of this approach is that a party intending to exclude categories of foreseeable loss, would be better off specifying what is included rather than specifying what is excluded. Examples are, loss of profit, overheads, additional costs required to bring the project back to the level contracted for, and loss of revenue. Note the outcome of *British Sugar*. By the wording of the clause, the parties had agreed that damages *not* flowing naturally and directly from the breach would be limited to the value of the contract. See also *Deepak Fertilisers*, where an explosion destroyed a low-pressure methanol plant. The losses suffered by the owner, in addition to the cost of reconstruction, included (i) fixed costs and overheads incurred from the date of the explosion to the resumption of commercial production (referred to as 'overheads') and (ii) increase costs due to reconstruction of the plant requiring more catalyst to operate the plant than to its original configuration (referred to as 'catalyst costs'). The court held that:

'Wasted overheads incurred during the construction of the plant, as well as profits lost during that period, are no more remote as losses than the cost of reconstruction. Loss of profits cannot be recovered because they are excluded by the terms of the contract not because they are too remote.'

This was a reference to the clause in the contract that stated that 'in no event shall [contractor] . . . be . . . liable for loss [of] anticipated profits'. As far as the extra catalyst claimed to enable the plant to operate safely and produce the amount of methanol in the quantities the plant was supposed to, this was held to be direct costs. '[It] could not be categorised as indirect or consequential loss or damage . . .' since it was a claim for any other costs such as additional price of plant or part, required to produce the intended result (i.e. the methanol plant contracted for).

Hotel Services Ltd v. *Hilton International Hotels (UK) Ltd* [2003] EWCA Civ 74 illustrates all the principles in a concise manner. Once again, the Court of Appeal had to construe the same type of clause. The appellants supplied Robobars (mini bars) in hotel bedrooms. These dispensed drinks which when removed from the mini bar were electronically registered to the account of hotel guests. The robobars leaked ammonia that corroded the equipment and also created a health risk to the guests. The problem could not be corrected and the defendant replaced them. The Hilton sued and was awarded damages in the Official Referee's court (now the Technology and Construction court). The amounts awarded included (i) £151,065 for the costs of removing and storage of chiller units and cabinets and (ii) £127,000 for loss of profit on the mini bars. The defendants appealed on the basis that a clause in their rental contract excluded liability for such damage. The clause read:

> 'The Company will not in any circumstances be liable for indirect or consequential loss, damage or liability arising from the any defect in or failure of the system or any part thereof or the performance of this agreement or any breach thereof by the company or its employees.'

In the Court of Appeal the issue was the meaning of the words 'indirect or consequential loss'. It unanimously held that:

1 The loss of profit and the storage fell within the meaning of the first limb of the rule in *Hadley* v. *Baxendale* and are therefore not indirect or consequential loss.
2 Direct loss is equivalent to the first limb of the rule in *Hadley* v. *Baxendale* whereas indirect or consequential loss is equivalent to the second limb of the rule. The Court of Appeal, adopted the statement of Rix J in *BHP Petroleum Ltd* v. *British Steel plc* [1991] LLR 387, in relation to the need for special knowledge 'the parties are correct to agree that authority dictates that the line between direct and indirect or consequential losses is drawn along the boundary between the first and second limbs of *Hadley* v. *Baxendale*' (at para. 18).
3 The losses suffered by Hilton were direct losses and not caught by the wording of the exclusion clause. At para. 20 the court concluded:

> 'We prefer therefore to decide this case, much as *Victoria Laundry* was decided, on the direct grounds that if equipment rented out for selling drinks without defalcations turn out to be unusable and possibly dangerous, it requires no special mutually known fact to establish the immediacy both of the consequent cost of putting it where it can do no harm and – if when it was in use it was showing a direct profit – of the consequent loss of profit.'

Deposits, advance payments and forfeitures

Advance payments

Parties frequently contract where one party is required to make advance payments ahead of performance by the other. In *Dies* v. *British Mining and Finance Corporation Ltd* [1939] 1 KB 724 a contract was entered into to buy Mauser rifles and ammunition at a price of £270,000. The contract provided that £13,500 would be forfeited by the buyer if the contract became impossible of performance (a *force majeure* clause). £100,000 was paid in advance but the buyer failed to accept delivery of the goods. The Mining Company elected to treat the contract as terminated. The buyer sued for the return of the deposit. Held: that the buyer was entitled to the return of the £100,000. The *Force Majeure* Clause only applied if the contract became impossible of performance. The Mining Company could however set-off against the £100,000 for the breach of contract by Dies. The court came to the opposite conclusion in *Hyundai Heavy Industries Co Ltd* v. *Papadopolous* [1980] 1 WLR 1126.

A Liberian company (the 'buyer') entered into a contract with the ship builder Hyundai, whereby the latter agreed to build and deliver a multi-purpose cargo ship for US $14,300,000. The price was to be paid in five instalments, the first two being 2.5 per cent of the total price (US $357,500). The first instalment was paid, but the buyer failed to pay the second instalment according to the timetable for payment. The ship builders elected to cancel the contract in accordance with the contract provisions. Papadopolous had provided a guarantee of the buyer's obligations, that provided that they would re-imburse the ship builders 'in case the buyer is in default of any such payment we will forthwith make the payment in default on behalf of the buyer.'

Held: the ship builders were entitled to the second instalment despite cancelling the contract. Papadopolous was liable under the express guarantee given in the contract.

The difference between the two cases

Dies is a sale of goods contract and did not require the seller to perform work or to incur any expense. The seller was a middleman who intended to buy the goods from a manufacturer and sell them to the buyer at a profit. In contrast *Hyundai* was not a discrete contract. It was equivalent to a contract that made provision for interim payments. Viscount Dilhorne said at p.1134:

> 'it was a contract "to build, launch, equip and complete" a vessel and to "deliver and sell" her. The contract price included "all costs and expenses for designing and supplying all necessary drawings for the vessel . . .". It was a contract which was not simply one of sale but so far as the construction of the vessel was concerned, "resembled a building contract".'

The practical effect of the rules about deposits is that the rules are the reverse of penalties. Deposits are often greater than the loss that would have been contemplated at the beginning of the contract. The difference between the two is strikingly illustrated where the payee is entitled to recover the deposit. In *Workers Trust and Merchant Bank Ltd* v. *Bojap Investments Ltd* [1993] 2 All ER 370 the Bank sold property in Jamaica at auction for $11,500,00. Clause 4 of the contract provided for payment of a deposit of 25 per cent and a deposit of $2,875,000 was paid. The contract required the balance to be paid within 14 days. The purchaser was unable to pay the balance within 14 days, although it tendered the full amount on the twenty-first day. The Bank claimed to be entitled to keep the deposit. The Privy Council held:

1 The deposit rule was not in general subject to the penalty rule. However this was only so where the amount was reasonable. Long usage had established that

10 per cent was a reasonable figure and any deposit above that was a penalty unless circumstances justified that level.

2   Had the Bank taken a 10 per cent deposit it would have been entitled to keep the deposit. As the 25 per cent the sum was a penalty. It could therefore, not keep the sum but had to repay it in full. However, since they had a legitimate claim for damages they would be allowed to retain some of the deposit as security for that claim.

Summary

1   Delay in a construction contract is controlled by the use of Liquidated and Ascertained Damages.

2   To retain the right to Liquidated and Ascertained damages the contact must contained an extension of time clause.

3   Such a clause enables the employer to retain the right to Liquidated and Ascertained damages for acts of prevention for which it is responsible.

4   In the JCT 98 contracts these acts are called relevant events. The provisions are comprehensively drawn and effectively transfer a great deal of risk from the contractor to the employer. Their use should enable the contractor to offer a more competitive price.

5   Delays caused by certain relevant events may enable the contractor to make a claim for direct loss and/expense. Whether it can do so depends on whether it is not re-imbursed by some other provision in the contract.

6   The rule in *Hadley* v. *Baxendale* is important in LAD provisions. A pre-estimate of damages is usually direct costs and falls into the first limb of the rule. Claims for loss and/or expense do so also. Consequential losses are more difficult. A clause limiting this is unlikely to be applied where the losses flow directly from the breach.

7   The rules about penalties are part of the historical jurisdiction of Equity. Equity would relieve a party from penal bonds in circumstances where it was unfair to hold the party to the agreement.

8   There are a number of reasons why parties would agree damages beforehand. It might not be worth suing at common law, the result is uncertain and costs can be heavy.

9   Penalties are quite readily undertaken by parties, who are not in the least terrorised by the prospect of having to pay them. Quite often they are not taken into account when tendering for work.

10 Deposits are normally a guarantee of performance. Deposits are still payable (forfeit) even if the money has not been paid. It makes commercial sense to have deposits, since in many contracts it may not be worth suing if one party fails to perform.

11 In the case of advance payments the claimant has the money and is usually suing for the rest. Whether it is recoverable depends on the particular contract and the purpose of the payment. *Dies* and *Hyundai* illustrate two different approaches.

9 Variations and the Right to Payment

Background

Most construction contracts make detailed provision for the employer to extend the time for completion of the work. The need to extend the construction period arises from either (i) events external to the contract or (ii) because changes in design or the construction method are required. These may arise as a consequence of extra work needed to accommodate changes made or required on behalf of the employer. For such a provision to be effective it has to permit a *variation of the work* itself rather than the contract, since the basic rule at common law is that one party cannot unilaterally vary a contract.

External events

The effect on the contract of external events depends on whether the parties have made provision for it. Thus events (called 'excepted risks' or 'specified perils') such as fire, strikes, shortages, riot, invasion, rebellion, terrorism and war are dealt with expressly in the standard forms of contract. See clause 22 (2) of *ICE 7th* and clauses 1.3 and 22.1.3 of JCT 98. There is also a category called *force majeure* (see, e.g., clause 25.4.1 of JCT 98) but its meaning in English law is unclear. Keating (2001, p.731) considers the words to have a restricted meaning because matters such as war, strikes, fire and bad weather are dealt with elsewhere in the standard forms. Whether an external event also amounts to the *frustration* of the contact depends on whether it falls within that category of event. *Frustration* is an *event outside the control* of the parties that excuses them from carrying out the contract. The contract is automatically brought to an end by a rule of law: see *Hirji* v. *Mulji* v. *Cheong Yue SS Co* [1926] AC 497. *Taylor* v. *Caldwell* (1863) 3 B&S 826 recognised that the parties could not always be held to their bargain. It decided that there was an *implied condition* that the hall, burnt down the day before a series of concerts began, *would exist*. As a result the contract was *frustrated* and the parties released from further performance. There are a number of situations where a contract may become frustrated:

1. The *subject matter* of the contract is destroyed, as a result performing of the contract becomes impossible. *Taylor* v. *Caldwell* (1863) is such an example.
2. An event central to the contract fails to occur. Many contracts were frustrated in 1903 when King Edward VII became ill the day before the state procession and the event was postponed. See also *Krell* v. *Henry* [1903] 2 KB 740.
3. The contract becomes illegal. Examples include the outbreak of war, which would make continuing with the contract 'trading with the enemy'. In the well-known case of *Fibrosa Spolka Akcyjna* v. *Friaburn Lawson Coombe Barbour*

Ltd [1943] AC 32, the German army occupied Gdynia in Poland shortly before the ordered machines could be delivered. Trading with the enemy of war became illegal and thus the contract was impossible to perform. Other examples are the imposing of sanctions or the compulsory purchase of property that is the subject of a contract.

The result of frustrating events

Where frustration occurs and the parties are excused performance of the contract, a number of consequences result:

1 The general rule was that loss lay where it fell, which meant that the parties bore their own losses when the frustrating event occurred.
2 The decision in *Fibrosa* led to a change in the rule. It was held there that there had been a *total failure* of consideration. As a result a deposit paid had to be returned because the paying party had *received nothing* in return. Where there was a partial failure and the paying party had received some benefit the loss still lay where it fell.
3 The Law Reform (Frustrated Contracts) Act 1943 was passed to regulate the effects of frustration. The Act extends the decision in the *Fibrosa*. Where money is paid before the frustrating event it can be reclaimed even though the consideration was partial. Money due but not paid before the frustrating event need not be paid. If a party has incurred expenses before the frustrating event two things can happen: (i) if money has been paid to him or money is due before the contract is frustrated, the court may award expenses up to that amount *and* (ii) if nothing was paid or due to be paid, he cannot recover expenses. Where a party had received a *valuable benefit* before the frustrating event, he may be required to pay for it.

The concept of a *failure of consideration* is now considered as *part* of the *law of restitution*. It had *nothing* to do with the *doctrine of consideration* needed as one of the elements of a valid contract. In the same way the Law Reform (Frustrated Contracts) Act 1943 is one of the statutes under which *benefits received* under a *failed contract* are returned.

Frustration and construction contracts

In *Metropolitan Water Board* v. *Dick, Kerr & Co Ltd* [1918] AC 119, the Ministry of Munitions suspended the building of a reservoir using wartime powers. Since the contract contained provisions for the granting of extensions of time for completion, performance of the contract after the war was *not impossible*. The House of Lords however, held that performance after the war was a *fundamentally different thing*.

The mere fact that that the contract had turned out to be *more expensive and difficult* than envisaged is not sufficient to amount to frustration. In *Davis Contractors Ltd* v. *Fareham Urban District Council* [1996] AC 696, the plaintiff contractor argued that as a result of a scarcity of skilled labour, the delay in completion amounted to frustration of the contract. A contract for the completion of 78 houses in a period of eight months at a fixed price took 22 months to complete. The contractor sought to escape the contract and claimed to recover the excess costs on the basis of a *quantum meruit*. Their claim failed. Dismissing it Lord Radcliffe commented:

> 'I am bound to say that, if this is the law, the appellant's case seems to me a long way from a case of frustration. Here is a building contract entered into by a housing authority and a big firm of building contractors in all the uncertainties of the post-war world. Work was begun shortly before a formal contract was executed and continued, with impediments and minor stoppages but without actual interruption, until the 78 houses contracted for had been built. After the work had been in progress for a time the appellants raised a claim, which they repeated more than once, that they ought to be paid a larger sum for the work than the contract allowed; but the respondents refused to admit the claim and, so far as it appears, no conclusive action was taken by either side which would make the conduct of one or the other a determining element in this case.'

Where the site is destroyed or made unworkable, the contract may be frustrated despite a clause allowing (i) an extension time and (ii) an express clause dealing with unforeseen circumstances: see *Wong Lai Ying* v. *Chinachem Co Ltd* (1979) 13 BLR 81 PC. The issue for the Privy Council was whether 24 contracts for the sale of flats had been frustrated. A landslip, unforeseen and unforeseeable and caused by circumstances beyond the control of the builders, obliterated the site where two tower blocks were being constructed.

The contract provided that time should be of the essence but allowed for time to be extended for various reasons. Clause 22 of the contract made provision for unforeseen circumstances beyond the vendor's control by which the vendor was unable to sell to the purchaser an undivided share and apartment. In such a case the vendor could rescind the agreement. The instalments of the purchase price paid would be refunded (but without interest or compensation), the agreement would become null and void and neither party would have any claim against the other.

Held: Dismissing the appeal, the following points were made:

1 The landslip was a 'frustrating event', as it made further performance uncertain and the character and the duration of any further performance radically *different* from that which the original contact contemplated.
2 Upon its true construction and upon the construction of the contract as a whole, clause 22 should not be read as making provision for the possibility of the particular unforeseen natural disaster which had occurred and, accordingly, the contract was frustrated.

See also *Codelfa Construction Proprietary Ltd* v. *State Rail Authority of New South Wales* (1982) CLR 33. A contract to construct an underground railway required the respondents to complete the contract in a fixed time, working 3 shift-days on a 7-day basis. Third parties obtained an injunction because of the noise, which limited evening and Sunday work. The appellants claimed in arbitration under the contract that there was an implied term in their contract that the employer should indemnify them against their additional costs. Alternatively they argued that the contract had been frustrated by the grant of the injunction. The High Court of Australia held that no term would be implied. By a majority they held the contract had been frustrated. Giving judgment Mason J said:

> 'I find it impossible to imply a term because I was not satisfied that in the circumstances of this case the term sought to be implied was one which the parties in that situation would necessarily have agreed upon as an appropriate provision to cover the eventuality which has arisen. On the other hand I find it much easier to come to the conclusion that the performance of the contract in the events which occurred is *radically different* from the performance of the contract in circumstances which it, construed in the light of the surrounding circumstances, contemplated.' (emphasis added)

The reluctance of the court to invoke the doctrine of frustration stems from two principal reasons (see McKendrick, 2000, p.302).

1 Parties should not be able to escape from contracts that had turned out to be bad bargains. In *Davis Contractors* Lord Radcliffe said that it was 'hardship or inconvenience or material loss itself which calls the principle of frustration into play'.
2 The future is uncertain and parties should make provision in their contracts to guard against increases in price, the shortage of materials, inflation and other uncertainties.

Design changes

In the traditional form of construction contract, provision is usually made for *errors* arising out of the proposed design. Clause 2.3 of JCT 98, for instance, requires the contractor to give notice of any divergence or discrepancy between any two or more documents. The documents listed include the contract drawings, the contract bills, instructions, drawings or documents. The contractor has no positive duty to actually look for such errors. Provision is also made for the giving of notice of any divergence between the statuary requirements and the drawings listed in Clause 2.3. Clause 6.1.3 requires the employer to respond within seven days of receiving such a notice by giving instructions to the contractor to deal with the error or discrepancy.

Variations

When the contractor makes a claim for payment for extra work carried out the primary question to be asked is whether the work is included in the contract or amounts to a variation? In order to be able to claim payment at common law the contractor must prove:

(a) the work was extra over the work actually agreed by the parties;
(b) there was either an express or implied promise by the employer to pay for the work done;
(c) any agent had authority to issue instructions for the work to be done.

What is a variation?

In the standard forms of contract there are express rules for the ordering of varied work. In deciding whether the work is extra, the *recitals* have to be examined. A second question to be answered is whether the variation goes to *root* of the contract. A variation must bear some relationship with the work of which it is a variation. If the work could not have been in the contemplation of the parties at the time they made the contract, it must fall outside the scope of the variation clause, therefore the less specific the work is at tender stage, the greater the scope for variations. It has to be stressed that the purpose of a variation clause is primarily to deal with the unexpected, rather than the consequences of poor planning or frequent changes of mind by the employer.

It is clear that a supervising officer, whether an engineer, architect or another construction professional, cannot order work which is wholly different from the original contract. In practice this principle is more complicated: see *Blue Circle Industries Plc* v. *Holland Dredging Co.* (1987) 37 BLR 40. During a dredging contract the claimant entered into negotiations with the defendant for the deposit of spoil which required the approval of the local authority. Its tender had allowed for deposit of spoil within the Lough. Following negotiation between the parties and the various agencies involved, the final solution was agreed. An artificial kidney-shaped island would be constructed to provide a bird sanctuary. The construction required the building of a bund-wall with ancillary works. Spoil from the dredging would be deposited in it. The defendant submitted a quotation for the erection of the artificial island. This was accepted with the stipulation that an official order was to follow. No order was ever sent. The works failed as the island was only visible at low tide and quite unsuitable as a bird sanctuary.

The claimant brought an action for misrepresentation amongst other claims. The defendant alleged that there was an arbitration clause in their contract which did not cover the action. The question for the court was what in fact the relationship between the two agreements was. The test proposed was whether the employer could have ordered the work as a variation *against* the wishes of the contractor? The court decided that it could not have done so, and therefore the work carried out *was not* a variation.

The scope of variation clauses

The variation clause considered in *Blue Circle* was the equivalent of clause 51 of the ICE (fifth edition) and entitled 'Alterations, Additions and Omissions'. It states that:

'51 (1) The Engineer
(a) shall order any variation to any parts of the Works as in his opinion
(b) necessary for the completion of the Works and
(c) may order any variation that for any other reason shall in his opinion be desirable for the completion and/or improved functioning of the Works.

Such variations may include additions omissions substitutions alterations changes in the quality form character kind position dimension level or line and changes in any specified sequence method or timing of construction required by the contract and may be ordered during the defects correction period.'

By comparison the FIDIC (fourth edition) defines variations more *closely* and *clearly*:

(a) increase (or decrease) in the quantity of work;
(b) omission of such work;
(c) changes in character, quality or kind of work;
(d) changes in levels, lines, positions and dimensions of any part of the work;
(e) execution of additional work of any kind necessary for the completion of the works;
(f) changes in any specified sequence of timing of construction of any part of the work.

Variations under the provisions of JCT 98

There are *four* mechanisms by which variations can be ordered by the architect under clause 13, variations (included in the *four* for convenience is performance specified work ordered under clause 42), and these will be examined in turn.

1 Under Clause 13.2.1 the architect is empowered to issue instructions requiring a variation. The list is comprehensive.
2 Clause 13.4.1.2 Alternative *A*: Contractor's Priced Statement *or* Alternative *B*.
3 Clause 13 A – contractor's quotation.
4 Performance specified work under clause 42.

Under Clause 13.2.1

Variation is defined in clause 13.1. The clause states that the term 'Variations' as used in these conditions means:

Clause 13.1.1 *'alteration or modification* of the design, the quality or quantity of the works including
Clause 13.1.1.1 "the addition, omission or substitution of any work"
Clause 13.1.1.2 the alteration of the kind or standard of any material or goods to be used in the works,
Clause 13.1.1.3 the removal from the site of any work executed or goods or material brought thereon by the contractor for the purposes of the works other than work material or goods which are not in accordance with the contract.'

The employer may also issue instructions under clause 13.1.2, which imposes obligations, restrictions or limitations on the contractor. These can involve changes of access or use of the site, limitations in working space or working hours and the execution or completion of work in any specified order.

Such instructions are subject to the contractor's right of reasonable *objection*. Clause 4.1.1.1 allows the contractor to refuse to comply with a variation within clause 13.1.2; where it does so the objection must be in writing. Any instruction given by the architect must be in writing. For the position where an instruction is not in writing see *Brodie* v. *Cardiff Corporation* (1918) AC 337. A contractor does not automatically lose his entitlement to a variation simply because it has not been formally ordered.

Clause 13.4.1.2 Alternative A: Contractor's Priced Statement or Alternative B

The Latham Report recommended that all variations should be priced in advance and that provision should be made for independent adjudication if the parties could not reach agreement. The JCT did not consider this idea to be practical, but in amendment no 18 to JCT 80 introduced an alternative to clause 13.2.1. above. Its purpose is to provide an alternative to the mandatory method of valuing variation by the quantity surveyor.

Clause 13.4.1.2 is headed 'Alternative A: Contractor's Priced Statement'. Under the provision the contractor can respond to either (a) a variation instructed by the architect or (b) where approximate qualities have included in the contract bills and the work has been carried out, or instructed to be carried out by the architect. The contractor may 'submit to the quantity surveyor his *Price Statement* . . . for such work' (paragraph A1). The price is based on the valuation rules in clause 13.5. To its price statement the contractor can attach two other things:

(a) an amount due for direct loss and/or expense,
(b) the extension of time required.

Paragraph A2 requires the quantity surveyor after consultation with the architect, to inform the contractor within 21 days in writing whether the Price Statement is accepted or not. The failure by the parties to agree means that either party can submit the matter to adjudication as a dispute or difference.

The alternative procedure to the above is outlined in clause 13.4.1.2 which is called 'Alternative B'. Where the contractor does not submit a priced statement or no priced statement had been agreed, then the valuation is to be made in accordance with clauses 13.5.1 to 13.5.7.

Clause 13 A – contractor's quotation

First introduced into the JCT 80 by amendment 13, this method requires a quotation by the contractor for carrying out varied work. The contractor has seven days in which to disagree with the application of the clause.

Sufficient information has to be given to the contractor to enable it to price the work. The information should be similar to that provided at tender stage including drawings, plans, bills, and so on. The contractor has seven days in which to request additional information. The quotation is to be submitted to the Quantity Surveyor not later than 21 days from receipt of the instruction, and remains open for acceptance for seven days from its receipt by the quantity surveyor.

What the quotation requires (clause 13A.2):

- the value of the adjustment to the contract sum
- the adjustment in time required for the completion of the works
- · the amount due for direct loss and/or expense
- a fair and reasonable amount for preparing the quotation
- (and) if required by the instruction, statements on the additional resources needed and a method statement.

Acceptance

If accepted, the employer shall accept it in writing, and the contractor is obliged to carry out the variation. The adjustment to the contract sum and the additional requirement contained in the quotation shall also be confirmed.

Non-acceptance

If it is not accepted, the architect shall either instruct the contractor to carry out the variation and it shall be valued under clause 13.4.1, or alternatively the architect may instruct that the variation is not to be carried out. In that case a fair and reasonable sum shall be added to the contract sum to cover the cost of preparing the quotation.

Performance specified work under Clause 42

Amendment 12 to JCT 80 issued in July 1993 first introduced the phrase Performance Specified Work (PSW) into the contract. The provision dealing with its operation is clause 42 of JCT 98. The amendment does not mention design but it is generally accepted that it covers simple elements of design to be carried out by the contractor after the award of the contract.

PSW must be identified in the appendix (Clause 42 requires each item to be provided by the contractor to be identified 'in the contract bills). The work cannot be introduced as a variation unless the parties are in agreement. Clause 42.12 states that no instruction under *clause 13.2 variation* can be given to carry out PSW. However, with the agreement of both parties, a *13.2 variation* to the work can be instructed and issued under clause 42.11.

Summary

Clause 13.2.1 instructions

This method frequently leads to dispute as the consequences of the variation are always assessed afterwards. The costs, additional time to be allowed and the effect on the rest of the programme is always judged by the architect and quantity surveyor *after* the work has been carried out.

Clause 13.4.1.2 Alternative A: Contractor's Priced Statement

Introduced by amendment 18 to the JCT 80 as a consequence of the recommendation of the Latham Report for the pre-pricing of variations. Although it achieves the aim of having the contractor provide the information on the costs, time and the effect on the rest of the programme, it does also provide for immediate adjudication where there is a failure to agree. This might be seen by discerning employers as not an altogether wise idea.

Clause 13 A – contractor's quotation

This method avoids the difficulties of Clause 13.2.1 instructions by introducing a method of assessing the impact of a variation *before* the work is carried out. It removes potential areas of dispute by having the contractor price the costs and time impact of the proposed variation before the varied work is carried out. Bearing in mind the time limits laid down, it is clearly not a method to be used for immediate variations.

Performance specified work under clause 42

Work carried out under this provision can only be done as a variation where the parties agree. It provides an alternative way of varying the work.

Variation problems

The starting point is to look at the contract and discover exactly what the contractor has agreed, in the first place, to do. It must be noted that just because work is not covered in the bills of quantities or specification does not necessarily mean that the contractor is entitled to be paid for it as 'an extra'. *Patman* v. *Furtheringham Ltd and Pilditch* (1904) decided that the employer is not bound to pay for 'things that everyone must have understood are to be done but which happen to be omitted from the quantities'. Note, however, that under the JCT 98 work omitted from the bill of quantities triggers an automatic variation.

Hudson (chs 5 and 8) observes that modern bills of quantities in fact transfer a substantial element of risk to the employer. See also the position where the *standard method of measurement has been incorporated*.

In *Malloy* v. *Liebe* (1910) 102 LT 616, the employer insisted that the contractor carry out work which, it claimed, was part of the contract. The contractor disagreed. The Privy Council upheld the arbitrator's award that there was an *implied promise* by the employer to pay for the work.

Pieter Kiewit Sons' Co. of Canada Ltd v. *Eakins Construction Ltd* (1960) 22 DLR (2d) 465 deals with the contractor's response to a variation instruction. A contractor who was engaged to build a bridge across Vancouver harbour sub-contracted the piling work. During the piling operations the employer's engineer insisted that the piles be driven to a greater depth than required by the contract. The sub-contractor complained to the contractor who insisted the work must be done. Afterwards both the contractor and engineer refused to pay for the extra work.

Held by a majority of the Supreme Court of Canada: The sub-contractor was not entitled to payment. Although the engineer's instruction was not justified, the sub-contractor should have refused to do the work without a variation order. Cartwright J. dissenting, considered this to be an unrealistic approach.

Can the architect use the wide definition of variation given in JCT 98 above as a means of making fundamental changes to the work? The answer is probably not, bearing in mind that the variation must relate to the original work.

What about *omitted work*? The architect cannot omit work and award it to another contractor. *Commissioner of Roads* v. *Reed & Stuart* [1974] 12 BLR 55 decided it could not do so. Similarly the contractor cannot be instructed to carry out work which has no connection with the contract by invoking the variation clause. In a shipbuilding case, *Russell* v. *Sa da Bandeira* (1862) CB (NS) 149, Earle LJ said that an architect/engineer's power to order extras applies only to: 'things not specified in, nor fairly comprised within the contract, but cognate to the subject matter and applied to the carrying out of the design'.

In *McAlpine Humberoak* v. *McDermott Int Inc.* (1992) 58 BLR 1 the relationship between frustration and a variation was considered. The tender for the construction of a tension leg platform for an oil field was based on 22 drawings and a completion period of 18 weeks. Work started based on a letter of intent. By the completion date 161 drawings had been issued. Considerable delays resulted from technical queries, which took a long time to be answered. More delays resulted from the requirement for the approval of the procedures needed in welding the free steel issued as part of the contract. The plaintiff brought an action to recover extra costs incurred and the defendant counter-claimed for additional costs incurred due to delays and defective work.

At first instance the judge decided that the contract was let before the design was complete. The variations as provided for by the contract were not variations but changes to the contract. The effect of the changes was to frustrate the contract by turning it into something completely different from what it was in the beginning.

McDermott appealed. The Court of Appeal allowed the appeal. It decided the following:

1 The revised drawings did not transform the original contract into a totally different contract. It remained a contract for the construction for four pallets. Frustration, which is dependent on upon unforeseen circumstances, cannot apply to circumstances known at the time of entering into contract.
2 The changes were within the variation clause.

Note that the *valuation* of a variation instructed under clause 13.2.1 of JCT 98 shall be in accordance with clause 13.5 unless the instruction states that the valuation of the variation shall be in accordance with clause 13A.

Clause 13.5.1 valuation rules

'To the extent that the valuation relates to the execution of additional or substituted work, which can be property valued by means of measurement . . . such work shall be measured and valued in accordance with the following rules:

1 the additional or substituted work is of similar character to, is executed under similar conditions . . . the Contract rates for the works set out shall determine the valuation;

2 the additional or substituted work is of a similar character to work set out in the contract bills but is not executed under similar conditions thereto and/or significantly changes the quantity ... the rates and prices in the bills shall include a fair allowance for such difference in condition and/or quality

3 where the additional work is not of a similar character to the work set out in the bills the work shall be valued at fair rates and prices.'

(For the construction of a similar clause under the ICE condition see *Henry Boot Construction* v. *Alstom Combined Cycles* [2000], discussed below.)

Wates Construction (South) Ltd v. *Brodero Fleet Ltd* (1993) 63 BLR 128 involved the construing of the phrase '*similar character* and *similar conditions*'. The arbitrator looked at a wide range of factors:

• pre-tender procedure
• negotiations which led to the bills (and prices)
• the conditions Wates had in mind when they priced the cost plan
• what knowledge they gained and what work changes were apparent
• the nature of the works, conditions relating to time, and other things.

He decided that clause 13 conditions were not only those directly connected with the bills but also the 'factual matrix surrounding the contract'. These included the tendering procedure and the negotiations. The contractor objected to this approach. They argued that the arbitrator had to look at the original conditions. The express words in clause 13, 'similar conditions thereto', referred back to the work 'set out' in the contract bills. The arbitrator should have looked at the contract documents and sorted it out from there. The Judge agreed with the approach advocated by contractor. In applying clause 13 the arbitrator had to look only at the contract documentation (i.e., it does not matter what the parties knew, hoped, expected or suspected).

The valuation of variations

This aspect was considered in *Henry Boot Construction* v. *Alstom* Combined Cycles [2000] EWCA Civ. The clause in question was clause 53 of the ICE (sixth edition). Its wording is similar to clause 13.5 of the *JCT 98*. The JCT uses the expression *fair rates and prices* and the ICE a *fair valuation*. It is submitted that the application of the phrases would result in similar valuations.

The appellant was employed under the ICE standard form of contract (sixth edition), to carry out civil engineering work for a new power station at Clwyd, North Wales. Work was required in three different areas: the turbine hall, the heat recovery steam generator area and, separated by a road, the cooling towers. Cool water pipes carried condensed water from the cooling towers back to steam generators, where it was turned into steam again by exhaust gases from the gas turbines in the hall. Initially the cooling pipes were to be constructed 4.45 m above datum line. After tender submission the employer changed the design and required the datum *to be lowered* to 3.35 m. This required the use of sheet piling for the temporary work involved in the installation of the pipe work, instead of the trenches with battered sides initially specified. The appellant responded to a request for a revised price by submitting a price of £250,880 as its additional costs arising from the change and confirmed that this price was only for the work in turbine hall.

Disputes arose between the parties which were referred to an arbitrator. Amongst other things the arbitrator held that the sum of £250,880 applied only to the work in the turbine hall. Work in the other two areas fell to be *valued as a variation*.

The main difficulty in doing so lay in the fact that, in calculating its prices, the appellant had mistakenly omitted to state that the figure of £250,880 also included the work in the generator area. It also did not give any credit for the change from trenches with battered sides to sheet piling.

Clause 53 (3) of the *ICE 7th* states:

'Failing agreement between the Engineer and the contractor under either sub clause (1) or (2) the *valuations of variations* ordered by the engineer in accordance with clause 51 shall be ascertained by the Engineer in accordance with the following principles and notified to the Contractor.
(a) where work is of similar character and carried out under similar conditions to work priced in the Bill of Quantities it shall be valued at such rates and prices contained therein as may be applicable
(b) where the work is not of a similar character or is not ordered under similar conditions . . . the rates and prices in the Bill of quantities shall be used as a basis for valuation so far as may be *reasonable* failing which a *fair valuation* may be made.'

The arbitrator's approach

The arbitrator took the following line.

1 He decided (*correctly*) that he had no power to correct the mistake.
2 Since there was a flaw in the make-up of the rates he decided that it was not reasonable to use prices arrived at by mistake or error.
3 He made instead a fair valuation. He awarded £74,460 for sheet piling in the heat recovery steam generator area and £500,474 in the cooling tower (a reduction of about £2 million in the appellant's claim).

The High Court

An appeal was made against the arbitrator's award on the grounds that he misinterpreted the word 'reasonable' in the phrase 'so far as reasonable' in clause 52 (1) (b). The judge agreed and remitted the award to the arbitrator with a direction to base the valuation of the variation on Rule 2 (see below). The employer appealed against the finding.

The Court of Appeal

The Court of Appeal in *construing* the clause decided that it consisted of *three rules*.

Where work was carried out of a similar character and carried out under similar conditions, the engineer was bound to apply the rates in the bills. *Rule 1* is *mandatory*. If the work had fallen into that category the rate derived from £250,880 would have applied however obvious the mistake or unreasonable the result. However, the arbitrator held it did not and that finding was not challenged on appeal.

Rule 2 is also mandatory. Where the work carried out *was not* of similar character *nor* carried out under similar conditions, the bills were used to arrive at the basis of the valuation. This valuation had to be a reasonable one. Lord Lloyd held that the sole function of the words "so far as it may be reasonable" is to permit the engineer to adjust rates as may be necessary to make adjustments in the rate. They call for a comparison between the work covered by the variation and the work covered in the bills of quantities.

Rule 3 applies where *Rule 2* was not applicable. The engineer is then bound to make a *fair valuation* instead.

The contractor had made an error in pricing its rates. This resulted in the variation instruction when given and valued under similar character/similar conditions, giving the contractor a *windfall profit* on the work carried out.

Application of Rule 3

Weldon Plant Ltd v. *The Commission for New Towns* [2000] EWHC Technology 76 also concerned the valuation of a variation. At issue was whether, when applied under Rule 3, a fair valuation included an allowance for overheads and profit in its calculation.

In a contract for the construction of a reservoir using the ICE (sixth edition), the contractor had to excavate material consisting of clay and gravel. Clay was to be removed to a tip on site but the contractor was allowed to sell any gravel it excavated. At its own risk it could excavate for gravel below the design level of the dam and sell that too. During the work the engineer issued an instruction requiring the contractor to remove all gravel below the design level and to backfill the dam up to its design level with clay.

The contractor disputed (amongst other matters), the valuation made by the engineers using the rates in the bills and the matter was referred to arbitration. The arbitrator decided the contractor had an option under the contract to remove the gravel rather than an obligation to do so. It was up to the contractor to do the work if it chose to. If the work proved to be uneconomical it could choose to stop it at any time. An instruction by the engineer to do the work amounted to a variation under clause 51 of the contract. In his award the arbitrator said:

> 'I consider here there was a particular situation. Normally there is no question but that the contractor is required to execute extra work ordered and he does not challenge the authority of the Engineer's Instruction. However the Instruction 17 did not order more work in the usual sense. In this case Weldon had the option not to carry out the work which was the subject of Instruction 17. That option was removed and consequently a fair valuation

is I consider appropriate and I accept the Claimant's argument that it should not be worse off as a result of having complied with the instruction. It would not be proper for Weldon to suffer financially because of the situation imposed upon them.'

The parties were unable to agree the value fair valuation and another hearing took place before the arbitrator. He awarded the contractor a sum that did not include additional site/office overheads and profit. The contractor appealed.

In the course of his judgment HHJ Humphrey Lloyd QC said at para. 15:

'Variations are performed by the contractor *involuntarily*, as it were, in the sense that the contractor does not tender to do the variation since it is of course unknown at the date of tender. A contractor offers to carry out variations ordered under clause 51 that may be required by the Engineer to meet the employer's needs, but the offer is not unqualified. First, the *variation* must be *within* the scope of the works. Secondly, the variation must be *valued* in accordance with *clause 52*, the provisions of which are clearly directed at seeing *that the contractor will not have to bear the costs of the variation,* except to the extent that, where rules 1 and 2 apply, the contract rates or prices were inherently insufficient, or to the extent that the costs incurred are not reasonably or (properly to be treated as forming part of the valuation). The *contractor* therefore takes the *risk* that that its rates or prices for the work may not cover its costs of carrying out the variation which is the same or comparable to the contract work, just in the same way the employer takes the risk that by having second thoughts more might have to be paid for the work than might otherwise have been the case had the need been known prior to the date of the contract and rates and prices for the work in question been obtained.' (emphasis added)

Giving judgment in favour of the contractor and remitting it to the arbitrator for reconsideration he held the following:

1 A fair valuation will normally consist of cost plus a reasonable percentage for profit.
2 Overheads required separate consideration.
3 Overheads that made a contribution to the fixed or running costs of a business could be recovered without proof that they had been actually incurred for the purposes of a fair valuation: 'In my view a valuation which in effect required the contractor to bear that contribution would not be a fair if the valuation did not include an element on account of such contribution.'
4 Proof that they had been actually incurred for the purpose of a fair valuation was not necessary. ('Although their approximate amount must of course be established, e.g., by deriving percentages from the accounts of the contractor including where appropriate associate companies that provided services or the like that quality as overheads': para. 19).
5 Time-related overheads need to be proven.

Variation in a period of delay

In *Balfour Beatty Building Ltd* v. *Chestermount Properties Ltd* (1993) 62 BLR 1, an appeal in the commercial court was made against an interim award made by an arbitrator. (Cases which raise questions concerning either (i) the operation of the Arbitration Acts or (ii) matters of general importance to contract law of a wider commercial context could at that time exceptionally be heard in the commercial court.)

By a contract under the JCT 80 conditions dated 16 June 1988, Chestermount ('the employer') awarded the construction of the shell and core of an office block together with elements of fit-out work to Balfour Beatty ('the contractor'). Work started on 18 September 1987 and completion was scheduled for 17 April 1989. A *special provision* of the contract allowed the employer to elect, by a certain date, to confine the work to the shell and core work only. In March 1988, the employer elected to *omit* the fit-out work. A grant of an extension of time fixed *a new completion date for 9 May 1989*. On 8 June 1989, the architect issued a certificate of non-completion (which entitled the employer to deduct LAD). By January 1990 the shell and core work had still not been completed, and in February the parties agreed to the reinstatement of the fit-out works as part of the construction contract. As a result the architect *issued instructions* for the carrying out of the work as a *variation*.

Practical completion of the shell and core work was certified on 12 October 1990 and that of the fit-out work on 25 February 1991. *Two* further extensions of time had been granted:

(a) one was on 18 December, which fixed the new completion date as *12 September 1989*;
(b) a second one on 18 December 1990 fixed the new completion date as *24 November 1989*.

This extension of time was given three months *after practical completion* of all the works. It resulted in the *completion date* for the project being *two and a half months before* the architect's instructions for the carrying out of the fit-out work.

As practical completion was not achieved until 25 February 1991, the architect issued a final certificate of non-completion. This entitled the employers to deduct £3.84 million as LAD. The contractors challenged the employer's entitlement in arbitration on two bases.

1 Where the employer issued a variation in a period when the contractor is in delay (culpable default), this *act of prevention displaces the contractual structure* for completion date, extension of time based on the occurrence of relevant events and LAD. The resulting obligation of the contractor was then to complete the work in a reasonable time. If it failed to do so the employer had a right only to *unliquidated damages for delay*.

2 The granting of the extension of time had to be carried out on a gross assessment. The re-fixing of the completion date had to *start on the date* the *instruction* was given. The time required to complete the instruction was calculated and projected forwards from that date by *adding on the additional time* required.

The arbitrator decided that clause 25 of the JCT 80 contract conferred jurisdiction on the architect to extend time for completion for a relevant event occurring in the period where the contractor was in delay. The calculation in the extension was correctly based on a '*net*' extension of time. The revised date was calculated by taking the date currently fixed for completion and adding a number of days, which the architect regarded *as fair and reasonable*. The contractor appealed against these findings.

Giving judgment, Coleman J said that there were two questions of law that arose out of the contractor's appeal.

The first question

The argument for the appellant was that an *act of prevention* by the employer would displace the completion date and liquidated and ascertained regime. There was no provision for the extension of time in respect of a variation instruction issued after the most recently set completion date. In the alternative it argued clause 25.3.3 was unclear as to whether the architect could take into account relevant events occurring before the most recently set completion date or take into account subsequent relevant events.

The judge accepted if these arguments were correct it would mean that the contractor would have been in culpable delay for 16 months at the time of the variation instruction, but not liable to pay the damages set at £16,000 a week.

In interpreting the provisions, the judge said there were *two distinct powers* to extend time to complete the works: first, under clause 25.3.1.1. This clause is triggered by the contractor giving notice that the *progress of the work* is being, or is likely to be, *delayed* by a *relevant* event. Only relevant events which occur *before* the original or a previously fixed completion date can be taken into account. The second power is under clause 25.3.3. Here the architect has a power of review exercised *after* the original or a previously fixed completion date (either before practical completion or no later than 12 weeks after that event). He has power to fix a new completion date if he regards it as fair and reasonable, whether upon reviewing a previous decision or otherwise and even if the contractor has not given notice of a relevant event. The judge concluded that the words 'or otherwise' opened up the circumstances which the architect could take into account *in addition* to those he had already taken into account.

Alternatively, using the power he had under clause 25.3.3.2, he could advance the completion date by fixing it *earlier or later* than the previously fixed date, 'having regard to the omission of any work or obligation instructed or sanctioned by the architect under *clause 13* after the last occasion on which the architect fixed a new completion date'.

On the question of whether the fixing of a new completion date to a date before the variation instructions was given 'flouted business commonsense', Coleman J said that the answer lay in examining the underlying *contractual purpose* of the completion date/extension of time/liquidated damage regime. The contractual code fixed the obligation of the contractor to complete within the contractual period, defined by start date and completion date (practical completion). Failure to complete resulted in the contractor paying liquidated damages between the completion date and practical completion. 'Superimposed on this regime is a system of allocation of risk.'

1 There are *non-contractor risk events* that entitle the contractor to an extension of time if the progress of the works is delayed.
2 There is also power to reduce the works by an *omission* instruction by the architect, where it may be fair and reasonable to reduce the contractual period. Because of the difficulty of predicting the impact of.
   (a) a non-contractor risk event and
   (b) a contractor risk event operating together,
   the architect is given the power of *retrospective adjustment* of the completion date.

In dismissing the contractor's arguments Coleman J used the example of a trivial variation representing one day's work. Following the contractor's argument this would result in the employer losing all rights to liquidated damages for the period the contractor was in culpable delay and be left to prove unliquidated damages for delay at common law.

Coleman J, holding that the arbitrator was entirely correct in his answer to the First question, said:

'in my judgement *the construction that the contractor* contends involves legal and commercial results which are inconsistent with other express provisions and with the contractual risk distribution regime applicable to pre-completion date relevant events such that, in the absence of express words compelling that construction, it *cannot be right*.'

The second question

The appellants argued for a 'gross valuation' of the period of the extension of time; in other words, the extension of time should be the period of time fairly and reasonably to be expected when the variation was ordered. The arbitrator by contrast held that the delay caused by the requirement to execute the instructions was to start with the existing completion and then to be extended by a period that was fair and reasonable.

The judge in his approach went back to the *purpose* of the architect's powers under clause 25. In assessing the period for extending time he had to be fair and reasonable. To do so he had to take into account *the effect the relevant event* had on the work. He had to take into account one of two things:

(a) Did the relevant event bring progress to a *standstill*; *or*
(b) did it merely *slow* the work?

If the variation of the work can be reasonably conducted simultaneously with the original works without interfering with their progress and is unlikely to prolong the achievement of practical completion, the architect might properly conclude that *no extension of time* was justified; he would leave the completion date as it was. This would leave the contractor to pay LAD for the amount of time he exceeded the original period of time for completion. His continuing liability to pay liquidated damages, while he is at the same time carrying out the variation works, does not reflect an assumption by him of the loss of time due to what would otherwise be an act of prevention. It merely reflects his breach of contract by failing to complete the original work within the original last-fixed contract period for completion.

The judge found support for the 'net' method of extension in the approach adopted by the Court of Appeal in *Amalgamated Building Contractors* v. *Waltham Holy Cross UDC* [1952] 2 All ER 452. The issue in the case was whether the architect could grant an extension of time retrospectively as well as prospectively. Denning LJ, in deciding the clause gave the architect power to fix a date for completion that had passed, gave some practical examples of when it would be necessary to fix time retrospectively. Taking the simple example of contractors being near the end of the work having overrun the contract time by six months without legitimate excuse, he stated the following:

1 A strike occurs when the work is uncompleted and lasts a month. The architect can extend the original completion date by a month.
2 Labour and materials are not available. The delay is continuous and operates partially because the work carries on. Only when the work is complete is the architect in a position to give an extension of time.

Coleman J answered the second question by concluding that it would be 'wrong in principle' to apply the 'gross' method and decided the 'net' method represented the correct approach. He *upheld* the arbitrator's award.

For further cases applying the principles laid down in *Chestermount* see *Balfour Beatty Construction Ltd* v. *The Mayor and Burgesses of the London Borough of Lambeth* [2002] EWHC 597 (TCC) at para. 17.

Instructions and variations

A question that often arises during construction is the status of an instruction given by the architect/engineer. Is an instruction needed, and when does it become a variation? In what circumstances must the architect/engineer give instruction to assist the contractor to complete the work?

Hudson is highly critical of the argument that the 'architect/engineer is the captain of the ship' (1999 p.349). He considers that the person primarily responsible, when seriously defective work is discovered, is the contractor. Three key factors contribute to the likely presence of the defective work, (i) cost-cutting by minimising expenditure on 'troublesome details of good practice and workmanship', (ii) the opportunity to do so without being detected and (iii) to counter this contractor's need to better supervisory staff and not expect the employer to install 'more elaborate and expensive systems of inspection and quality control.'

If, in response to such a refusal, the employer employs other contractors to carry out the work, the contractor is obliged to let them do so. Should the contractor refuse to allow these contractors to enter the site to carry out the work, the employer may obtain an injunction: see *Bath and North East Somerset District Council* v. *Mowlem Plc* [2004] EWCA Civ 115). The contractor argued that on the 'balance of convenience' the employer would be adequately compensated by LADs for the subsequent delay and therefore the order granted at first instance should be overturned by the Court of Appeal. The employer successfully argued that LADs would not adequately compensate it since its likely losses, as public authority would exceed its claim for LADs no matter the length of the delay. The court accepted this argument, but reserved the position on whether LADs represented the only claim for damages available in the circumstances.

Architect's instructions

Clause 4 of JCT 98 empowers the architect to issue instructions to the contractor 'with regard to any matter in respect of which the architect is expressly empowered by the conditions to issue instructions.' Such instruction has to be in writing but clause 4.3.2 makes comprehensive provision if it is not.

1 The contractor can confirm the verbal instruction in writing.
2 The architect can also do so and, where neither party does so, and the contractor does comply, the architect has power to confirm the instruction in writing at any time up to the issue of the final certificate.

The power to give instruction is not unlimited. Clause 4.1.2 provides for a challenge to the instructions. The architect *must forthwith* comply with a request from the contractor to specify in writing the provision in the contract that empowers the issue of the instruction. If neither party challenges the instructions (by invoking the dispute resolution procedure), then the instruction shall be deemed to be to be empowered by the provision in the condition specified by the architect, in reply to the contractor's request (clause 4.2). Failure by the contractor to comply with the instructions within seven days of receipt enables the employer *to employ and pay others to do the work*. Clause 4.2 further specifies that such costs are either to be deducted from monies due or can be recovered as a debt from the contractor.

In *Holland Hannan & Cubitts (Northern) Ltd* v. *Welsh Health Technical Services Organisation and others* (1981) 18 BLR 80, design and installation work was carried out by the nominated sub-contractor. The sub-contractor accepted liability but the architect *refused to issue instructions*, amounting to a variation, to deal with the problem. The judge held that where unexpected difficulties are encountered (in this case the windows leaked), the architect may have to issue changed drawings or details in order for the work to proceed. HHJ Newey QC said at p.118:

'I think the second part of clause 3 (4) required [the architect] to issue drawings or details to enable [the contractor] to get on with the building the hospital, when it would otherwise have been impractical for them to have done so. Inability to proceed might be due to a variety of circumstances, including site conditions proving different from those anticipated or specified materials being unobtainable. In fact it was due to design falling below standard to such a degree that water penetration would have damaged internal partitions or the like, which were fitted.'

Later on he said that the architects were under a duty in accordance with clause 3 (4) to provide drawings or details to overcome defects of design. There was also an implied term – which he held to be wider than clause 3 (4) – to issue a variation under the then clause 11(1) or to take appropriate action if the building of the hospital was brought to a stop. *Note* that the implied term was to *do all things necessary* to enable the contractor to carry out the works.

Variations under the ICE

The engineer under clause 51 can order variations to the works. The clause is headed 'Alterations, additions and omissions.'

51 (1) the engineer
(a) shall order any variation to any part of the works that is in his opinion necessary for the completion of the works and
(b) may order any variation that for any other reason shall in his opinion be desirable for the completion and/or improving functioning of the works.

Such variation may include additions substitutions alterations changes in quality form character kind positioning dimension level or line and changes in any specified sequence method or timing of construction required by the Contract and may ordered during the defects.

Correction period

The ICE guidance notes to the seventh edition emphasises that clause 51(1) specifically permits the ordering of variations *during* the defects correction period. It this respect it differs from the JCT 98 which has no corresponding provision. Clause 17.3 does, however, permit the issue of instructions for *necessary work*. Such an instruction does not amount to a variation.

Eggleston (2001) makes two observations about the nature of civil engineering variations. First, such variations usually arise from *necessity* rather than simple changes of mind by the employer; second, being a re-measurement contract, there is less certainty about the final price than in lump sum contracts (such as *JCT 98*). At p.255. he lists a number of reasons for the use of variations clauses:

• to authorise the engineer/contract administrator to order variations
• to entitle the contractor to a variation if the specified work proves impossible to construct
• to define the scheme for the valuation of variations

- to prescribe procedures to be followed to recover payment
- to preclude payment for unauthorised variations
- to set financial limits on unauthorised variations.

In addition to the three-tier method of valuing variation, the ICE seventh edition introduces contractor quotations for varied work in *clause 52*. This was a response to the Latham Report's recommendation for pre-priced variations.

Clause 52 (1) 'if requested by engineer the contractor shall submit his quotation for any proposed variation and his estimate for any consequential delay. Whenever possible the value and delay consequences (if any) of each variation shall be agreed before the order is issued or before work starts'.

Where no request is made or the parties fail to reach agreement under Clause 52(1), clause 52(2) deals with the consequences. As soon as possible after receipt of a variation the contractor shall submit to the engineer:

- a quotation for the work using any rates or prices included in the contract
- an estimate of any resulting delay
- an estimate of the cost of that delay.

The engineer has 14 days to either accept the submission or negotiate with the contractor to resolve the matter. If the parties fail to reach agreement, the valuation of the variation ordered by the engineer will be made using the three-tier method specified clause 52 (3) (see the examples above under the valuation of variations).

Giving instructions

Neodox Ltd v. *Borough of Swinton & Pendlebury* (1958) 5 BLR 34, the contractor argued that it was an implied term that it would receive 'all necessary instructions and details . . . in sufficient time to complete in an economic and expeditious manner and/or sufficient time'.

Held: A reasonable time does not depend only on the convenience or financial interest of a contractor; it also depends on the interests of the engineer, its staff and other relevant factors, such as when it was needed.

The specification read: 'excavations shall be timbered with suitable timber or alternative forms of sheeting . . . to the satisfaction of the engineer'. The arbitrator found that the engineer had acted unreasonably in:

(a) ordering the contractor to use timbering in sections where excavation by machinery with battering of the slopes was practicable; *and*
(b) exposing the contractor to additional expense.

Diplock J decided the engineer's decision was honest and his requirements were conclusive. The specification gave the engineer power to determine the manner of working. In the absence of a specific method of carrying out the work *the instruction did not amount to a variation.*

Hudson injects a note of caution on the wording of the specification (at p.502) and stresses that in the absence of express power the architect or engineer:

(a) can only identify defective or potentially defective work;
(b) can require it to be done again/or replaced;
(c) has no right to control or interfere with the contractor's method of working.

If the engineer does so the *instruction may amount to a variation*. Hudson also points out (at p.892) the pitfalls of instructions to assist a contractor having difficulty in completing the contract to the original design. However, the employer has an interest in ensuring the long-term suitability of the permanent work. It is therefore in its own interest to assist the contractor by making design changes if required. He cautions the architects or engineers to use permissive language only when authorising such an alteration.

In *Pearce* v. *Hereford Corporation* (1968) 66 LGR 947, under clause 13 of the ICE contract (fourth edition), the contractor was required to comply with and adhere strictly to the engineer's instructions and direction on any matter. Whilst driving a sewer heading under a road, the contractor ran into an old sewer, which collapsed into and flooded his heading. A solution was agreed at a site meeting. The old sewer would be stanked off at a point off-site, and a shaft would be sunk on the far side of the road and work back to the crossing, which would be open cut (referred to in Hudson at p.895). In refusing the contractor's claim for payment for the extra work involved, Paull J held the work was in the nature of a joint decision as to the best way of doing the work. Even if there was an instruction under clause 13, they did not create any financial liability for the work which the contractor was bound to do in order to complete the contract.

Legally or physically impossible work

Clause 13(1) of the *ICE* conditions provides for the giving of instructions where the contractor is unable to complete the works due to the effect of *legal or physical impossibility*. Clause 13(3) deals with the consequences of the instruction given by the engineer in those circumstances. Where instructions result in delay the contractor is entitled to an extension of time and loss and expense incurred.

In *Yorkshire Water Authority* v. *Sir Alfred McAlpine* (1985) 32 BLR 114, the contractor's method statement was incorporated as a contract document. When the work was found impossible to perform, it was held to be a matter for the decision of the arbitrator whether there was an impossibility to perform.

Whether the work was physically impossible arose in *Turriff* v. *Welsh National Development Authority* (1985) 32 BLR 117. Tolerances for precast sewer segments proved impossible to overcome and the contractor abandoned the work on the ground that it was impossible to carry out.

The judge held that the works were impossible in the ordinary commercial sense. Judge William Stabb QC said: 'But the real issue is as to the construction of the words "physically impossible" in the light of those authorities, which establish the common law position. [Counsel for the employer] urges me to construe it as having an absolute meaning, so as to give it certainty.' *Holland Dredging (UK) Ltd* v. *Dredging and Construction Co. Ltd* (1987) 37 BLR 1 establishes that impossibility can apply to both supervening and pre-existing events.

Design liability and impossibility

Norwest Holst Construction Ltd v. *Renfrewshire Council* (1996) had a special clause whereby the contractor was responsible for the design of piled foundations for a railway under-bridge. The contractor, together with the piling sub-contractor, produced a design which met the contract specification, but proved impossible to construct. One of the matters the court had to decide was whether the engineer was contractually responsible for resolving the issue of impossibility. Giving judgment in favour of the contractors the judge said:

> 'Even though the fact that the piling could not be constructed emerged at the time the piling was being designed, in my view that is irrelevant. The reality is that Norwest were telling the engineer that part of the work that had been specified in the contract could not be constructed and the reason for that lay in the specification of the abutment which was not a part of Norwest's design responsibility. In that situation, because of the terms of clause 13(1), Norwest would come under no contractual obligation to construct the piling and they can have no responsibility for devising a solution to the difficulties, although, in fact, along with their sub-contractors they put forward a number of possible ways round the problem. The reality was therefore, if (the employer) wanted the contract works complete, they or the *engineer* acting on their behalf *were required to devise a solution and to give appropriate direction or instruction*. Moreover in terms of *clause 51* the engineer has responsibility to order "any variation to part of the works that may in his opinion be necessary for the completion of the works".' (emphasis added)

Summary

1 Frustration is an external event outside the control of the parties that brings their contract to an end. Such events are rare in construction contracts.
2 Where the contract is found to be frustrated the Law Reform (Frustrated Contracts) Act 1943 regulates the effects. This is now considered to be part of the law of restitution based on a principle of unjust enrichment.
3 Design changes are usually dealt with expressly in the standard form contracts by a provision for variation to the contract. A variation must bear some relation to the contract of which it is a variation.
4 The scope of variation clauses in the standard form contracts is broad and wide-ranging. Complicated provisions are made for the valuing of variation.
5 The ICE valuation rules consist of three rules: *Rule 1* – where work was carried out of similar character and carried out under similar conditions, the engineer was bound to apply the rates in the bills. Rule 1 is *mandatory*. *Rule 2* is also mandatory. Where the work carried out *was not* of similar character *nor* carried out under similar conditions – the bills were used to arrive at the basis

of the valuation. *Rule 3* – where *Rule 2* was not applicable the engineer was bound to make a *fair valuation*.

6   Under the JCT the phase 'fair rates and prices' is used instead of 'fair valuation and prices'. In other respects the clauses are very similar.

7   Where a variation is ordered when the contractor is in culpable delay, the variation does not displace the contractual structure. The provisions under the standard forms are broad enough to cover the position. The liquidated and ascertained damages provisions are not displaced and replaced with a claim for unliquidated damages for delay.

8   Under the JCT the contractor can be given a variation when in delay that fixes a new completion at a time before the variation was given.

9   Care has to be taken in giving instructions to the contractor to assist it in doing work that it was bound to do under the contract. This is to prevent the contractor claiming the instruction as a variation.

10 Where the work is legally or physically impossible to carry out, clause 13 of the ICE provides for instruction to be given by the engineer.

10   Payment and Certification

The right to payment

The main *obligation* of the employer is *to pay* the contractor for work and material supplied. It used to be said that the contractor's right to payment depended on the wording of the contract, and that within the limits of legality the parties were free to make arrangements they chose. These could be grouped under *three broad heads* (Keating, 2001, p.90):

1 A lump sum contract – the right to payment only arose after the work was carried out.
2 An express contract other than a lump sum (common under the standard form contracts).
3 A claim for a reasonable sum frequently called *quantum meruit*. This arises in two situations:
    (a) either under an informal contract where it is classified as a 'contractual' *quantum meruit* (see *Clarke & Sons* v. *ACT Construction Ltd* [2002] EWCA Civ 972);
    (b) or where a variation is made outside the contract. For an example see *Costain Civil Engineering Ltd and Tarmac Construction Ltd* v. *Zanen Dredging and Contracting Co. Ltd* (1996) 85 BLR 85.

In *Lubenham Fidelities* v. *South Pembrokeshire District Council and Wigley Fox Parts* (1986) 33 BLR 39, the contractor suspended work because the interim certificate issued by the architect was incorrect. The court held that the only right under the then JCT 63 (now *JCT 98*) was to have the certificate corrected in the *next interim certificate* and, failing that, to go to arbitration. By suspending work the contractor was in breach of contract. *No right* to *suspend work for non-payment* existed in English law. In the light of the HGCRA 96 *both propositions are wrong* as regards contracts falling within the legislation.

1 There is now a *statutory right to interim payments* and, if the contract does not contain such a provision, the 'Scheme for Construction Contracts' (a statutory instrument) applies.
2 The contractor has a *statutory right to suspend work* for non-payment.

These quite fundamental changes in construction contract law were the result of the Sir Michael Latham's Final Report, *Constructing the Team*. Only some of his many recommendations for legislative intervention have been carried out. Amongst the major changes that resulted were the implementation of a *payments regime* and the introduction of a system of *adjudication* for resolving disputes during the construction period.

The obligation to pay

In any contract the parties are free to stipulate both:

(a) how much is to be paid;
(b) when payment is to be made.

They may also stipulate what conditions have to be met before the obligation arises and how payment is to be made. Payment need not always have to be in money (e.g., land can be exchanged in return for services), and payment need not be in sterling or in the United Kingdom. Most construction contracts regulate the matter in detail. The introduction of the HGCRA 96 has complicated payment provisions and is discussed further below.

The difference between debt and damages

A *debt* is money claimed under a party's contractual obligations, whilst *damages* are compensation claimed for a breach of contract. The advantage in claiming for payment of a debt is that the claimant has only to establish the following:

1 That the *sum is due*, and any *condition precedent* has been met.
2 Interest is payable at the commercial rate from the date the debt is payable to judgment date. There is now a statutory right to interest on late payments and the standard forms of contract regulate the position. *JCT 98* sets it at 5 per cent above the base rate (see clause 30.1.1.1). The *ICE 7th* allows for it in clause 67. Interest on late payments is to be compounded monthly at 2 per cent above the base rate. In practice certifying engineers and employers tend to resist claims for the payment of compound interest on outstanding balances.

When does right to payment arise?

The obligation depends on the agreement and the payment structure agreed under it. The sum payable may be called:

• the tender sum
• the contract sum
• the contract price.

The construction industry 'uses terminology, which is *vague* and often *downright inaccurate*' to describe how the contract due under the contract is calculated; see Murdoch and Hughes (1992). There are a number of ways of describing the means by which the price is determined:

1 Fixed price – only the method of calculation is fixed.
2 Lump sum – does not mean a single payment but *specified* work is to be done at a *quoted price*.
3 Firm price – no guarantee that the price given does not mean that there will be no claims for extra payment. It means only that there is no provision for the

contractor to pass on increases in the cost of work and materials to the employer.

Principles used in calculating the cum

There are only *two bases* for calculating payment under a construction contract, a fixed *price* and cost *reimbursement*.

Fixed price items

This consists of a *pre-determined estimate* of the likely cost of the work plus an allowance for the *risk* involved plus an *allowance* for the contractor's *situation*. This will depend on the contractor's workload, the contribution to overheads required, how much is the job wanted (or needed), and whether there is the possibility of follow-on work. The employer pays the sum calculated *regardless* of the contractor's actual costs.

Cost reimbursement

This is paid on the basis of the *actual cost* of the works. The contractor provides an *estimate* beforehand but gives no guarantee that the sum will be the same or not increase. The *choice* between these methods will fundamentally *alter the risk* between the parties. Payment regime as described are not mutually exclusive and it is unusual to find a contract using only one method. Both methods are normally used and how the contract is described depends on which method predominates.

* * *

Building contracts usually have a different price structure from civil engineering. The differences in the pricing structure reflect this. In the JCT 98 form with quantities, the *tender is based on* work exhaustively described and quantified in contract bills and specifications and includes variations. By contrast, in both the ICE 7th and JCT 98 (with approximate quantities) the contractor's tender *prices* are purely for tendering purposes. The contractor is paid at quoted rates but for measured quantities only. Civil engineering usually carries greater risks of the unknown than building contracts, because in the construction is of a road, railway, dam, bridge, tunnel, drainage or laying pipelines, the 'as built' quantities are usually difficult to predict with accuracy. The larger the scale of the operations, the greater the spread of boreholes or trial pits made during the site investigation. As a result the extent and location of quantities for large operations which are quite straightforward e.g., the building of a motorway or the construction of a dam), must inevitably be provisional. For this reason civil engineering contracts have always used re-measurement to arrive at the contract price. By comparison, building sites are much more compact and the amount of work below ground correspondingly represents a much smaller part of the final total. The pre-planning of the superstructure can be prepared in greater detail beforehand and the quantities are less likely to change. For this reason the priced bills of quantities will more accurately reflect the final price.

Contractor cash flow

Contractors make a profit from the installation and assembly of materials and components to meet a specified performance within a given time period. They will pay their suppliers and sub-contractors at differing intervals depending on the nature of what has been provided. Suppliers to the contractor usually invoice at the end of one month for payment at the end of the following month. In practice reputable contractors enjoy substantial periods of credit after the materials have been incorporated into the works. Typically this may be up to four to six months. Sub-contractors usually have similar waits for payment.

The contractor has to pay the directly employed staff, and its balance sheet reflects the income from work certified for payment. Most certified sums have to be paid within 14 days where the standard forms of contract are used. It is *critical* for the contractor's operations to hang on to monies received from the employer for two reasons. *Little* or *no mark-up* needs to be added to services (which makes it more competitive when tendering) and it can *operate with limited capital* (it can also gain interest on sums due to suppliers and sub-contractors). Since the contractor's pricing stricture is very *sensitive* to changes to payment regimes on projects, the provisions of the HGCRA 96 have made a significant change to how they operate.

Progress payments

There is no *implied* right to payment on account *unless* the contract says so.

1 Staged payments may be due under the contract geared to the completion of separate stages. An alternative is a provision for periodic payments usually made monthly. Most construction contracts, whether made under a standard form or a non-standard one, adopt one of these methods.
2 *Progress payments on account.* These are *not really instalments* since they are paid when the price is not yet due. The courts will, it appears, readily imply an obligation to make progress payments. Hudson (at p.493), states that there is a class of contracts for work and services in which, from the nature of the work, it is necessary to imply a term for reasonable payments on account of the work, at reasonable intervals (see also *DR Bradley (Cable Jointing)* v. *Jefco Mechanical Services* (1989) 6 CLD-7-21). A dispute over non-payment arose between the main contractor and the sub-contractor. The sub-contract made no express provisions as to the time or method of payment. Rich QC sitting as Official Referee said:

> 'Having heard their evidence as to the custom of the industry, I conclude that it was an implied term of a contract that application could be made for payment not more frequently than monthly. Payment would then be due in a reasonable time of such application, for the work done and any unused materials on site, valued in each case in accordance with the contract between the parties. I announced in the course of the hearing my intention

to hold that I construed a reasonable time in all the circumstances as 30 days with a period of grace to allow 42 days in all, and I now hold that 42 days was a reasonable time for payment.'

3 *Interim certificates.* The general rule is that where there is no certificate there is no entitlement to payment. Where a certificate of payment has been issued the contractor cannot be paid more than that sum even if it is incorrect. The only right the contractor has is to request a correction in the next certificate. Where the contract specifies it, this is a ground for immediate arbitration. See *Lubenham Fidelities* v. *South Pembrokeshire DC* (1988).

Interim certificates merely represent the contract administrators' view of work and services supplied. They reassure the employer that sufficient work to that value has been done. Any certificate can be 'opened up and reviewed' by an adjudicator, arbitrator and by the court.

There are a number of grounds on which a certificate can be challenged, including where the proper procedure has not been followed, or it has not been issued by the right person, or it is not in the correct form and at the right time, or where there is fraud or collusion, or the employer has interfered in the issue or the certifier has died or become incapable.

Where the certification regime under the contract is brought to an end (for example, by an unlawful repudiation of the contract), recovery can be made despite the absence of the certificate: see *Scobie & McIntosh Ltd* v. *Clayton Bowmore Ltd* (1990) 49 BLR 119. In this case, the employer informed the main contractor that their own client had informed them that the contract had been 'frustrated and must be regarded as terminated'. They accepted this as a *repudiation* and terminated the sub-contract under the provisions of the contract. The sub-contractors issued a writ claiming that the termination was a repudiatory breach of contract of the sub-contract and claiming damages, or alternatively a *quantum meruit* payment. The main contractor challenged the court proceedings on the basis that the contract provided for arbitration. In addition, the right to payment only arose when the employer certified for payment the work carried out by the sub-contractor. Since this had not been done, no right to payment had arisen. The judge summarised this argument as follows:

'The essential question was whether the . . . scheme of payment meant that the contractor's obligation to pay was conditional on the employers certification of all the works: i.e. that the contractor would be obliged to pay only if the employer paid him.'

The judge dismissed this proposition for two reasons. First, each contract was an entire one, separate and distinct from each other. Second, there was no privity of contract between the employer and the sub-contractor. The machinery for payment only existed while both contracts existed. Once the main contract was repudiated, the contractual machinery for payment fell with it. The sub-contractor was therefore entitled to payment of its claim for an interim payment less the amount of money that was in dispute.

The formal requirements

The certificate must be clear and unambiguous and the use of the word 'certify' is not essential. The general advice is to avoid any ambiguity or uncertainty arising by following closely the wording of the clause under which the power to certify arises.

Payment clauses usually require the certificate to be issued and delivered. JCT 98 clause 5.8 requires 'any certificate shall . . . be issued to the employer with a duplicate copy to the contractor'. Whether it had been so issued arose in *Camden London Borough Council* v. *Thomas McInerny* (1986) 3 Const LJ 293, a contract under the then JCT 63. Under the contract 291 dwellings were to be constructed including a high rise block with brick cladding. The Director of Architecture was the architect under the contract, although the day to day supervision was done by architects on site.

The final account was agreed and interim and final certificates were prepared by a technical assistant in the quantity surveying department. These showed the amounts due to McInerny. During routine inspections by the site-based architects, cracks were discovered in the brickwork. A note was attached to the certificates saying 'do not pay' and signed by the chief quantity surveyor. The certificates came to light just before trial. There were three preliminary issues to be decided at trial:

1 Were the signed certificates valid final certificates that the work had been properly carried out and certified?
2 Even if not issued, did the certificates represent the architect's view that the work had been properly carried out? If they did, the court had no power to open up the certificate. (Note: in the light of the *Crouch* decision only an arbitrator could exercise that power).
3 The interim certificate showed the total value of the work that had been properly executed. Every interim certificate represented that opinion. Only an arbitrator could revise it.

Judge Lewis rejected these novel arguments. The circumstances in which the disputed certificates were prepared showed that those concerned with certification clearly intended that McInerny should not receive any further money once the alleged bad workmanship had been discovered.

In deciding that the certificates had not been issued, the judge said: 'I have some difficulty in thinking that there would be sufficient compliance with (the contract) if the architect certified in writing and then locked the document away and told no one about it.' Mere signing of the certificates was not enough to comply with the clause; the certificates had also to be issued. Neither did the mere signing of the final certificate amount to the architect's opinion that the work had been properly carried out: 'I see no reason why the architect should not change his mind after signing but before issuing it. He was not irretrievably committed to that opinion merely by signing it.' The judge also refused to accept the contractor's argument about the status of the final certificate. For the status of the final certificate, see Chapter 6.

Status of an interim certificate

The issue of the certificate is no proof that works was properly done and is merely the mechanism by which stage payments are made. Certificates can be opened up and revised by an adjudicator and an arbitrator or, where the parties have allowed for it, in legal proceedings by the court.

What the employer must pay

The certificate reflects the opinion of the certifier that it is safe for the employer to make advances, since enough work to cover the value of the certificate has been done. They are not binding with regard to quality or the amount, and are subject to adjustment in later interim certificates. Where there is an *undervaluation*, it was held in *Lubenham* that the only right the contractor had was to request an adjustment in the following certificate. Failing that, the dissatisfied contractor could initiate immediate arbitration where provided for in the contract. This is now subject to the provisions of the HGCRA 96.

Entire contracts and substantial completion

The rule only applies these days to small informal contracts, since most formal contracts deal with payment expressly. The Law Commission found that the rule was of great benefit to homeowners since they often have great difficulty in getting small contractors to complete the work. Essentially the rule allows the contractor to be paid the contract sum less the cost of curing any defects (i.e., the cost of putting things right). If, however, the contractor abandons the work there is no entitlement to any payment. The application of the rule is a matter of degree in all the circumstances. If sued, the contractor can offset the value of the work against damages awarded if the work proves to be of value.

When payment is due

Payment is due at the time stipulated in the contract, if not stated, within a reasonable time in all the circumstances. Where payment is only due on the occurrence of some event, such as the issue of a certificate or payment being received from a third party, the general rule is that payment is only due on the occurrence of that event.

Note that in general, the time of payment is not of the essence. Delay in payment does not entitle the other party to treat the contract as at an end. However, an *express* clause that '*payment to the contractor is due 14 days after the issue of a certificate*' is a breach of condition which allows the contractor to bring his employment under the contract to an end. Such a provision is usually found in standard forms of contract.

Excuses for non-payment

Set-off

Historically the common law made no provision for dealing with *cross-claims between* parties in the same proceedings. From the eighteenth century onwards statute and later *equity* intervened by allowing a defendant in an *action for money due* certain monetary cross-claims which could also form a *separate action by the defendant*. The defendant could *set off* (as a defence):

- mutual debts
- defective work in the price claimed
- cross-claims closely connected with the original claim.

Cross-claims from *another contract* cannot normally be set off. It is possible to argue in court that the two contracts are so *closely connected* that it would be *unfair* and *inequitable* for judgment to be enforced until the other claim has been settled. A case where such a claim failed was *Anglian Building Products Ltd* v. *W&C French (Construction) Ltd* (1992) 16 BLR 1. Pre-stressed concrete beams were supplied by Anglian to French on three separate motorway contracts. On being sued for their failure to pay, French raised a counter-claim alleging that beams supplied for one motorway had been defective. Judgment was given against them on two of the contracts. They sought a stay of execution of these contracts until the matter of the third contract had been settled. The court refused their request. See also *AB Contractors Ltd* v. *Flaherty Brothers Ltd* (1990) 6 Const LJ 141.

Set-off is a *true defence* which reduces a claim or extinguishes it. Recently in *Bim Kemi* v. *Blackburn Chemicals Ltd* [2001] EWCA Civ 457, Potter LJ said:

'The judgment of Morris LJ has been described as "authoritative" . . . and . . . "a masterly account" of the subject and it has subsequently been generally accepted as the starting point for modern consideration as to the scope of the defence of equitable set-off.' (p.10)

Morris LJ in *Hanak* v. *Green* [1958] 2 WLR 755 gave what is considered the modern analysis of set-off in its historical context. He summarised the position as follows.

1   There can be set-off of mutual debts in bankruptcy or liquidation: section 323 of the Insolvency Act 1986 and rule 4.90 of the Insolvency Act Rules 1986.
2   In certain cases, there may be matters of complaint which, if established, reduce or even extinguish the claim raised by way of defence in that they affect the value of the plaintiff's claim.
3   There can be an equitable set-off which can be used as a matter of defence.

A right to payment on certificate

The whole matter of set-off against interim certificates was raised by Lord Denning in *Dawnays Limited* v. *F G Minter and Trollope and Colls* [1971] 1 WLR 1205. He wanted there to be no right of set-off in respect of payment on certificate in the construction industry. Cash flow he said was the lifeblood of the industry. The FASS green form used in conjunction with the JCT 63 standard form of contract at that time contained *no express* provisions dealing with set-off. In *Dawnays* the main contractor received payment from the employer, but refused to pay it over to the sub-contractor, claiming the right to deduct from it damages for delay resulting from their defective work. The Court of Appeal held that they were only entitled to deduct LAD under the provisions of the sub-contract.

In *Gilbert-Ash (Northern) Ltd* v. *Modern Engineering (Bristol) Ltd* [1973] 3 WLR 421, the House of Lords held that the contractor was entitled to set-off unless there were clear words excluding it. It rejected the idea that 'cash flow' in the construction was a special case. Viscount Dilhorne stressed in his judgment that the contractor's only entitlement was payment for work properly done. He was not entitled to payment for defective work. It is generally accepted that *Gilbert-Ash* abolished the *Dawnays* principle. This decision caused great uncertainty in knowing when and in what circumstances set-off could be used. Because of the importance to sub-contractors and contractors of cash flow, express provisions for set-off and the procedures to be followed were introduced in the standard forms after 1976. The problem continued in domestic sub-contracts where main contractors used their own standard forms. This greatly disadvantaged sub-contractors.

Jones *et al.* (Const LJ, p.101) comments: 'in very recent times, it has become apparent that, insofar as domestic sub-contracting is concerned, contractors can, by challenging the valuation rather than seeking to claim set-off for delay and disruption, out-maneuver the sub-contractor, with the cash flow advantage moving strongly in favour of the contractor'. See, however, the changes made by the HGCRA 96 below. Its provisions could be described as a statutory re-enaction of Lord Denning's belief in the honouring of interim certificates of payment.

Counter-claim

A counter-claim is a *procedural convenience*. It need not be related to the original claim, it can be for any amount and additional defendants can be joined. A counter-claim is not a defence (although it may involve set-off). The distinction is vital as far as cash flow is concerned. Where the counter-claim is in law also a set-off, the party raising it can withhold the set-off of monies properly due to the other party. If the counter-claim does not amount to a set-off the party raising the counter-claim has to pay the amount in full. The claim must be paid in full and the counter-claim must be by a separate claim. Its existence is not affected by the withdrawal of the claimant's claim (Jones, p.84).

Abatement

This is an old common law remedy based on *Mondell* v. *Steel* (1841) 8 M&W 858. At that time civil procedure did not allow claim and cross-claim in the same proceedings. In *Mondell* v. *Steel*, the cross-claim was for *defective work*. A ship delivered under a shipbuilding contract contained certain defects. The technical *distinction* between set-off and common law abatement (which applies only to contracts for the sale of goods and contracts for works and materials) was explained by Park B, who said at pp.871–2:

'It must however be considered, that in all these cases of goods sold and delivered with a warranty . . . the rule which has been found so convenient is established; and that it is competent for the defendant . . . not to set-off by proceedings in the nature of a cross-action, the amount of damages which he has sustained by breach of the contract, but simply to defend himself by showing how much less the subject matter of the action is worth, by reason of the breach of contract; and to the extent that he obtains, or is capable of obtaining, an abatement of price on account, he must be considered as having received satisfaction for the breach of contract, and is precluded from receiving in another action to that extent; but no more.'

The *Judicature Acts 1873–75* changed procedure so that claims, whether at common law or in equity, could be brought in the same court. *Abatement* is a useful remedy available to an employer where the work has proved defective and there is no contractual mechanism to deal with it. It is a defence to a claim for payment where the work was not done properly. It is not a claim for breach of contract and the *employer cannot get more than the contract sum.* In *Raymond Slater* v. *C.A. Dunquemin Limited* (1993) CLJ 64, the late Judge Newey QC, Official Referee, used the law of abatement on a JCT standard form contract.

Remember that with abatement the employer is not arguing that the contract sum is not due, only that the work was done so badly that the *contract sum is not justified.* In *Investors Compensation Scheme Ltd* [1998], Lord Lloyd of Berwick said that the technical meaning of abatement 'in connection for contracts for the sale of goods and provision of services' simply meant a reduction.

In *Mellowes Archital Ltd* v. *Bell Projects Ltd* [1997] EWCA Civ 2491, the Court of Appeal considered the *relationship* between *set-off* and *abatement.* Mellowes was the sub-contractor employed by Bell under the standard form of sub-contract, DOM/1. The contract provided for interim payments to a set timetable calculated by (i) taking the gross value of the sub-contract work and (ii) deducting any retention (clause 21.5) and any set-off permitted (clause 23). This required a clear calculation of the sum subject to set-off and advance notice of any sum. In an application by Mellowes for payment under (the then) order 14 of the Rules of the Supreme Court, Bell argued that they were entitled to rely on the defence of abatement for additional costs for delay which would extinguish the sum claimed. The official referee held that clause 23 precluded sums by way of set-off but Bell could rely on abatement if the delay claim fell within the rule in *Mondel* v. *Steel.*

The Court of Appeal recognised that Bell was only driven to the law of abatement since clause 23 did not exclude a defence or plea of abatement. It held the law of abatement could not apply to claims for delay. Hobhouse LJ put the matter clearly at p.9:

> 'The case for the defendants is that the plaintiff's delays caused them serious losses through prolongation of the head contract, the disruption of their own contractual works and those of other sub-contractors. Thus, the defendants' case is based upon financial losses which they say have been suffered as a consequence of the plaintiffs' breaches of their obligations of timeous performance. Subject to the terms of clause 23, those losses can be relied on to support an equitable set-off but cannot justify the legal defence of abatement of the price.'

This conclusion is in line with the views expressed by the Court of Appeal in the case of *Acsim*. There Ralph Gibson at p.67 and Slade LJ at p.80 expressly agreed with the decision of Judge Hawser QC and the concession of counsel that the losses suffered as a result of delay in performance of the contract cannot be relied upon to abate the price. In *Acsim (Southern)* v. *Danish Contracting and Development* (1989) 47 BLR 47, the clause limiting set-off was similar to clause 23. It was held this did not affect the right of the contractor to defend a claim for interim payments by showing that the value of the work was less than that claimed.

Remedies for non-payment

Before the introduction of the HGCRA 96 a contractor who was owed money had a number of options available. Chief amongst these were an application under the *Rules of the Supreme Court* for either *summary judgment* or an *interim payment* in the High Court. Similar rules applied to claims in the County Court. A substantial body of law developed as to the circumstances in which summary judgment was available. The advantage of the procedure was to speed up the payment of monies due, in circumstances where disputes arose later about the value of work carried out. Two factors have affected this body of rules: first, the changes brought about by the Woolf Report and the consequent introduction of the Civil Procedure Rules (CPR), and second the sanction by the courts of the procedure to enforce the payment of awards made in adjudication.

The Latham Report

Sir Michael Latham in his report made a number of recommendations for a *Modern Contract*. Amongst these were:

1 The introduction of 'express provisions for assessing *interim payments* by methods other than monthly valuations i.e., mile stones, activity schedules or payment schedules. Such arrangements must also be reflected in related sub-contract documentation. The eventual aim should be to phase out the traditional system of monthly measurement or re-measurement but meanwhile provision should still be made for it.' (Item 5.18.8 at p.37)
2 'Clearly setting out the *period within* which interim *payments* must be *made* to all participants in the process, failing which they will have an automatic right to compensation, involving payment of *interest* at a sufficiently *heavy rate* to deter slow payment.' (Item 5.18.9 at p.37)
3 'While taking all possible steps to avoid conflict on site, to provide for *speedy dispute resolution* if any conflict arises by a pre-determined impartial adjudicator/referee/expert.' (Item 5.18.11 at p.37)
4 A *Construction Contracts Bill* to exclude unfair conditions. 'Any attempt to amend or delete the sections related to times and conditions of payment including the right to interest on late payments', such as pay when paid clauses and the use of set-off without notifications or justifications. (Item 8.9, pp.84–5)

The provisions were introduced by the Housing Grants, Construction and Regeneration Act 1996 part II (HGCRA 96) supplemented by the Scheme for Construction Contracts.

Both apply to construction contracts entered into after 1 May 1998. *Building Magazine*, 17 April 1998, called it 'the industry's *most important piece* of legislation since the war (1945) . . . [it] presages a *fundamental* change in the nature of contractual relationships'. A *substantial critique* of the Latham proposals and subsequent legislation can be found in Contemporary Issues in construction law Volume II, *Construction Law Reform: A plea for sanity*, which is a collection of essays by well-known academics, construction lawyers and construction professionals published by the Construction Law Press in 1997.

The Act *defines a construction contract* as an agreement to carry out construction operations, or arranging for others to carry out construction operations under a sub-contract, the provision of one's own labour or the labour of others for carrying out construction operations. It also covers professional services, architectural, surveying, engineering, decoration and landscaping.

Construction operations have been *widely defined*. Operations covered include all normal building and civil engineering activities including site preparation, new work, alterations, services, cleaning and decoration of the internal or external part or whole of a building or structure.

There is no minimum value for contracts that fall under the Act. Contracts with a residential owner are excluded. Since entire contracts are common in this sector and are usually made informally, the courts may well extend the Scheme to them in the future. Note that the provisions of the HGCRA 96 only apply to *contracts in writing, which again, is widely defined* (see Chapter 16) for detailed discussion.

The Housing Grants Construction and Regeneration Act 1996 provisions

Section 109 (1) entitles 'a party to a construction contract to payment by instalment, stage payments or other periodic payments for any work done under the contract', except where the work was to be for less than 45 days. The parties are free to agree the amount payable and the intervals when payments are due. If they failed to agree this, the payments provisions of the Scheme for Construction Contracts apply.

Section 110 (1) provides further that the 'construction contract shall (a) provide an adequate mechanism for determining what payments become due under the contract and when, and (b) provide for a final date for payment in relation to any sum which becomes due'.

'The parties are free to agree how long the period is to be between the date on which a sum becomes due and the final date for payment.' Section 110 (2) requires the paying party to give notice by a party not later than five days after the date on which payment becomes due from him under the contract, or would have become due if:

(a) the other party had carried out his obligations under the contract; and
(b) no set-off or abatement was permitted by reference to any sum claimed to be due under one or more other contracts;
(c) specifying the amount (if any) of payment made or proposed to be made and the basis on which the amount was calculated.

If or to the extent that a contract does not contain such a provision as is mentioned in sub-section (1) or (2), the relevant provisions of the Scheme for Construction Contracts apply.

Incorporation

All the standard forms have been re-issued since the passing of the act. The JCT 80 did it via amendment 18. This is now reflected in JCT 98 clause 30.1.1. The clause kept the old provision that payment was to be made 14 days after the issue of an interim certificate. It then grafted on to it the HGCRA 96 provisions.

Application

The following steps are required:

1 Interim certificates shall *state* the *amount* due.
2 The final date for payment is *14 days* from the date of *issue* of an interim certificate.
3 Not later than five days *after* the issue of an interim certificate, the employer is to give *notice* of the *amount* of the proposed payment and the basis on which it is *calculated*.
4 Not later than five days *before* the final date of payment, the employer may give *notice* specifying any *amount* to be *deducted/withheld* from the amount due and the *grounds* for such deduction/withholding.
5 If the employer fails to pay by the final date for payment and the failure continues for *seven days* after the contractor has given *notice* of his intention to *suspend* his performance, the contractor may suspend performance.
6 Under the new clause 26.2.10, the contractor has *a right to claim loss and expense caused by the suspension which was not frivolous or vexatious.* (Compare the position with the Act, which makes *no* such provision.) In *John Jarvis* v. *Rochdale Housing Association* (1986) 36 BLR 48 the word 'vexatiously' was held to mean ulterior motive, to oppress, harass or annoy.

The Act also makes provision for compulsory *adjudication*. This can be initiated by either party to the contract and covers *any dispute* between the parties. After practical completion the *JCT 98* now provides the parties with a choice between arbitration and litigation. The various modes of dispute resolution, including adjudication, are discussed further in Chapter 16.

Rupert Morgan Building Services (LLC) v. *David Jervis and Harriet Jervis* [2003] EWCA Civ 1563 laid to rest the issue of the role and significance of a withholding certificate as required by s.111 (1) of the HGCRA 96. Prior to this case it was a matter of great concern in the administration of the Act's requirement when a certificate should be issued, and in what circumstances the absence of one could be used in adjudication as a defence to a claim.

Withholding payment

Initially a substantial amount of case law developed around notices of intention to withhold payment. Two issues were raised:

(a) If no withholding certificate had been issued could the defence of set-off or abatement be raised in adjudication and
(b) could the defence be raised that the sum claimed was 'not due' to be paid under the contract and therefore, no withholding notice was required?

Furthermore, if the paying party has not made payment nor issued a with-holding notice, is the sum claimed fixed by the failure to observe the section? Can the paying party resist the enforcement of the award on the grounds that the sum is not due? In *S L Timber Systems Ltd* v. *Carillion Construction Ltd* [2001] BLR 516, Court of Sessions, Lord Macfadyen concluded at para 20 that the absence of 'a timeous notice [under s 110 (2)] of intention to deduct does not relieve the party making the claim, from the ordinary burden of showing that he is entitled under the contract to receive the payment he claims.' He went on say that if the claimant discharges the burden on him, one of two things result:

1 If he proves that the sum is due, he is protected by the absence of a notice to withhold all or part of the sum due, from any attempt on the part of the losing party to withhold any sum due.
2 If he does not prove this, then the sum claimed is subject to the defence that the sum is not due. 'Without the benefit of authority, I would have been inclined to say that a dispute about whether the work in respect of which a claim was made had been done, or about whether it was properly measured or valued . . . went to the question whether the sum claimed was due under the contract, therefore did not involve an attempt "to withhold . . . a sum under the contract" and therefore not require the giving of a notice of intention to withhold payment.'

Allied London and Scottish Properties Plc v. *Riverbrae Construction* [1999] BLR 346 was another decision of Outer House, Court of Session. In adjudication the contractor was awarded payment on four contracts carried out in England and Scotland. The employer argued that the adjudicator should delay enforcement until cross-claims from other contracts for liquidated and ascertained damages had been ascertained. In addition, they argued that the monies should be placed on deposit in the joint names of the parties. On appeal, the Court of Session decided that there was no logical basis for the adjudicator to lawfully make an order postponing payment. There was no error of law.

In *Rentokil Ailsa Environmental Ltd* v. *Eastend Civil Engineering Ltd* (1999) CILL 1506 the Sheriff's Court refused to allow the use of cross-claims against the adjudicator's award. The losing party paid the monies owing but immediately issued an arrestment order against the assets of the winning party, reclaiming the money paid. Note an arrestment order is a protective measure available in Scotland. Funds in the hands of third parties can be frozen pending the outcome of the arrestment proceedings. Held: its use during the enforcement adjudication award was an abuse of the process by the court. The decision of the Court of Appeal in *Parsons Plastic (Research and Development) Ltd* v. *Purac Ltd* [2002] EWCA Civ 459 also concerned rights to reduce the sum payable at common law and seemed to suggest that the sum due could be challenged. What must, however, be appreciated from the outset was that the contract for 'odour control' was not a 'construction operation'. It was not therefore subject to the Act, but by their sub-contract the parties had made provisions that mirrored it. The appeal arose out of proceedings to enforce the adjudicator's award. The contract allowed for notice to be given of any amounts subject to set-off or withheld. The Court of Appeal decided that the parties had by another clause retained their common law rights of set-off despite not following the procedures laid down. They were entitled to use those rights to dispute the amount due.

The Court of Appeal in *Rupert Morgan* clarified the position with regard to both issues. Giving the leading judgment, Jacob LJ (with whom his fellow Lord Justices agreed) said that the purpose of s. 111 (1) was to protect cash flow. It did not seek to make any certificate issued conclusive. All that it decided was that in the absence of a withholding notice, the amount certified had to be paid. With regard to the sub-section, the two situations as identified in *S L Timber* had to be considered. Both arose from the provisions of the contract under which the certificate was issued. One was where a third party certified the sum due and the other was where the claim (whether in the form of an invoice or the builder's bill itself) was itself the claim for payment.

Third party certification

This was where a third party (an architect, engineer or other construction professional) is appointed to perform the certifying role under the building contract. Note that this is the usual position under the standard forms of contract. He adopted the analyses of Sheriff Taylor in *Clark Contracts* v. *The Burrell Co (Construction Management) Ltd* [2002] SLT 103. Where the contract provides that interim certificates of payment shall be issued by the architect (whether the sum is wrong or not), the employer has to pay the sum certified unless a withholding notice has been given. In *Rupert Morgan* the contract also made provision for certificates of payment to be made by the architect. In contrast to *Clark*, which was the Scottish version of the standard form of building contact, the architect is certified from the builder's bill. However, the result was the same. Jacob LJ left open the issue of whether there was a duty owed by the architect to advise the client to issue such a certificate. He also left open the matter of whether an action lay in negligence against the architect for over-certifying where the contractor subsequently became insolvent. Note that such an action does exists under the JCT standard form of contract: see for instance the leading case of *Sutcliffe* discussed in ch 5.

No third party certification

The case of *S L Timber* can be explained on this ground. The application for payment was an invoice submitted by the contractor. No provision had been made for independent certification. In the subsequent referral to adjudication the employer could defend the claim by arguing that 'the sum was not due' under the contact. The contractor's invoice did not make the sum due. What made it due was that the work had been done. Work not done was not due for payment and required no withholding notice. See also *Shimsu* where the sum claimed was held to be not due until a VAT invoice or an authenticated VAT receipt was delivered. A withholding certificate could be issued against the invoice and the adjudicator's award.

Withholding payment and common law defences

Section 111 (1) provides that a party may not withhold payment after the final date for payment due under the contract, unless he has given an effective notice of intention to withhold payment. The notice mentioned in section 110 (2) may suffice as a notice of intention to withhold if it complies with the requirements of this section. To be effective the notice must do two things. Section 111 (2) states it must specify (i) the amount to be withheld and (ii) the ground for withholding the payment. Where there is more than one ground, it must be given separately, together with the amount resulting due to it. This notice must be given not later than the prescribed period before the final date of payment
The parties are free to agree what that prescribed period is to be. In the absence of such agreement the period shall be that provided by paragraphs 4 to 7 of the Scheme for Construction Contracts.

Although *Rupert Morgan* clarified the general position, the questions posed in many applications for the enforcement of the adjudicators' awards was this. Where no notice of withholding was given, can the losing party plead the defence of set-off or abatement either during adjudication proceedings, or when defending the enforcement of the award?

Set-off

Northern Developments (Cumbria) Ltd v. *J & J Nichols* [2000] EWHC Technology 176 held that the adjudicator cannot take set-off into account where the other party has not issued a valid notice to withhold. Can the losing party raise the defence of set-off after the adjudicator's decision but before payment has to be made? *VHE Construction Plc* v. *RBSTB Trust Co. Ltd* [2000] EWHC Technology 181 held it could not do so. Where a party failed to comply with section 110 in raising a notice of intention to withhold payment, it could raise the matter when first responding to a notice of adjudication. This is however not the case where there is a certification process involving a third party.

Abatement

An argument that the losing party could use abatement to reduce the amount ordered by the adjudicator failed. See *Whiteways Contractors (Sussex) Ltd* v. *Impressa*

Castelli Construction UK [2000] EWHC Technology 67. HHJ Bowsher QC decided the Act made no distinction between set-off and abatement. A failure to give the requisite notices under sections 110 and 111 meant that the law of abatement could not be used to reduce the amount due. Despite that decision in *Woods Hardwick Ltd* v. *Chiltern Air Conditioning* [2001] BLR 23 it was held that s. 111 was not a matter caught by abatement. The paying party was entitled to adjust the amount payable to reflect the value of the work. HHJ Thornton QC pointed out that: 'It was a significant feature of the adjudication that *Chiltern* had not served any appropriate notices under s. 111 and was therefore not entitled to withhold payment to *Woods Hardwick*.' However, he went on to say that although the defendant had put forward a set-off and cross-claim for damages, its principal defence was that the money was not yet due. This was because the project was incomplete, fees due for payment fell to be abated by virtue of the breaches of contract and in any case the claimant had not carried out any additional work.

Millers Specialist Joinery Co Ltd v. *Nobles Construction Ltd* (2001) CILL 1770, also concerned the failure to give withholding notices. Three arguments were raised against the application for summary judgement to enforce the adjudicator's award.

1   The contract was not a written one; this matter was disposed of by the judge who said 'it would in my judgement be a remarkable result if a document which the parties accept is a correct record of what has been agreed was not be regarded as written evidence of what they have agreed'.
2   No notice of intention to deduct was required because the sums were not due:

> 'the circumstances that previous overpayment may operate under the general law by way of an equitable set-off and thus technically a defence or that it may perhaps be characterised as an abatement which technically in law prevents the amount claimed from ever becoming due does not in my judgement obviate the need for the paying party if he wishes to rely on a right to deduct previous overpayments to give the requisite notice under s. 111.'

3   Defences it wished to raise were a good reason for not disposing of the matter by summary judgment.

HHJ Gilliand QC held that s.111 was wider than the purposes of adjudication. As no certificate had been given, he allowed summary judgment in favour of the claimant.

The right to suspend

This new right is provided for in s.112 (1). A party has the right to suspend performance of the work where (i) the sum due has not been paid in full by the final date for payment and (ii) no effective notice to withhold payment has been given. This right is without prejudice to any other right or remedy that they may have. The right may not be exercised without first giving the party in default at least seven days notice of intention to suspend performance, stating the ground or grounds on which it is intending to suspend performance. The right to suspend performance ceases when the party in default makes payment in full of the amount due.

'Any period during which performance is suspended in pursuance of the right conferred by this section shall be disregarded in computing for the purposes of any contractual time limit the time taken by a party exercising the right or by a third party, to complete the work directly or indirectly affected by exercising of the right.

Where the contractual time limit is set by reference to a date rather than to a period, the date shall be adjusted accordingly': section 112 (4).

In *Ferson Contractors Ltd* v. *Levolux AT Ltd* [2002] EWCA Civ 11, the Court of Appeal considered:

(a) the issuing of notices to withhold money;
(b) the right to suspended the work under the Act;
(c) the provisions of the contract for termination for wrongly suspending the work.

Levolux entered onto a sub-contract with Ferson for the supply and fitting of brise soleil and louvre panelling. A first interim payment was received after an application for payment. A second application for an interim payment of £56,413 was made. Ferson issued a notice of intention to withhold payment so that only a balance of £4,753 was left outstanding. That payment was claimed by Levolux to be three weeks late. They suspended work on site. Ferson issued a notice requiring work to be restarted, failing which they would terminate the contract under the provisions of clause 29.6.2, which included amongst its grounds the wrongful termination of the work. Levolux issued a notice referring the matter to adjudication and Ferson terminated the contract.

The adjudicator decided that Levolux was entitled to the second interim payment of £56,659 (including VAT) plus interest and costs. One of the main findings was that the notice to withhold payment did not comply with s. 111 (2) (a) in that it did not specify the grounds for withholding payment.

In proceedings to enforce the adjudicator's award, the judge held that Levolux had been entitled to suspend the work. That meant that the purported determination based on wrongful suspension had no contractual effect. Ferson appealed, arguing that the judge had wrongly interpreted the adjudicator's decision. Its case was that the adjudicator was not required to decide whether there had been a valid determination of the contract. In rejecting it, Mantell LJ at para. 25 said that that 'argument [was] ingenuous at best'. He accepted that the judge had been right to recognise that implicit in the adjudicator's decision was the finding that there had been no valid determination of the contract.

Other provisions

Section 113 (1) deals with what was commonly known as 'pay when paid' clauses in sub-contracts. This is discussed further in Chapter 11. The section provides that a provision making payment conditional on the payer receiving payment from a third person is ineffective except in insolvency.

The JCT 80 have incorporated the provisions of the HGCRA 96 with amendment no 18 issued in April 1998. Parties to the contract can also decide whether they wish to incorporate amendments arising out of the Latham Report into their contract. The amendment has 60 pages and is longer than the actual JCT 80. One of the report's prime recommendations was that a modern contract should be in language that was easily understood. Soon afterwards the Joint Contracts Tribunal issued JCT 98. It is more comprehensive than JCT 80 and contains all the amendments issued in the past 18 years. The nature of a contract that requires substantial amendment every year opens itself to the criticisms levelled at it that it is ineptly, obscurely and badly drafted.

Section 115 of the HGCRA 96 requires the parties to agree the manner of service of any notice under the Act. Clause 1.7 of JCT 98 deals with the giving of such notices, which should be to an agreed address or, if no address is agreed, the last known address.

Section 116 of the HGCRA 96 provides a means of reckoning periods of time. The new Clause 1.8 specifies that where something is to be done within a certain period of days after and from a specified date, the period begins immediately after that date. Public holidays are excluded from the calculation.

The employer can provide the contractor with an 'Information Release schedule'. The sixth recital states that the schedule shall state what information the architect will release and when. If not provided, the requirement can be deleted.

If not deleted, the architect is obliged by clause 5.4.1 to ensure that the Information Release schedule is released at the time stated in the schedule. Any delay is a new relevant event (clause 25.4.6).

Summary

1 At common law the parties are free to decide how and when payment will be made. The HGCRA 96 has, however, fundamentally altered these basic common law rules.

2 The terminology used to describe the way price is calculated is complex in the standard forms. The agreed sum can be called the tender sum, the contract sum or the contract price.

3 Contractors are very sensitive to changes in the payment regime, so the introduction of the HGCRA 96 has resulted in many changes to sub-contracts.

4 The means by which deductions from price could be made has caused a great deal of litigation in construction contracts over the years. Excuses for non-payment, such as set-off, abatement and counter-claims have all resulted in complex rules.

5 The value of a set-off may not only extinguish the sum owed but can give rise to a further claim for the cost of delays and the remedying of defective work.

6 Abatement means simply a reduction in price. It is limited to the value of the cost of the work.

7   In theory, excuses for non-payment should now be much more transparent because of the notice provisions of the HGCRA 96.

8   Two issues have emerged: one is where there is a third party provision for certifying interim certificates. A withholding notice is required to deduct any disputed sums. The second is where there is no such provision. The sum claimed then can be resisted on the grounds that the payment is not due.

9   The value of set-off or abatement claimed must be specified in a withholding notice if the paying party wishes to rely on it.

10 The HGCRA 96 has not abolished the old common law rules. It has merely clarified the application, and by adjudication made the position clearer at an earlier stage.

11 Sub-contracting

Although the construction contract is made between the employer and the contractor, sub-contractors will do most of the work. As a general rule, however, the contractor always remains liable for the work carried out by its sub-contractors (i.e., liable for the defaults of the sub-contractor). In law it does not matter what sort of sub-contractor you are. This general rule may be altered by (i) the circumstances surroundings the procedure of 'nomination' and (ii) the express terms of the main contract with regard to the contractor's usual implied obligation.

Vicarious performance

This is where the contractor delegates performance of either all or some of his obligations to a third party. This is usually done under a contract to which the employer is not a party. Whether the contractor is allowed to do so depends on the terms of the contract. *Vicarious* performance of the work and whether it is permitted is usually decided on two alternate bases: either it does not matter who does the work, or the contract expressly prohibits it. Whatever the basis, the contractor remains *responsible* for the resulting work carried out.

Where the work is of a *non-specialist nature* the right of the contractor to employ a sub-contractor is normally *implied*. *British Waggon Co.* v. *Lea* (1880) 5 QBD 149, is usually quoted as the authority for that proposition. The original contractor contracted to keep railway carriages in repair over a number of years. On liquidation the benefit and burden of the contract was assigned to *British Waggon*. Lea could not refuse performance despite the assignment. The basis of the decision was clearly spelt out by Cockburn CJ:

> 'All that the hirers, the defendants, cared for in this stipulation was the wagons should be kept in repair; it was indifferent to them by whom the repairs should be done. Thus if, without going into liquidation, or assigning these contracts, the company entered into a contract with a competent party to do the repairs, and so procured them to be done, we cannot think that this would have been a departure from the terms of the contract to keep the wagons in repair.'

Where the contractor has been chosen for its specialist skills, reputation and ability to perform complex operations efficiently, the rule in *British Waggon* may not apply (see *Southway Group Ltd* v. *Wolff and Wolff* (1991) 57 BLR 33).

The contractor's liability for the defaults of the sub-contractor may arise in either tort or in contract. Amendment 8 to the JCT 80 made implicit what was formerly implied. Changes to clauses 19.2 and 19.5 have the effect of making the contractor responsible for carrying out the work whether it is sub-contracted or not. This responsibility is carried regardless of whether the sub-contractors are domestic or nominated.

The standard forms recognise three types of sub-contractors:

(a) domestic;
(b) named;
(c) nominated.

Domestic sub-contractors

The JCT 98 states in clause 19.2.1 that 'a person to whom the contractor sub-lets any portion of the work . . . is in this contract referred to as a domestic sub-contractor'. In order to sub-let part of the work clause 19.2.2 requires the contractor to obtain the *written consent* of the architect. Such consent is not to be unreasonably delayed or withheld. The contractor can challenge a refusal of consent in adjudication under clause 41 as *dispute or difference*. Failure to comply is a ground for the determination by the employer of the contractor's employment under clause 27.2.1.4. The remainder of the sub-clause reiterates that the contractor remains wholly responsible for carrying out the work in accordance with clause 2.1 regardless of whether part of it has been sub-let or not.

The ICE 7th has a more relaxed attitude to sub-contracting. However, the seventh edition reintroduces certain restrictions which were abandoned in the sixth edition, where sub-contracting was allowed without restriction. Clause 4 provides that the contractor 'shall not sub-contract the *whole of the work* without the written prior consent of the engineer'. The guidance notes to the seventh edition state that the clause recognises the practice of the construction industry concerning the use of sub-contracts and self-employed labour. Accepting, as a general principle, that the contractor should be free to choose its own sub-contractors, it adopts the device of providing in the appendix restrictions on work which cannot be sub-contacted with the permission of the engineer.

For the work not excluded in the appendix, the contractor under clause 4(2) must notify the engineer of the names and addresses of any sub-contractor in writing. This information is to be given as soon as practicable and in any event no later than 14 days after entry on site or, where it involves design, on the appointment. Within seven days the engineer may *object* to the employment of the sub-contractor in writing, giving *good reasons*. The guidance notes states: 'caution should be exercised in the manner of objection to a specified sub-contractor so that legislation on procurement procedures are not infringed'.

Where labour only sub-contractors are employed no notification need be given according to clause 4(3). Clause 4(4) reiterates that the contractor is fully responsible for all the sub-contracted work and for the acts, defaults and neglects of the sub-contractor.

Named sub-contractors

Under the standard form of JCT 63 the practice grew of naming a single sub-contractor to undertake certain work. This was a response to the greater risks that nomination posed after *Bickerton* v. *NW Hospital Board* [1970] 1 WLR 607.

However, it left the position on liability unclear as the status of the sub-contractor was uncertain.

Using the naming provision of JCT 98 makes the position is much clearer. Under the contractual procedure, the named sub-contractor *remains* a *domestic* sub-contractor. Clause 19.3.1 provides that where in the contract bills certain work is to be carried out by named persons listed in the bills, only persons listed will be selected by the contractor. The list must comprise at least *three persons* and both the contractor and the employer can add to the list at any time prior to the execution of a binding sub-contract. Where the list contains fewer than three persons, either of the parties may add to the list by agreement. Failing agreement, the work may be carried out by the contractor who may sub-let it under clause 19.2.

Nominated sub-contractors

The object of nomination is to ensure that specialist parts of the work are carried out by the employer *selected* sub-contractor and not by the contractor. The reasons are that it gives the employer and its advisers *control* of the specialist auction. It can balance the *price* of work against the quality it requires. There are also circumstances where the specialist will provide a *design* input before the main contract is let.

Clause 35.1.4 of JCT 98 states that:

'[where] the architect, whether by use of a prime costs sum or by naming a sub-contractor, reserved for himself the final selection and approval of the sub-contractor who shall supply or fix any materials or execute work, the sub-contractor so named or to be selected and approved shall be nominated in accordance with clause 35 and a sub-contractor so nominated shall be Nominated Sub-contractor for all the purposes of this contract.'

The ICE 7th deals with nomination in clause 59. In clause 1(1)(m) 'Definitions and interpretation' it defines a nominated sub-contractor as:

'Any merchant tradesman specialist or any other person firm or company nominated in accordance with the contract or to be employed by the contractor for the carrying out of the work or the supply of goods materials or services for which a prime cost or a provisional sum has been included in the contract.'

The JCT treats a nominated sub-contractor differently from a nominated supplier. Clause 36.1 defines the latter as a supplier who is nominated by the architect to supply materials and goods which are to be fixed by the contractor. Clause 36.5 gives the contractor the benefit of any restrictions, limitations or exclusions contained in the sub-contract. In contrast, the ICE 7th treats suppliers as nominated sub-contractors.

Provisional and P-C sum items

Hudson (at p.1,352) points out that the peculiar characteristic of traditional English nomination draftsmanship is the procedural technique, whereby nearly all work intended to be sub-contracted by selected specialists is designated in the specification or bills of quantities as *prime cost* or *provisional items*.

In the ninth edition it suggested that 'provisional sum' was work which had not been finally determined at the time of contracting whereas 'P-C' referred to determinable items whose price was uncertain. Normally, provisional sums would apply to work alone or work and materials. P-C items would usually apply to materials or fittings which needed to be purchased by the contractor, once selected. This view has been modified by various definitions contained in the Standard Forms of Measurement, introduced since then. Hudson further produces a number of permutations that need to be considered in interpreting particular descriptions of work contained in construction contracts:

1   Is the contractor allowed to carry out the reserved work, itself?
2   Is that decision reserved for the employer through its agents?
3   Is the work itself 'provisional' in the sense that it may never be carried out?
4   Is it intended that only the specialist will carry out the work?
5   How is the work to be paid for? Does the provisional sum include mark-ups for the contractor's handling of the work? Will it include allowances for profit and 'cash discount' for prompt payment to the specialist?

In the light of the HGCRA 96 a further clarification has to be added to this permutation. The contractor will need to ensure that the provisions of the Act are closely followed. Unlike the domestic sub-contractor, a specialist is likely to be highly conversant with its provisions. Emden (at p.272) agrees with Hudson that it is difficult to describe accurately at the time of contracting everything that has to be done under it. Instead, the contractor is directed by the specification to provide and fix an article as a 'pc' or prime cost. The Standard Method of Measurement defines 'provisional sums' and 'prime costs' as follows:

1   The term 'provisional sum' is defined as a sum provided for work or costs which cannot be entirely foreseen, defined or detailed at the time the tendering documents are issued.
2   The term 'prime costs' is defined as the sum provided for the work or services to be executed by a nominated sub-contractor, a statutory authority or public undertaking or for materials or goods to be obtained from a nominated supplier. Such a sum shall be deemed to be exclusive of any profit by the general contractor and provision should be made for the addition thereof.

Keating deals with provisional sums in relation to clause 13.3 of JCT 98 which provides for instructions for the expenditure of such sums included in the contract bills or in a sub-contract. Clause 1 defines a provisional sum as including a sum provided for work whether or not identified as being defined or undefined work and a provisional sum for performance specified work (see clause 42.7). A provisional sum included in the contract bills can be converted into a prime cost sum to enable the architect to make a nomination (at p.660).

Links with employer

There is no privity of contract between the employer and the sub-contractor. Details of the difficulties caused by this rule and the exceptions created are dealt with in detail in Chapter 13. The creation and use of collateral warranties and direct warranties are explained as well. See also the Contracts (Rights of Third Parties) Act 1999 which stipulates that a third party may enforce contract a term in a contract if:

(a) it expressly *states* that the third party may enforce the term; or
(b) the term purports to enforce a *benefit* on him.

In the absence of these contractual links, the result of the restrictions on suing for economic loss in tort resulting from *D & F Estates*, together with *Murphy* is explained there as well. Claims in tort can only be brought for injury to persons and damage to other property. There is a further category of torts explained below.

Orders

Where the employer gives a direct order to the sub-contractor to carry out work, the employer is liable for the costs. The architect has no implied authority on the employer's behalf to order the sub-contractor to carry out work. However, the employer may later adopt the work and agree to pay for it.

Liability in tort

The general rule in tort is that the employer is not liable for the torts of the independent contractor. An aspect of *D & F Estates* which is often neglected is its confirmation of the rule: 'It is trite law that the employer of an independent contractor is, in general, not liable for the negligence or other tort committed by the contractor in the course of execution of the work.'

In this case the claim was made in tort alleging that the main contractor had not adequately supervised its sub-contractors. The House of Lords held no such duty existed in tort. *Haseldine* v. *CA Daw & Son Ltd* [1941] KB 343 illustrates the extent of any claim on the main contractor. It was held not responsible for the negligent installation of a lift by independent contractors. This was because the independent contractors had been properly selected, and it was *reasonable* to expect they were competent.

Rowe v. *Herman and ors* [1997] 1 WLR 1390 confirmed the general rule: 'the starting point must be the basic principle that an employer was not liable for an independent contractor's negligence'. There are, however, a number of exceptions.

First, the sub-contractor was *not* carefully and competently selected. This duty was not discharged in *Saper* v. *Hangate Builders Ltd* (1972). A large builders skip hired from (and delivered by) a plant hire company was inadequately lit by the contractors and its servants who had hired it. A car ran into the skip. The hire company was held to owe a duty of care to third parties to see that the skip was adequately lit. They did not discharge this duty by merely having left the skip in the control of the company and its servants: 'They did nothing about lighting the obstruction, they simply left the skip there in the road with a man whose capacity, experience and equipment they knew nothing about, and they never said a word about lighting.'

Second, there is the issue of non-delegable duty. Where the work is inherently *dangerous* work (near the highway, or explosive substances being used) or likely to cause *damage*, the contractor cannot avoid liability by simply employing a sub-contractor. In *Alcock* v. *Wraith* (1995) 59 BLR 20, the employer was held liable for defective work to a roof by a now bankrupt builder. During the replacement of slate tiles by concrete tiles, the builder filled the gap between the different tiles by using newspapers and filler. This poor workmanship later caused damage to the premises next door. In contrast, the result in *Rowe* v. *Herman and Ors* [1997] TLR June 9 was different. The employer was held not liable for injury to a road user who was injured by metal plates inserted in a pavement by contractor during the building of a garage.

Third, public or statute law imposes duty. In the *Pass of Ballater* (1942) PC 112, some ship repairers employed a firm of consulting engineers. When certifying that the vessel was free from gas, they omitted to mention that a certain part of the ship had not been properly inspected. An explosion resulted when the ship repairers used an oxyacetylene burner in it. The ship repairers were held liable. It was held that the duty was not only to take care, but a duty to provide care is taken where implements or substances dangerous in themselves, such as flame-bearing instruments or explosives are necessarily incidental to the work themselves (at p.11).

In *Murphy* the local authority's duty to approve and inspect a property for compliance with building regulations could not be delegated to an outside consultant. The duty was imposed on the council no matter from whom it sought and obtained expert advice.

Contractual liability of the sub-contractor

The general principle is that a contractor remains liable for any default of the sub-contractor. The question of whether work can be sub-contracted depends on the wording of the contract. JCT 98 clause 19.2/3 allows the contractor to sub-contract but requires the consent of the architect, 'such consent [to sub-contracting] not to be unreasonably withheld'. It further requires in JCT 98 clause 20.2 an express indemnity against liability for the default of the contractor's servants or agents.

Contractual liability for sub-contractor exceptions implied by law

Reliance

This situation arises where the employer has not relied upon the contractor (who accepted the other for the work) but on some *third party*. An example would be a sub-contractor to the main contractor selected by the architect. In an ordinary contract for work and materials the contractor is not liable for defects in the materials if (i) the employer selected them and (ii) the contractor was required by the employer to use a particular material. No warranty is then given that that material is reasonably fit for its intended purpose, merely that materials supplied have been obtained from a selected source. (Note: see *Young & Marten* v. *McManus Child* discussed in Chapter 6. Roofing tiles were chosen by the employer and supplied and fixed by the nominated sub-contractor. It was held that there was an implied warranty of quality despite obtaining them from specified supplier.) See also *Rotherham* v. *Haslam*, which illustrates the principles mentioned above. The contractor was held not liable for the defective material supplied. There was no reliance upon the contractor's skill and judgement.

Specialist design

Where this carried out by a nominated or independently selected sub-contractor, liability depends on the circumstances. Was there, to the knowledge of all parties involved, reliance by the employer on the sub-contractor for some element of specialist design? In *University of Warwick* v. *Sir Robert McAlpine* (1988) 42 BLR, for example, it was held that there was no reliance on the main contractor. Epoxy resin was injected to remedy problems found in white ceramic tile cladding, but the resin itself caused further cracking of the cladding. The resin was (i) recommended by architect and (ii) supplied and injected by a sub-contractor selected by them. The main contractor was not involved in the selection and expressed reservations about method chosen.

Garland J held the following:

1 The employer relied on the skill and judgement of a selected sub-contractor rather than on the main contractor.
2 No term warranting the fitness for purpose of resin would be implied in the main contract.

Main contractor employs specialist design sub-contractor

Express imposition

In this situation, the main contract contains, (however briefly or incidentally) an express imposition of design liability. Where there is an express imposition of design liability as for example in a JCT contract with a design supplement, or a design and build contract, the contractor's price will reflect this. If it chooses to sub-contract the design work to a specialist, it still remains liable. The situation differs from where there is an 'incidental' allocation of design liability. Here, the main contractor may be held responsible for the design, (even though the design was performed in direct consultation with the employer or with its consultants) and there was no practical involvement by the contractor. This happened in *IBA* v. *EMI and BICC* (1980) 14 BLR 1, where the main contractor was held to have accepted responsibility for the design. The design for a television mast turned out to be defective and it fell over. The sub-contractor carried out the design, but the main contractor was held liable because it had adopted the sub-contractor's work as its own.

Chain of liability broken

In this situation, the employer is unable to recover damages from the main contractor. This is because the contractual structure may lead to the defaulting sub-contractor by its actions, breaking the chain of liability. The contractor, and ultimately the employer, may be unable to recover its damages from the sub-contractor, even though under the terms of the sub-contract it may the liable. An illustration of this principle is the case of *John Jarvis* v. *Rochdale*. The nominated sub-contractor for piling work started late, failed to carry out the work properly and withdrew from the site leaving the work defective and unfinished. The architect instructed the contractor to postpone the work while it attempted to find another sub-contractor (for the problems of nomination see below). One month later the contractor gave notice of termination, as it was entitled to, under clause 28.1.3.4 of the then JCT 80. Under clause 23.2 the architect is entitled to suspend the work for a month unless caused by some negligence or default of the contractor.

Held by the Court of Appeal:

1 The clause did not refer to any breach of contract.
2 Its language was directed at practical men on the ground – whose fault was it that the instruction to postpone was given?
3 The contractor did not lose the right to rely on the clause just because the nominated sub-contractor's failure to perform put the contractor in breach, despite lack of actual fault. In other words 'negligence or default is not the same as a breach of contract.'

Note the suggestion of the editors of Building Law Report on their commentary on the case. The JCT should amend the word 'contractor' to include the sub-contractor and supplier. Note too that this interpretation is regarded as one the factors which led to the decrease in the use of nomination.

Sub-contractor forms of contract

Up to the late 1980s, main contractors were reluctant to use the domestic standard form of contract; instead, they produced their own bespoke sub-contract form. These were heavily biased in favour of the main contractor. Such terms in a contract, whilst onerous and sometimes unfair, are readily accepted in the offshore oil industry. Oil companies have a general policy that the contractor is in business to make a profit. By contrast, the terms produced by main contractors tended to promote their own interest at the expense of the sub-contractor.

Typical provisions

Pay when paid clauses

The essence of such a clause is to pass the risk of late or non-payment to the sub-contractor. Typical examples are: 'Our liability for payment to you is limited to such amounts as we ourselves have *actually received* from our Employer in respect of work under this order by virtue of Certificates received from the Architect.' Another version runs:

'Notwithstanding the provisions of clause 21 of (DOM/1) Standard Form there shall be no obligation upon the Contractor to make payment to the sub-contractor of any sums under the Sub-Contract unless or until the Main contractor shall have received payment from the Employer as defined in the main contract in respect of such sums.'

Such clauses can be construed in two ways:

(a) timing – payment is *only due* to the sub-contractor when the contractor *receives payment* from a third party (the employer).
(b) entitlement – the sub-contractor is *entitled* to payment only if the contractor has received payment (and only in most cases *only that amount*) from the employer.

There is limited case law on the subject. Cases such as *Schindler Lifts (Hong Kong)* v. *Shui On Construction Co.* [1985] HKLR 118 and *Brightside Mechanical & Electrical Services Group* v. *Hyundai Engineering & Construction Co.* (1988) 41 BLR 110 suggest that the court would enforce such clauses if construed as only relating to the timing of payment.

Three grounds have been suggested on which such clauses can be attacked: the first is the *contra proferentum* rule, there is an ambiguity in the purpose of the clause and the court should hold it against the paying party.

The second ground relates to *notice*, and this is where the payer has not brought a particularly onerous clause to the attention of the other party. Where the contract has been executed this would be difficult to show. (See *Interfoto Picture Library* v. *Stiletto Visual Programmes* [1889] 1 QB 433.)

The third ground concerns *The Unfair Contract Terms Act 1977*. It has often been argued that s.3 of the Act would make ineffective such a clause as being unreasonable. First, the contract would be on the payer's 'written standard terms of business', and second the party cannot by 'reference to any contract term (unless the term satisfies the requirement of reasonableness) claim to be entitled in respect to the whole or any part to render no performance at all'. However, a carefully drafted clause that states that the obligation only arises until or unless payment is received from a third party would not be capable of challenge. This is because no contractual obligation to make payment has arisen on which the Act can bite.

In *Durabella Limited* v. *J. Jarvis & Sons* (2001) CILL 1796, HHJ Lloyd QC declined to find as unreasonable under the Act a 'paid when paid' clause. In passing, he commented that there were two reasons for such clauses: first, the work has to be financed until payment is received; such a clause extends the period for financing until payment is received and relieves the contractor of an obligation to do so until payment is received. The second reason is that the risk of insolvency is shared proportionately.

Outlawing pay when paid clauses

Section 113.-(1) of the HGCRA 96 now makes such provisions ineffective except where there is insolvency.

> 'A provision making *payment* under a construction contract *conditional on the payer receiving payment from a third person is ineffective, unless* that third person, or any other person payment by whom, under the contract (directly or indirectly) is a condition of payment by that third party, *is insolvent.*' (emphasis added).

Discount

Most sub-contracts contain revision entitling the contractor to deduct a certain amount as a discount. In *Team Services Ltd* v. *Kier Management and Design Ltd* (1993) 53 BLR 76, it was decided that this deduction was not dependent on prompt payment.

No right to arbitration

The deletion of any reference to arbitration in these sub-contracts was striking; especially since the main contract usually provides for arbitration. The sub-contractor therefore had to embark on litigation to resolve any dispute. This was likely to be expensive as well as protracted. Sub-contractors would be reluctant to antagonise the main contractor, especially when they held the sub-contractor's money.

Programme

An express clause would require one of two things:

(a) the sub-contractor to provide a programme (again a striking requirement when the main contract did not treat any programme as a contractual document);
(b) a requirement that the sub-contractor would adhere to the main contractor's programme and be liable for any delay resulting from its failure to do so.

The problem with nomination

For a number of essential reasons, nomination distances the employer contractually from (i) contractor default and also (ii) sub-contractor default. It achieves this by requiring the contractor to enter into a contract with the sub-contract specialist. The contractor is responsible for co-ordination of the specialist work into the main works programme. If the contractor defaults on its obligations to the specialist, the employer is not liable. Similarly, if the specialist defaults, the employer is free from liability. This *commercial reality* was understood by Court of Appeal as early as 1901, observes Duncan-Wallace (1986, p.329).

In *Leslie & Co. Ltd* v. *Metropolitan Asylums District* (1901) 68 JP 86, it was held that the contractor could not sue for delays caused by the specialist. The converse case where the nominated sub-contractor sought to sue the employer on the insolvency of the contractor was dealt with by the House of Lords in *Hampton* v. *Glamorgan CC* [1917] AC 13. They found that there was no liability *on the employer* once the contractor had placed an order with the specialist.

Contractor pressure on the standard forms

This led to *two* significant *changes* in the standard form of contract: first, contractors were given the right of *reasonable objection* to specialists; and second, in the JCT forms, they were given the right to an extension of time for *delay caused* by the specialist. See *Jarvis Ltd* v. *Westminster Corporation* [1970] 1 WLR 637 for the difficulties caused by such a provision.

Repudiation by the nominated sub-contractor

Bickerton v. *NW Hospital Board* [1970] 1 WLR 607 was the first case to come before the English courts on the question of a *'second nomination'*. Before *Bickerton* the idea that the employer was liable to compensate the contractor for the defaults of the nominated sub-contractor was unthinkable.

Where the specialist *repudiates* the contract it leaves the parties with:

* the question of responsibility for the unfinished work
* increased costs resulting from such issues as inflation
* the cost of remedying any defective work
* consequent disruption of the main contractor's works programme, and so on.

In *Bickerton*, the Court held that the employers had a duty to renominate and to pay a new specialist's account even if it was more than the original price. *Percy Bilton Ltd* v. *GLC* [1982] 2 All ER 623 confirmed that the duty of the employer was to renominate in a *reasonable time*. The contractor was entitled to an extension of time only where there was unreasonable delay. It must be noted that the contractor still has the right of reasonable objection if the new specialist cannot complete within the contract period as a whole, and this position was confirmed by *Fairclough Building Ltd* v. *Rhuddan DC* (1895) 30 BLR 26.

Practical consequences of Bickerton

It must be remembered that the issue is not one of 'death, sickness, disappearance' or 'involuntary dropping out' of the first specialist. It is usually *deliberate repudiation* by or on behalf of the specialist, such as a solvent specialist abandoning an unprofitable or difficult contract for a more profitable one or, in the case of insolvency, the calculated decision of the trustee in bankruptcy or the liquidator to disclaim the contract rather than perform it (Duncan Wallace, 1986, p.335).

The Bickerton position in the JCT standard form

The JCT 98 position

Clause 35.24 deals with the situation where a renomination is necessary. The clause places the risks of nomination on the employer. The provisions are complex and comprehensive in scope. For ease of understanding they have been classified under different grounds that trigger the application of particular provisions.

Ground I   Clause 35.24.1 requires the contractor to inform the architect if it is of the opinion that the sub-contractor has made default (as outlined in clauses 7.1.1 to 7.1.4 of the *Condition* NSC/C).

Clause 7.1.1 – 'without reasonable cause he wholly or substantially suspends the carrying out of the sub-contract work'

Clause 7.1.2 – 'without reasonable cause he fails to proceed regularly or diligently'

Clause 7.1.3 – 'he refuses or neglects to comply with an instruction of the architect issued to the sub-contractor . . . , to remove any work, materials or goods not in accordance with this sub-contract and by such refusal the or neglect the works are materially affected or to remove defective work.'

Clause 7.1.4 – breach of the undertakings in clause 3.13 and 3.14, not to assign or sub-let without consent.

Note that Ground I replicates almost word for word the provisions in Clause 28.2 of JCT 98 which enables the employer to determine the contractor's employment under the contract.

Ground II   Clause 35.24.2 [if] the contractor informs the architect that the nominated sub-contractor has made default in respect of:

Clause 7.2.1 of Condition NSC/C

'If the sub-contractor makes a composition or arrangement with his creditors, or becomes bankrupt, or being a company makes a proposal for a voluntary arrangement for composition of debts etc, or has provisional liquidator appointed or a winding-up order made or passes a voluntary winding-up' *or*

Clause 7.2.3 'a provisional liquidator or trustee in bankruptcy is appointed or a winding-up is made, the employment of the sub-contractor is automatically determined' *or*

Ground III   Clause 35.24.3 – 'the nominated sub-contractor determines its employment because of contractor default under clause 7.7.' One ground is that the contractor without reasonable cause wholly or substantially suspends the carrying out of the works. A second is that without reasonable cause he fails to proceed with the works so that the reasonable progress of the sub-contract work is seriously affected *or*

Ground IV   Clause 35.24.4 – the employer *has* required the contractor to determine the employment of the nominated sub-contractor under clause 7.3 of conditions NSC/C and has done so. The condition includes bribery or corruption etc. *or*

Ground V   Clause 35.24.5 – the refusal of the nominated sub-contractor to take down or re-execute or re-fix or re-supply, work executed as result of instructions by the architect to the contractor. These arise in JCT 98:

(a) under clause 7 – errors arising from inaccurate setting out of levels to be corrected at contractor's expense'

(b) under clause 8.4 – 'removal of works and goods that do not comply with the contract';
(c) under clause 17.2 – the making good of defective work contained in a 'snagging' list;
(d) under clause 17.3 – necessary instructions issued during the defect liability period.

The procedure

Clause 35.24.6 requires the architect to instruct the contractor to issue to the nominated sub-contractor, a *default notice*. It may state that the contractor must obtain a further notice before determining the employment of the nominated sub-contractor. On confirming to the architect that the employment has been determined, the architect shall make such *further nomination of a sub-contractor* to 'supply and fix the materials and goods or to execute work and to make good and re-supply or execute as necessary any work executed by . . . [the defaulting nominated sub-contractor]'.

Clause 35.24.7 deals with the situation in two ways. If the nominated sub-contractor is bankrupt, *renomination is automatic*. Where the defaulter makes a composition or arrangement and to the reasonable satisfaction of the architect and contractor demonstrates it can meet its liabilities, the architect may refuse his consent to the determination.

The architect shall also make a further renomination where default has occurred in accordance with Grounds III to V. It shall be made in a reasonable time having regard to all the circumstances.

Duncan-Wallace is highly critical of the adoption of the *Bickerton* principle in the civil engineering forms of contract. The problem arose, he asserts, out of the particular interpretation of the express wording of the clause in that case. As such it only represents judicial interpretation of that particular clause and is not a principle of law. Yet the ICE drafting committee incorporated the decision into its contracts 'as if it were a principle of law' (1986, p.351). Sir William Harris, chair of the Joint Contracts Committee, said of the policy of the ICE fifth edition that the committee was of the opinion that the chain of liability should not be broken. Where it was, the results should be clearly spelt out so that engineers and employers should be clear about the risks involved.

> 'The committee had to decide upon whom the risk should rest: upon the employer who chose the sub-contractor, and ordered the contractor to employ him or the contractor who was obliged to employ him. The committee decided that, despite the power of objection given to the contractor, he who called the tune should pay the piper and the clauses are designed so that if a loss should ultimately be suffered as a result of his chosen specialist's fault or bankruptcy, that loss should be on the employer.'

Direct warranty

The employer may require a direct warranty by the nominated sub-contractor allowing the employer to sue the nominated sub-contractor directly. For the difficulty that can be caused by such a warranty see *Greater Nottingham Co-op*. The owner obtained a direct warranty from a piling sub-contractor in the recommended RIBA/JCT form, by which Cementation gave a duty of care warranty as to the design and selection of materials for the sub-contract work. The sub-contractors drove their piles negligently, the adjoining property was damaged, and the employer suffered loss as a result of delay to the main contract. The Court of Appeal held that the warranty only required that the piling contractors use proper materials, and was silent as to *how* the work was to be carried out. The warranty was of *no use* to them.

Summary

1 The contractor remains liable for the default of his sub-contractors.
2 There is no requirement in tort for the contractor to supervise its sub-contractors but this is subject to some exceptions.
3 The standard form JCT 98 classifies sub-contractors as domestic, named or nominated.
4 Where design is involved the contractor is not liable unless the contract expressly provides that the contractor is responsible for the sub-contractor design.
5 Generally, the contractor is not liable in tort for not supervising its sub-contractors. In certain circumstances it will be liable in tort for the activities of the sub-contractor if they are dangerous.
6 Prior to the HGCRA, contractors used their own sub-contractors forms of contract to impose onerous conditions on their sub-contractors.
7 The decline in nomination can be traced to the decision in *Bickerton*.
8 The provisions in the standard forms, especially JCT 98, are complex. In essence the employer takes the risk in using nomination to procure the specialist parts of the work.

12 Supply Contracts

Background

Whether the construction contract is of the traditional type or a more modern variation, there will be many other contracts to support the construction activities. This chapter concentrates on the supply of goods and with related issues of quality, title and exclusion or limiting clauses in those contracts. As discussed in Chapter 1, a construction contract is one for the supply of *work and material*. Until 1954 it was very important to distinguish between a contract for *work and materials* and one for the *sale of goods*. Up to that time s. 4 of the Sale of Goods Act 1893 required written evidence or part performance of contracts of £10 or more. The Law Reform (Enforcement of Contracts) Act 1954 repealed this section in contracts. There is still *a practical* difference between these categories of contracts. The distinction matters where *title to goods* used in the construction of buildings is disputed. This is reflected in the case of *Dawber Williams* v. *Humberside CC* (1979) 14 BLR 70.

The sub-contractor supplied roofing tiles under a supply and fix contract. The contractor had been paid for the tiles but had not in turn paid the sub-contractor. Subsequently the contractor went bankrupt. The tiles were stored on site but had not yet been fixed. The employer refused to allow the sub-contractor access to remove them. It later paid other contractors to fix the tiles.

Held: When sued by the sub-contractor in conversion, it was held that the employer had no title to the tiles. In fixing them to the roof it was liable to the sub-contractor in conversion (a tort).

In countries governed by a *Civil Code*, such as France, Germany and Italy, this distinction between the two types of contract is still very important because of the way their codes deal with latent defects. English law treats the question of latent defects in goods and issues of fitness for purpose in an identical manner in *both* types of contract. See *Young & Marten* v. *McManus Child* [1969] 1 AC 454 and the Supply of Goods and Services Act 1982 for the effects of the development of both statute and common law.

Other types of contract

Below are listed some other types of contract:

(a) sub-contracts (both nominated and domestic) for the supply of services, work and materials (these were dealt with in Chapter 11);
(b) sub-contracts (both nominated and domestic) for the supply of goods or materials;

(c) hire contracts, which envisage short-term use of equipment, sometimes with an operator, and the return of the item transferred.

It is common to describe some contracts for goods and equipment such as computers, cranes, cars and trucks as 'leased'. Unlike leases of land there is no separate legal regime for leased goods. They are really simply contracts *of hire*. In many leases of goods, the 'lessor' is not the supplier but the bank or finance house. The supplier sells the goods to the bank who in turn leases them to the buyer. The resulting triangular arrangement can lead to problems when the goods are defective. A recent illustration is *Anglo Group PLC* v. *Winter Browne and BML (Office Computers) Ltd* [2000] EWHC Technology 127.

Winter Browne (WB) purchased a computer system from BML (Office Computers) Ltd (BML). Finance was then arranged through Anglo Group PLC (Anglo). Anglo claimed that under the leasing agreement dated 15 April 1996, they had agreed to lease the computer system to WB for a 60 month period, at £4,485 plus VAT per quarterly instalment. WB had defaulted in making quarterly payments and had unlawfully repudiated the agreement. Acting on that repudiation Anglo now claimed the outstanding amount of £67,000 plus interest.

In their defence WB claimed that there were many defects in the system supplied and that Anglo were in breach of the leasing agreement. WB also claimed against BML alleging that it had made a number of *representations to induce* them to purchase or hire the system. The effect of the representations were that BML would be able to operate a computerised tele-sales, order processing, stock control and accounting system which, as well as being easy to use, would provide comprehensive management information.

At the heart of this case was the extent to which the 'Charisma' software worked. This was a standard system supplied by BML. The judge found that in the contract between BML and WB there was an *implied* term: 'the design and installation of a computer system requires the *active co-operation* of both parties'. Since WB had not employed an outside consultant to advise them of their requirements, this active co-operation was of crucial importance. 'A customer must accept where possible, reasonable solutions to problems that have arisen.'

Held:

1 The defects, though many, were of a minor nature. A year after repudiating the contract WB was still using the system. In addition they had unreasonably refused to change their working practices and adapt their systems to Charisma. Since they had bought a standard system and not a bespoke one the onus to change was on them. They had no case against BML.
2 WB was liable to Anglo. There was no evidence to show that they had relied on the skill and judgement of Anglo in leasing the equipment. Since there was no reliance there was no implied term that the system would be fit for its purpose.

WB had defaulted on its agreement, so Anglo was entitled to judgment in the sum of £67,006 plus interest (a total of £110,008).

Bailment

This is the delivery of goods by one person called the bailor to another, the bailee, so that they might be used for some specified purpose. The contract of bailment is based upon a condition that they shall be redelivered by the bailee, or in accordance with the specified instructions of the bailor or kept until he reclaims them.

Any person is to be considered as a bailee who (otherwise as an employee) receives:

(a) possession of a thing – a thing being defined as a personal chattel (jewellery, bricks, cars, lorries, concrete mixers, etc.) from another; *or*
(b) consents to receive or hold possession of a thing from another, upon an undertaking to keep, return and deliver the specific thing according to his instructions.

The person who receives the thing is the bailee and the owner is the bailor. When the owner *transfers possession* to the bailee, the bailee has all the rights of the owner subject to the right of the owner to call for the thing. Bailment can be by contract, though this requirement is not essential.

Failure to return the chattel

The bailee is under an obligation to return the chattel to the bailor at the end of the period of bailment. Under the Tort (Interference with Goods) Act 1977, a refusal to allow the bailor to enter the bailee's premises to reclaim the chattel is *conversion*. If the bailee (i) wrongfully parts with the chattel or (ii) loses it through negligence, it is no defence that he is unable to return it. Any loss or injury to the chattel puts the onus on the bailee to show it was an accident. If the chattel has been lost or destroyed the onus of proof is on the bailee. On being sued by the bailor, the court has discretion to order the return of the chattel or the payment of its value.

Conversion of the chattel by the bailee

The bailee has a duty not to convert the chattel. Conversion is an intentional act, which is inconsistent with the bailor's right to the property. Where the bailee pledges it or offers it for sale, bailment terminates the right of possession and re-invests it in the bailor.

Types of bailment

Chitty classifies bailment as consisting of two kinds (1999, p.102, vol.2): the first, gratuitous bailment, is where a chattel is deposited with the bailee without reward. It must be returned on demand and cannot be used for personal advantage. The second, involuntary bailment, means that where a chattel is left with the bailee without consent his consent, the bailee is liable for its loss if he has not taken reasonable care of it. If he has, he is not liable for its loss.

Valuable consideration

This is bailment for reward. Where goods are delivered to the bailee to be taken care of by him in return for remuneration to be paid by the bailor, the contract is one for custody by reward. The bailee must take reasonable care of the chattel.

At common law, *hire* gives the bailee the following obligations:

1 Both possession and the right to use the chattel are given in return for valuable consideration (i.e., a remuneration to be paid by the bailee).
2 An obligation to return the chattel at the end of the fixed period of hire and the costs of returning it.
3 The Supply of Goods and Services Act 1982 implies terms into the contract of bailment. The bailor has the right to transfer the goods. A warranty for quiet enjoyment is also given for the hire period. There is also no encumbrance known to the bailee. Terms of quality and fitness are also implied.
4 At common law terms of fitness are also implied into the contract. The chattel will be reasonably fit for the purpose for which it was hired. In *Hadley* v. *Droitwich Construction Company* (1967), a mobile crane was hired to the builder. Two months later it collapsed and injured workmen. The clearance between the rear roller and the bottom of the base had been incorrectly adjusted. The builder had failed to service it. An express agreement was made to put a competent man in charge and service it. The owner had, however, given no warranty that the crane would be fit for its purpose.

Note also the Unsolicited Goods and Services Act 1971. Section 6 (1) defines unsolicited goods as goods being sent without prior request. If the recipient sends notice to the sender that he does not want the goods, the sender must collect them within 30 days. If he does not, the recipient becomes the owner of the goods. Where no notice is sent and the goods are not collected within six months, the recipient also becomes the owner of the goods.

Contracts for the supply of goods

The contractor will purchase materials and goods either directly from the manufacturer or through a supplier, who may in turn purchase them from the manufacturer. It may also employ a sub-contractor who may purchase them from a supplier or the manufacturer. In that case the contract is one for works and materials and the implied terms of *Young & Marten* apply to the contract. In the first case the contract of purchase will be subject to the Sale of Goods Act 1979 (as amended).

In 1893 the common law on the sale of goods was codified. The decisions of the courts in many cases over the years were stated in statutory form. The draftsman of the Act, Sir Mackenzie Chalmers, 'was not trying to change the law but to state it clearly and accurately' (Furmston, 2000). In the context of the construction industry terms dealing with title, possession, quality and fitness are of vital importance. Liability for these may be excluded by clear drafting in commercial contracts.

Section 2 (1) of the Sale of Goods Act 1979 describes a contract for the sale of goods as: 'A contract by which the seller transfers or agrees to transfer property in the goods to the buyer for a money consideration called the price.' This definition *excludes* contracts for the hire of goods, for services, and for the supply of goods and services. The main characteristic of the sale of goods contract is that the contract passes title in the goods to the buyer. Title (more commonly described as ownership) is the greatest right a person can have in a chattel. Without it, (i.e., if the seller has not transferred title to the buyer), the goods cannot be sold to anyone else.

Section 12 (1) states that there is an implied condition that the *seller has the right to sell* goods or will obtain the right when title is due to pass to the buyer.

In addition to the right to sell, the seller gives two *warranties*: under s.12 (2) (a), that the goods will be free from any charge or encumbrances. What this means is that there is not a charge or lien in favour of a creditor of the seller. Section 12 (2) (b) deals with the second, which is *quiet possession* in the goods. Where there is limited title or an encumbrance on it, the seller has to disclose this to the buyer. *Rowland* v. *Duvall* (1923) 2 KB 500 illustrates the working of the section.

The buyer of a car found out that it had been stolen. As a result he had to return it to the true owner. He sued the seller on an implied condition that the seller had the right to sell. He recovered the full price despite having the use of the car for four months and the car being now worth less than he had paid for it.

Section 13 provides that where goods are sold by description, there is an implied term that the goods shall correspond to their description. In *Beale* v. *Taylor* (1967) 1 WLR 1193, a private motorist advertised his car for sale as a 'Herald Convertible, white, 1961, twin carbs.' Another private motorist bought it. Later he discovered that it was in fact two cars welded together. Only the rear half fitted the description given. The seller was unaware of the position.

Held: The buyer had relied upon the description, it was a sale by description and the seller was liable under the section.

The specification

Quite often the specification will rely on the description of goods using names commonly used in the trade. If some words have acquired a special meaning in a trade, there may be no breach of the section if they conform to that meaning. An example is *Grenfell* v. *E.B. Meyrowitz Ltd* [1936] 2 All ER 1313. There it was proved that 'safety glass' had a special meaning understood in the trade. The sellers were not in breach of the section in supplying 'safety glass' goggles which corresponded to the special meaning. Care needs be taken when the purchase order is drawn up, so that those trade names are included or excluded as the case may be.

Note that sections 12 and 13 also apply to private sales between two consumers.

Quality and fitness

Section 14 (1) retains the *caveat emptor* rule by providing that except where provided in sections 14 and 15, there is no implied term about quality or fitness for any particular purpose of goods supplied under a contract of sale. Section 14 implies two conditions as to quality and fitness for purpose provided the sale is made 'in the course of a business'.

Section 14 (2) dealt with *merchantable quality* in the following manner. 'Where the seller sells goods in the course of a business, there is an implied condition that the goods supplied under the contract will be of *merchantable quality*, except there is no such condition (a) as regards defects specifically drawn to the buyer's attention before the contract is made; or (b) if the buyer examines the goods before the contract is made, as regards defects which the examination ought to reveal.'

The words *merchantable quality* have been amended by the Sale of Goods (amendment) Act 1994, by the words *'satisfactory quality'*. The effect is discussed below. Because the words *merchantable quality* are central to understanding of the implied warranties of *Young & Marten*, some explanation of the old law is given. A statutory definition of *merchantable quality* was introduced in 1973 by the Supply of Goods (Implied Terms) Act 1973. This became s. 14 (6): 'Goods of a kind are of merchantable quality . . . if they are fit for the purpose or purposes for which goods of that kind are commonly bought as is reasonable to expect having regard to any description applied to them, the price (if relevant) and all the other relevant circumstances.'

In *Bartlett* v. *Sidney Marcus* [1965] 1 WLR 1013, the buyer, having been told that the clutch of the car he was intending to purchase was defective, chose the option of a reduction in the price of £25, rather than having it repaired. After driving 300 odd miles the clutch had to be repaired at a cost of £45 more than initially anticipated. The car was held to be merchantable despite the defect proving more serious than initially thought.

In *Kendall* v. *Lillico* (1969) 2 AC 31, one of the ingredients in animal feeding stuff which the plaintiff bought to feed his pheasants was contaminated. Brazilian groundnut extract proved poisonous to pheasants. It was held that the feed was not unmerchantable because the extract was perfectly suitable for animal feed. If the extract had been sold as poultry feed it would not have been merchantable. As animal feed, it was perfectly suitable as feed for most animals. If the goods as described in the contract have a number of purposes, it would be merchantable if it fitted one of the purposes for which such goods are usually used. See also *Aswan Engineering* v. *Lupdine* [1987] 1 All ER 135; [1987] 1 WLR 1. The Court of Appeal rejected the argument that the wording of the then s. 14 meant that goods were not of merchantable quality unless they were fit for all the purposes for which such goods are commonly bought.

Satisfactory quality

Section 14 (2A) makes clear that what is satisfactory is to be judged *objectively*. Goods need to meet the standard that a *reasonable person* would regard as satisfactory. Account has to be taken of the following:

(a) fitness for all purposes for which goods of the kind in question are commonly supplied;
(b) appearance and finish;
(c) freedom from minor defects;
(d) safety;
(e) durability.'

Fitness for purpose

Section 14 (3) reads:

'Where the seller sells goods in the course of business and the buyer expressly or by implication makes known

(a) to the seller, or
(b) [blank]

any particular purpose for which the goods are bought, there is an implied condition that the goods supplied under the contract are reasonably fit for its purpose, whether or not that is the purpose for which such goods are commonly supplied, except where the circumstances show that the buyer does not rely, or that it is unreasonable for him to rely, on the skill and judgement of the seller.'

In *McAlpine* v. *Manimax* (1970) fire extinguishers sold by Manimax were used to extinguish a fire in a site hut. The extinguishers were defective and made the situation worse. The extinguishers were held to be not fit for their intended purpose.

In *Venables* v. *Stephen Wardle and others* [2002] EWHC (TCC), s. 14 and the 1994 amendment were considered by the court. Copper piping installed in a large domestic dwelling house corroded and developed leaks within three years of being installed. The claimant alleged that all the pipes would have to be replaced and since the pipes were concealed behind expensive floor, wall and ceiling finishes, the work required would be expensive and involve substantial upheaval. The defendant joined the builder's merchants and their suppliers as part 20 defendants and claimants. The issues before the court were whether:

(a) the pipes fitted were of satisfactory quality and/or reasonably fit for their purpose;
(b) the pipes sold were of satisfactory quality and/or reasonably fit for their purpose.

The single expert explained that copper pipes have a life of 20 years with hot water and 25 with cold. Copper pipes require a sufficient flow of water during the first few months of service. This is critical because during this period a sufficient flow of water is required to build-up a layer of cuprous oxide on the inside of the pipes. If there is low or intermittent use, the failure to build up that layer allows instead the build up of slime or bio film leading to pitted corrosion. This condition was caused by the low or intermittent use of the house by the claimants and their practice of shutting off the mains when they left the house. This made the problem worse because half-full pipes would increase the likelihood of microbiological growth.

As some of the pipes were supplied before 3 January 1995, the judge had to apply the old s. 14 (2) containing the words 'merchantable quality' and the new s. 14 (2) containing the implied term of 'satisfactory quality'.

Held:

1   'Merchantable quality': intermittent use had prevented the build up a protective layer of cuprous oxide. Misuse of the pipes by the claimants in shutting off the mains had resulted in the pipes not acquiring satisfactory protection. Since there was no reason why copper pipes could not be used at this property again, provided water flow was maintained and not allowed to stagnate in it, the pipes were merchantable.
2   Satisfactory quality: the test now is the standard that a reasonable person would regard as satisfactory, taking into account all of the relevant circumstances. Even on that test the pipes were of satisfactory quality for normal use in a domestic environment.
3   Fitness for purpose requires reliance on the skill and judgement of the seller. The judge said at para. 36: 'the pipes were durable with normal use. It was abnormal or idiosyncratic, use, which caused the problems.' There had been no reliance on the skill and judgement of the defendant.

Exclusion clauses

There are two types of clause that exclude liability for what would otherwise be a breach of contract. One excludes liability altogether, and is known as an exemption clause. The other limits liability to an agreed sum or the replacement value of a defective item or service. It may place a financial limit (e.g., up to £1,000 per occurrence) or it may state 'our sole responsibility is to repeat the defective work up to the value of our contract'. This is called a limiting clause. In the celebrated case of *Photo Production* the House of Lords had to construe the meaning of the following clause in a service contract. The particular clause read:

'Under no circumstances shall the company . . . be responsible for any injurious act or default by any employee of the company which could have been foreseen and avoided by the exercise of due diligence on the part of the company . . . nor in any event shall the company be held responsible for any loss suffered by the customer through burglary, theft, fire or any other cause except insofar as such loss is solely attributed to the negligence of the company's employees.'

The employee of the company who was employed to guard the factory, started a fire instead. The fire spread and destroyed the factory. It was accepted in the House of Lords that the fire had not been started negligently. The court had therefore to decide whether the clause protected the company. In construing the clause the House of Lords decided it was a fair and reasonable one. The contract had been negotiated with lawyers advising both sides. Therefore it was a fair allocation of the risk. In accepting the clause, the owners of the factory gained the services provided at a much lower rate. This left it to them to arrange their own insurance of their premises.

Unfair Contract Terms Act 1977

Despite its name, the Unfair Contract Terms Act 1977 (UCTA), deals with exclusion clauses and not with terms which are unfair. Where the purchaser is a consumer, any exemption clause is rendered void by the Act. Where UCTA applies in commercial contracts the position is dealt with by s. 1 (3).

'In the case of both contract and tort sections 2 to 7 . . . applies only to business liability, that is liability for breach of obligations or duties arising:

(a) from things done, or to be done by a person in the course of business (whether his own or another's); or
(b) from the occupation of premises used for business purposes by the occupier.'

Section 2 provides that liability for death or injury resulting from negligence cannot be excluded by a term in a contract or by a notice that is non-contractual. Other loss or damage (examples are damage to property or loss of profit) excluded must, as far as the term or notice is concerned, satisfy the requirements of reasonableness. Section 3 applies between contracting parties where (i) one deals as a consumer *or* (ii) the contract is in the other party's written standard terms of business. A number of restrictions are contained in s.3 (2). A person cannot use a contract term to (i) excuse or restrict his liability when in breach or (ii) render a substantially different performance or not perform at all 'unless the term satisfies the requirements of reasonableness'.

Guidelines for assessing reasonableness

Section 11 (1) requires of the term of reasonableness that it should have been a fair and reasonable one to be included, having regard to the circumstances which were, or ought reasonably to have been known, or in the contemplation of the parties when the contract was made. Circumstances to be taken into account include: (i) the resources available to meet the liability and (ii) the availability of insurance cover.

Section (11) 2 provides further that regard should be had to the guidelines given in Schedule 2 in assessing reasonableness. The following factors are to be borne in mind.

1 The strength of bargaining position of parties relative to each other, taking into account (amongst other things) alternative means by which the customers requirement could have been met.

2 Whether the customer received any inducement to agree to the term, or in accepting had an opportunity of entering a similar contract with other persons, but without having to accept a similar term.
3 Whether the customer knew or ought reasonably to have known of the existence and extent of the term (having regard, amongst other things to any custom of the trade and any previous course of dealing between the parties).
4 Whether the term excludes or restricts any relevant liability if some condition is not complied with, whether it was reasonable at the time of the contract to expect that compliance with that condition would be practicable.
5 Whether the goods were manufactured, processed or adapted to the special order of the consumer.

Note: see below: s. 6 (3) states that liability for s. 14 of the Sale of Goods Act can be excluded by a contract term if it satisfies the requirement of reasonableness.

Relevant examples of exclusion clauses

Rees-Hough Ltd v. *Redland Reinforced Plastics* (1985) 27 BLR 136 provides a useful example of the application of the guidelines. Rees-Hough Ltd called (RH) were tunnelling and pipe-jacking contractors for the construction of a tunnel and pumping station at the Heysham/Esher sewage works. In January 1979 they awarded the contract for the supply of concrete pipes to Redland. The specification required that the pipes should be able to bear a maximum load of 400 tonnes and withstand a deflection of 1.5 degrees.

The pipes failed during the tunnelling work and RH had to complete the contract using alternative methods. They refused to pay for the pipes and Redland sued for breach of a contract of sale. RH counter-claimed for their additional costs resulting from the defective pipes supplied. In their defence to the claim Redland relied on clauses 1 and 10 of their standard terms, excluding responsibility for the quality of their pipes.

The relevant terms were as follows: Clause 1 was headed application of terms and stated that 'all sales of goods made by the company shall be on these terms. In any contract these terms shall apply to the exclusion of all others . . . even if contained in the customer's order.' Clause 10 dealt with 'quality and description.' In it the company warranted the workmanship of the goods and materials and promised to repair or replace any defective goods on being notified. However 'all terms of any nature, express or implied, statutory or otherwise, as to correspondence with any particular description, fitness for purpose or merchantability are excluded.' Note that this clause restates in contractual form the provisions contained in UCTA. Section 6 (3), which provides that:

'As against a person dealing otherwise than as consumer, the liability specified in s. 2 i.e. liability breach of for ss. 13, 14 & 15 (sellers implied undertakings as to conformity with description or sample, or description, or to their quality or fitness for a particular purpose) of the Sale of Goods Act, can be excluded by a contract term if it satisfies the requirement of reasonableness.'

The issue for the Court was whether their terms were fair and reasonable in the circumstances. In applying the guidelines the court took into account the fact that the companies were of similar size. They had dealt with each other over many years and both were aware of these terms.

Held: In past dealings between the parties, Redland had always repaired and replaced pipes. They had never in the past relied on the clauses to escape liability. This was the clearest indication that the terms were unreasonable.

The case of *Pacific* was not concerned with the reasonableness of the clause. The Court of Appeal decided the effect of the clause in the context of the professional services and whether the clause gave immunity to a third party (note that professionals can exclude liability for tort by a suitably worded clause in their contracts). The relevant clause provided that:

> 'Neither any member of the employer's staff not the engineer nor any member of his staff, nor the engineer's representative shall be in any way personally liable for the acts or obligations under the contract, or answerable for any default or omission on the part of the employer in the observance or performance of any of the acts matters or things which are herein contained.'

Puchas LJ in dealing with the effect of the clause on parties outside the contract said at p.1,022:

> 'The presence of such an exclusion clause whilst not directly binding between the parties, cannot be excluded from general consideration of the contractual structure against which the contractor demonstrates reliance on, and the engineer accepts responsibility for, a duty in tort, if any, arising out of the proximity established between them by the existence of the very contract.'

In essence, all three judges agreed that the disclaimer contained in a contract, to which the defendant (the engineer) was not a party, negated any liability which the defendants owed to the plaintiffs. In short, the plaintiff entered the contract in full knowledge and accepted the risk structure contained therein. Part of the decision was based too on the fact that the court considered that the contract provided for arbitration as a method of resolving any claims the contractor had against the employer.

St Albans City and District Council v. *International Computers Ltd* [1996] 4 All ER 481 demonstrates also the application of the Sale of Goods Act to software contracts. In December 1988, ICL contracted with St Albans for a poll tax system based on its Comcis software package. Bugs in the Comcis system used by the council resulted in an overestimate of 2,996 people, leading to the setting of lower levels of tax. The council sued for the recovery of their losses and ICL relied on their limitation of liability and exclusion clause limiting any damages to a £300,000.

Held: The clause was unreasonable under UCTA. In doing so it made a number of rulings.

1 Software. Software when sold as a package application is goods and subject to statutes such as the Sale of Goods Act 1979 and UCTA.
2 Software sold as part of an application of a development contract is not a good. However, there will be an implied term that the software will be fit for the purposes for which it is intended.

It is clear from the case that in a contract between a supplier and an end user, a liability limitation clause which has been taken without a substantial change from the supplier's standard terms and conditions is subject to UCTA. It can be tested for reasonableness even though freely accepted by the end user in negotiations. To avoid UCTA applying the clauses must be negotiated.

In *Pegler Ltd* v. *Wang Ltd* [2000] EWHC Technology 137 the parties were involved in long negotiations. Pegler's main business is the manufacture of engineering and plumbers' brassware including domestic taps and mixers, radiator valves and plumbing fittings. In 1990, realising that they were heading for extinction because of poor delivery and service, they implemented radical remedies. Part of their restructuring included replacing their out of date computers and replacing them with a new integrated computer system.

On 8 August 1991 in response to their tender, Pegler informed Wang that in principle it intended to contract with them. A written contract was entered into in December 1991 to supply computer hardware, software, bespoke programming and services for a total price of £1,198,130 plus £235,000 per annum for three years for the maintenance and business process management. Commenting on the performance of this contract, HHJ Bowsher QC said at para. 20 that 'Wang's performance was disastrous. The parties have agreed by December 1995 Wang ceased to offer any relevant performance'. At trial, the parties agreed that Wang had abandoned the contract and the only live issue was the amount of damages. Pegler claimed damages estimated at £22,898,472 including amongst other things lost sales, lost opportunities, ongoing expenditure and wasted management time. One of Wang's defences was their exclusion clauses. By their clause 5.15.1 they excluded all 'warranties, conditions, guarantees or representations as to description, merchantability or fitness for a particular purpose or other warranties, conditions, guarantees or representations whether express or implied by statute or otherwise, oral or in writing'.

Clause 5.15.2 promised to 'replace or repair defective hardware or correct non-conforming software in accordance with the warranties given'. If it was unable to do so, the customer could reject non-conforming equipment 'and upon its return to Wang's premises is entitled to recover the purchase price of the hardware or the initial licence fee for the software, as appropriate'. The judge applied the guidelines mentioned earlier in this chapter in deciding whether the clauses were reasonable. He reached the following conclusions:

1 Pegler was a substantial company and therefore there was no imbalance in its negotiations with Wang. It however 'burnt its boats' by allowing work to start before standard terms were discussed.
2 All computer companies contract on the same terms.

3 Pegler was advised by solicitors throughout their negotiations, and was well aware of the terms on which they were contracting.
4 Pegler had every reason to be confident that the system they were buying was suitable.
5 Wang acknowledged that the system sold to Pegler required much more bespoke programming than they had let on.
6 Wang portrayed their system as a 'low risk' solution.

He decided that it was one thing for Wang to exclude liability in their standard terms intending to exclude liability for some unforeseeable lapse on their part. This was not so when they had so misrepresented what they were selling that breaches of contract were not unlikely. In the circumstances, it was unreasonable to impose their standard terms on Pegler, who had no choice but to accept them. Wang's exclusion of liability clauses was held unenforceable by reason of UCTA and Pegler was awarded damages of £9,047,113.

Unfair terms and adjudication

The HGCRA 96 excludes contracts made with residential owners. However, some professional bodies have in their standard forms for use in such contracts, incorporated (i) the payment provisions of the Act and (ii) provided for adjudication as a means of resolving disputes. In the case of *Rupert Morgan* for example, the written contact was one prepared by the Architecture and Surveying Institute (which also naturally provide for the employment of an architect or surveyor). In *Picardi* the dispute between the parties was whether the appointment of the architect was made under the RIBA Conditions of Engagement/99. This included a right adjudication, which the court held to be an 'unusual' term. It had therefore, to be specifically drawn to the attention of the 'lay person' in order to be relied on. The judge held it was an unfair term at common law. He said at para. 125 'The common law rule is that where there is a tender of a contract with conditions which are particularly onerous or unusual . . . must be brought properly and fairly to the other parties attention.' No adjudication clause was incorporated and therefore the adjudicator had no jurisdiction to make the award.

A note of caution

Recently the Court of Appeal in two decisions emphasised that in property drafted commercial contracts, the court should be slow to find clauses unfair. In *BHP Petroleum Ltd* v. *British Steel Plc* [2001] 2 LLR 277 the court had to interpret an exemption and limitation clause. It decided that where the clause is entirely clear and unambiguous, a contractor or a supplier could limit the period in which the claim is brought. A clause limiting claims for defective work to 2 years after the delivery of steel pipes was held to be effective. Under this clause British Steel had promised to remedy at its own expense defects in the work due to faulty design, material or workmanship which appeared within 24 months of delivery. A second clause limited their liability to 15 per cent of the purchase price. *Watford Electronics Ltd* v. *Sanderson* [2001] EWCA Civ 317 was another case of the installation of a computer software system that failed to deliver on its promises. HHJ

Toulmin QC decided the clauses contained in the three sets of documents that contained the three contracts were unreasonable. Sanderson appealed against these findings. Chadwick LJ outlined the approach to be taken by an appellate court in reviewing an original decision as to what was 'fair and reasonable': 'It must follow, in my view, that, when asked to review such a decision on appeal, an appellate court should treat the original decision with utmost respect and refrain from interference with it unless satisfied that it proceeded upon some erroneous principle or was plainly and obviously wrong.' All three judges were unanimous in holding the *clauses reasonable*. Chadwick LJ said at 55:

> 'Where experienced businessmen representing substantial companies of equal bargaining power negotiate an agreement, they may be taken to have regard to matters known by them. They should, in my view be taken to be the best judges of the commercial fairness of the agreement they have made; including the fairness of each term of that agreement. They should be taken to be the best judge on the question whether the terms of the agreement are reasonable. The court should not assume that either is likely to commit his company to an agreement which he thinks is unfair, or which he thinks includes unreasonable terms, unless satisfied that one party has, in effect, taken unfair advantage or that term is so unreasonable that it cannot properly have been understood or considered – the court should not interfere.'

Gibson LJ said that where parties have agreed the allocation of risk, the price must be taken as reflecting that allocation, and there might be thought to be little scope for the court to unmake the bargain of commercial men. He adopted the observation of Forbes J in the *Salvage Association* v. *CAP Financial Services* [1995] FSR 654 at p. 656:

> 'Generally speaking where a party well able to look after itself enters into a commercial contract and, with full knowledge of all relevant circumstances, willingly accepts the terms of the contract which provides for the apportionment of financial risks of that transaction, I think that it is very likely that those terms will be held to be fair and reasonable.'

Disclaimers of liability

Surveyors often include disclaimers in their valuation and survey reports. These expressly disclaim any responsibility for the consequences of relying, in whole or in part, on the report. This is a common feature in building society valuations, which will typically point out that:

(a) they are not full structural surveys *and*
(b) no warranty or representation is made to the purchaser in respect of the various statements made by the surveyor to the building society.

Effect of disclaimers

There is no doubt that an appropriately drafted disclaimer can exclude responsibility for statements made by professionals. The clearest example is *Hedley Byrne*

v. *Heller* itself. The House of Lords held that although statements could be made negligently, in that case it was given *without responsibility* and the phrase was therefore effective to exclude responsibility for the statement. In *Smith* v. *Eric Bush: Harris* v. *Wyre Forest District Council* [1990] 1 AC 831, two appeals were held together before the House of Lords. Each plaintiff applied for a mortgage (one to a building society and the other to the local authority). In the first case, a standard valuation survey was carried by an independent surveyor. In the second, a surveyor employed by the local authority did so. Following the survey, mortgage moneys were advanced to both applicants.

Both properties later developed serious defects. In *Smith* the report mentioned that a number of chimney breasts had been removed, but no inspection was made above them. A year and a half later a chimney breast collapsed because it was not supported and caused considerable damage. In *Harris*, the property subsided and the repair cost more than the property itself.

The House of Lords found the surveyors owed a duty of care under *Hedley Byrne* v. *Heller* and both disclaimers were *subject to reasonableness* under the UCTA.

Other situations where liability may be excluded

Misrepresentation

Under s. 3 of the Misrepresentation Act 1967 (see now s. 8 of UCTA 77), liability cannot be excluded for misrepresentation unless it satisfies the requirement of reasonableness. In the context of construction contracts it is common to exclude liability for statements made *before* entering into the contract. In *Cremdean Properties Ltd* v. *Nash* (1977) 241 EG 837, the invitation to tender gave particulars of the dimensions of the property to be built as well as the amount of lettable space they would contain. These statements proved to be false. The invitation to tender contained a statement that tenderers were to check the accuracy of the information not to rely on them.

Occupiers' liability

The occupier owes a duty of care to his lawful visitors under the Occupiers Liability Act 1957. The duty contained in s.1(2) is to ensure that visitors, including those 'who would have been treated at common law as invitees or licensees', are safe from dangers posed by the premises. The occupier owes a common duty of care to ensure that the visitor is safe in all the circumstances of the case that are reasonable. Premises are given a wide meaning and are not limited to immoveable property such as land, houses, railway stations, bridges and construction sites. The definition has been extended to ships, gangways, scaffolding and even ladders.

The occupier owes the duty of care. It is not necessary for the occupier to have an estate in land, or even to occupy it exclusively. The contractor of a construction has a licence to occupy and is therefore the person in control. It may share that control with other occupiers such as sub-contractors and suppliers and their sub-sub-contractors. In such circumstance, note the case of *AMF International Ltd* v. *Magnet Bowling Ltd* [1968] 1 WLR 1028. In the case of technical work the occupier may have to employ a competently qualified professional person to check the work of independent contractors. In *Keally* v. *Heard* [1983] 1 All ER 973, Mocatta J held that the negligence of the architect or other supervisor would not by itself involve the occupier in liability for otherwise, in technical cases, the common duty of care would become equivalent to the obligation of an insurer (Winfield and Jolowicz, 2002, 9.14).

Unfair terms in consumer contracts

The Directive on Unfair Terms in Consumer Contracts was adopted by the Council of Ministers in 1993. Member states were required to implement these by 31 December 1994. As a result the Unfair Terms in Consumer Contracts Regulations were laid before Parliament on December 1994 and came into force in July 1995. They were then replaced in October 1999 by the Unfair Terms in Consumer Contracts Regulations 1999.

It is unclear what the relationship between the two pieces of legislation, UCTA and the Regulations is. The regulations are based on the German Unfair Contracts Act in that they require both parties to act in good faith. The role of policing the regulations has been given to the Director-General of the Office of Fair Trading. Powers include the right to seek an injunction to prevent a trader using unfair terms. The Office of Fair Trading issues regular bulletins. These indicate that in practice, many traders abandon the offending term without application to the court. The Regulations apply to consumer contracts and only where standard forms of contract are used. A term deemed unfair is struck out and in principle the rest of the contract is left in place. Leaving the offending term out could of course leave the rest of the contract without any meaning or sense. Any unfair term is deemed by clause 8 (1) not to be binding on the consumer. Where the offending term can be struck out without affecting the existence of the contract, the contract is binding on the parties per clause 8(2). Clause 5 (1) defines an unfair term as one that causes a 'significant imbalance' in the rights and obligations of the parties which operates to the detriment of the customer. An assessment of the fairness of term excludes the definition of the main subject matter of the contract or the adequacy of the price or remuneration as against the goods or services supplied. However clause 6 (2) requires this to be in plain and intelligible language.

Terms not covered by regulations

The Office of Fair Trading has on its website (www.oft.gov.uk), a list of standard contracts whose terms are not covered. The exceptions are those that:

1 reflect provisions that by law have to be included in contracts
2 have been individually negotiated
3 are contracts between businesses
4 certain contracts that people do not make as consumers (e.g., in employment or the setting up a business).
5 contracts entered into before 1995.

The first major interpretation of the act has come in *Director of Fair Trading* v. *First National Bank* [2001] UKHL 52. The Director sought an injunction restraining use or reliance on the grounds that a term in a credit agreement with the bank was unfair. The House of Lords considered the Regulations and came to the conclusion that the terms were fair. The importance of this case from a point of view of construction contracts can be seen from the House of Lords' conclusions on the nature of good faith in the Regulations. Giving his opinion Lord Bingham of Cornhill said para 17:

> 'The requirement of good faith in this context is one of open and fair dealing. Openness requires that the terms should be expressed fully, clearly and legibly, containing no concealed pitfalls or traps. Appropriate prominence should be given to terms which might operate to the disadvantageously to the customer. Fair dealing requires that the supplier should not, deliberately or unconsciously, take advantage of the consumer's necessity, indigence, lack of experience, unfamiliarity with the subject matter of the contract, weak bargaining position or any other factor listed in Schedule 2 of the regulations. Good faith in this context is not an artificial or technical concept, nor, since Lord Mansfield was its champion, is the concept wholly unfamiliar to British lawyers. It looks to good standards of commercial morality and practice.'

The requirement that prominence be given to terms which are disadvantageous to the customer, is reminiscent of Lord Denning's comments in *Spurling* v. *Bradshaw* [1956] 2 ALL ER 121: 'Some clauses I have seen would need to be printed in red ink on the face of the document with a red hand pointing to it before the notice could be held to be sufficient.' Note that with the exception of (d) dealing with treating the customer fairly and equitably, these factors are identical to those factors in schedule 2 of UCTA. For further discussion on the Regulations and adjudication: see the cases of *Lovell Projects as* well as *Picardi* discussed further in chapter 16.

Retention of title clauses

Section 28 of the Sale of Goods Act provides that the parties can agree that delivery and payment shall not take place at the same time. To protect its property, the seller requires a clause in the supply contract that enables it to retain ownership of the goods until paid. In the event of the buyer becoming insolvent after delivery, the seller is then entitled to retrieve the goods. Goods (and services) are usually provided in commercial contracts on credit terms. Companies finance their operation by borrowing money to support their cash flow, usually by having overdraft facilities. The bank providing it will in turn, require security and this will usually be in the form of a general floating charge. This is a charge 'that "hovers" over a changing fund of assets, including assets acquired by the debtor after the creation of the charge': see Seally and Hooley (2003, p.1079) for further explanation of this complex area of law. The company can however deal with its assets without interference until some event occurs. For example, should the company default on its loan, the bank can step in and take over the company's assets (usually by appointing a receiver to enforce its security by, if necessary, selling up the company's assets). When this happens, the supplier, under the supply contract, may find that its buyer is unable to pay for the goods after their delivery. It then ranks as an unsecured creditor who may receive very little money when the company is eventually wound-up. Meanwhile, the receiver (or a liquidator) can dispose of the goods to the benefit of the secured creditors. To protect its interests in the property, the seller has to include a clause reserving title to the goods until paid. This is known as a 'reservation of title clause' or a *'Romalpha'* clause, so called after the case which first recognised the legality of the device in English law. Where the buyer becomes insolvent a number of consequences result. In the case of a company, the court may make an administration order, a holder of a charge may appoint a receiver or the company be wound up. Note that where a petition is being made to the court to appoint an administrator or a moratorium is in force, the administrator can sell goods subject to a retention of title clause with the leave of the court. Where a receiver is appointed, the goods can be used or sold, but the receiver must give an undertaking to pay for its use or for the price sold.

Examples of clauses

Simple examples

'The ownership of any goods delivered by us shall remain with us until paid'; this does not specify when the goods can be retrieved and gives no access to the property where the goods are stored. This can be varied to: 'Notwithstanding delivery the property in the goods shall remain in the company until the customer has paid in full therefor'. However, it suffers from same defect in the earlier example of not giving access to retrieve the goods. It also does not specify how the goods are to be identified. Clearly when the seller retrieves goods they should be identifiable as belonging to him: see clause (b) in the *Romalpha* clause below, which tries to ensure this.

More complex clauses

'Notwithstanding delivery of the goods or any part thereof the property in the goods shall remain in the seller until the purchaser has paid the purchase price in full as well as any payments due to the seller whether hereunder or in respect of any other liability to the seller whatsoever'; this clause requires the purchaser to pay previous bills outstanding as well as current ones.

Admixtures and tracing

The simple and more complex clauses above provide protection where the goods do not change shape, appearance or are mixed with other material to produce a new product. In such a case the seller may be made a trustee in the product where admixtures are sold on to another buyer. Such a clause can read:

'If any of the goods are processed into other goods before payment in full for the goods have been received by the company (the Seller), the goods including all other goods aforesaid shall be the property of the company, and the customer hereby declares itself trustee of such goods for the company until such payment is made, and the customer shall hold such goods and any proceeds of sale of such goods and any right arising from any sale thereof as trustees for the company.'

In *Aluminium Industry Vaasen BV* v. *Romalpha Aluminium* [1976] 2 All ER 552, the contract of sale was between a Dutch company and English buyers. Aluminium foil supplied was to be used in manufacturing processes in the English factory. The buyers took possession of the foil and mixed it in various processes. Full payment was never made and the manufacturers later became insolvent. At the time of the appointment of a receiver the buyers still had some of the unmixed foil. Some of the foil had been mixed in manufacture and some sold unmixed. The Dutch buyers claimed proprietary rights entitling them to priority over the buyer's other creditors. The relied on the following clauses:

(a) that ownership of the foil was to be transferred only when the buyer had met all that was owing to the seller;
(b) the buyer was required to store the foil in such a way that it was clearly the property of the seller until it had been paid for;
(c) that articles manufactured from the foil were to become the property of the seller as security of payment and until such payment had been made the buyer was to keep the articles manufactured *as fiduciary owner* for the seller and if required was to store them separately so that they could be recognised.

Held: Regarding the unmixed foil, by a combination of the clauses above, the Dutch sellers retained ownership in the foil. Ownership passed when the parties intended it to. No payment had been made to the sellers so ownership did not pass to the buyers. As regards the sale of unmixed foil, this foil belonged to whosoever had bought since (c) expressly authorised the sellers to sell it. However, the sellers were entitled to the proceeds of the sales by the buyers. By imposing a fiduciary duty the sellers could trace the proceeds of sale. Thus they had priority over both secured and unsecured creditors.

As regards the sale of mixed foil, clause (c) because it said so, gave the sellers property rights over the mixed i.e., the manufactured goods. (This, however, is not because title in the mixed goods is reserved to the sellers). It is possible to reserve title only in goods which one has title to start with i.e., in the unmixed goods. In *Re Peachdart* [1983] Ch. 131, the sellers of leather, which the buyers then made into handbags could reserve title only in the leather they supplied. They could not reserve title in the handbags, which were newly created goods. The contract of sale stated that the buyers granted the seller ownership in the mixed goods. This property right was void unless registered as charge under s. 94 of the Companies Act 1948. See also *Re Bond Worth Ltd* [1979] 0 WLR 629, where the floating charge was held ineffective because it was not registered. A similar outcome was the result of *Stroud Architectural Systems Ltd* v. *John Laing Construction Ltd* [1994] 2 BCLC 276. Suppliers of glazing units inadvertently created a floating charge. The use of the words 'equitable and beneficial ownership' meant that the floating charge that resulted had to be registered.

Separation of the goods

As indicated in the *Romalpha* case the mixing of the goods with other goods creates problems in drafting an effective retention of title clause. In *Borden (UK) Ltd* v. *Scottish Timber Products Ltd* [1979] 3 WLR 672, the resin supplied was mixed with sawdust to make chipboard. The Court of Appeal decided that the reservation of title clause was ineffective once the resin had been mixed, a result similar to one held in *Re Peachdart* (1983) above. The fact that the purchaser has reworked the material will not always prevent the clause from being ineffective. The fact that work was done on the steel strips supplied did not prevent the clause from biting in *Armour* v. *Thyssen Edelstahlwerke AG* [1991] 2 AC 339. Similarly in *Hendry Lennox (Industrial Engines) Ltd* v. *Graeme Puttcik* [1984] 2 All ER 152, the supplier of diesel engines was held to have effective title in them. This was despite the engines being installed into generators. They could, however, be easily removed and it was held that they had not been irretrievably mixed with other goods.

Unfixed material and goods

A basic rule of English law is that things fixed to the land become part of the land. Work and materials supplied and affixed to a building or structure become fixtures and thus part of the land. The process of fixing things to the land transfers ownership to the owner of the land. This is so even if the goods were stolen or the supplier (and fixer) never paid. This problem arose in *Dawber Williams* v. *Humberside CC* (1979) 14 BLR 70. The sub-contractor, under a supply and fix contract, supplied roofing tiles. The contractor had been paid by the employer but had not yet paid the sub-contractor. Subsequently the contractor went bankrupt. The tiles were on site but not yet fixed. The employer refused to allow the sub-contractor access to the site to remove them. Other contractors were paid by the employer to fix the tiles.

Held: When sued by the sub-contractor in conversion, it was decided that the employer had no title to the tiles. In fixing them to the roof it was liable to the sub-contractor in conversion (a tort).

Problem of unfixed goods

The JCT 63 (now reflected in *JCT 98*), amended the standard form to deal with this situation. Clause 16.1 deals with *unfixed* material and goods on-site. It provides that where the value of materials unfixed on-site is included in a certificate, property in the materials is vested in the employer. Clause 16.2 further states that *off-site* material and goods are property of employer. This is reinforced by the provisions in clause 19. Clause 19.4 states that the sub-contractor shall not deny Employer' right to ownership and that the contractor shall require the sub-contractor to provide similar provision in the sub-contract.

These provisions try to reverse the outcome of *Dawber Williams*. The problem with it is that no one can give better title than they have. If the contractor does not have title a mere contractual provision will not do so. Only a person with title can transfer it.

The *nemo date* rule

This Latin maxim restates the rule that *a person cannot give better title than they have*. Where the supplier claims the right to remove unfixed materials from the site, the employer may have a defence under s. 25 (1) of the Sale of Goods Act 1979. It states:

> 'where a person having agreed to buy goods obtains, with the permission of the seller, possession of the goods . . . , the delivery and transfer by that person . . . of the goods . . . under any sale, pledge or other disposition thereof, to any person receiving the same in good faith and without notice of any . . . right of the original seller of the goods [passes a good title to the goods].'

The purpose of the section is to protect (i) the purchaser and (ii) to facilitate trade. This however has to be balanced against the right of the seller to protect his property in the goods until paid. It is clear, however, that the section can override a retention of title clause. The employer has to beware of a number of points that could affect the application of the operation of section.

1 The employer and his professional advisers may have prior knowledge and therefore *notice* of the supplier's rights under the contract with the contractor. This is especially so where the supplier is nominated under the provisions of the main contract.
2 The contractor may not 'have bought or agreed to buy' but may instead have entered into a supply and fix contract.
3 Where interim payments are made under clause 30(2) of JCT 98, this may not invoke the protection of the section. There must be clear provision in the contract that the employer is buying the goods from the contractor.

The *nemo date* rule was applied in *Archivent Sales & Developments Ltd* v. *Strathclyde RC* (1985) 27 BLR 111. The employer succeeded in the face of a retention of title clause. Lord Mayfied in the Court of Sessions (Outer House) accepted that the clause: 'Until payment of the price in full is received by the company in the goods supplied by the company title in the goods shall not pass to the company', meant that the suppliers had *retained* property in the goods.

The seller supplied ventilators to the contractor under a contract of sale. These were to be incorporated into the building of a primary school. The ventilators were delivered to the site but were unfixed when the contractor went into receivership. Having paid for the goods in an interim certificate the employer disputed ownership. While accepting that the retention of title clause was effective, the judge held that title had passed under s. 25 (1) of the Sale of Goods Act 1979. Even though the ventilators had not been in the employer's control, possession was acquired when their surveyor measured the goods and did not reject them. The contractors had ostensible authority to deal with the goods. Nothing in their conduct suggested that they were not acting as mercantile agents.

This result on similar facts differed from *Dawber Williams* v. *Humberside CC* (1979). That was a contract for the supply and fixing of the tiles and therefore the Sale of Goods Act 1979 did not apply. *Archivent Sales & Developments Ltd* v. *Strathclyde RC* was a supply contract and s. 25 (1) applied and was capable of passing title. In both contracts the issue was who took the risk of the buyer becoming insolvent (a situation now dealt with by JCT 98, clause 16.1).

Summary

1 Besides sub-contracts, a number of other contracts support the construction operations.
2 These include contracts for leases of goods and for the supply of goods.
3 Similar obligations as to fitness for purpose and satisfactory quality (called merchantable quality in pre-1994 cases) are implied into contracts for work and materials and contracts for the sale of goods.
4 *Venables* v. *Stephen Wardle* provides the most recent example of the application of the rules in a construction setting.
5 The modern approach to the interpretation of exclusion clauses in contracts is to start from the proposition that terms agreed by businessmen are 'fair'. When applying the provisions of UCTA to commercial contracts, the courts should not without good reason interfere with their allocation of risk.
6 At the lower end of the market disclaimers of liability by surveyors are subject to the test of reasonableness laid down by UCTA.
7 The use of reservation of title clauses was approved of in *Romalpha*.
8 Problems arise when the goods have been incorporated into the building. Whether title has passed depends on whether it was under a supply and fix contract or one for the sale of goods.

13 Design Liability of Professionals and Contractors

Introduction

English law distinguishes between those who contract to supply a product and those who provide a service. In the case of the former there is a duty to provide a product which is reasonably fit for its intended purpose; in the latter, the duty is only to take reasonable care in providing the service. Since *Henderson* it is clear that this duty can arise in both contract and tort.

The 1970s saw a rapid expansion in tortious liability starting with *Dutton* and culminating in *Murphy* v. *Brentwood* in 1990 (these cases are discussed in greater detail in Chapter 14). The effect of cases such as *D & F Estates* and *Murphy* has been to reinforce the *primary distinction* between the law of contract and the law of tort.

In *Esso Petroleum Co.* v. *Mardon* [1976] 2 WLR 583, the Court of Appeal held that where the same set of facts produced both a breach of contract and a tort, the plaintiff could choose in which action to frame his claim. This produced a number of advantages to prospective claimants. One major advantage was that the *limitation period* in tort was longer, since a tort action could be brought when the limitation period in contract had expired. An example of such a case was *Pirelli General Cable Works Ltd* v. *Oscar Faber and Partners* [1983] 1 All ER 65. A chimney had been constructed at the plaintiff's factory and its construction was completed in 1970. In 1977 during routine maintenance, cracks were discovered in the chimney. Since the limitation period of six years in contract had expired, Pirelli brought their action against their designers in the tort of negligence, the builders being in liquidation. In the event they lost their case. The House of Lords held that they were *out of time* for starting their action. Time began to run when the *cracks first occurred*. This had happened in 1970 when the chimney was completed and thus they had started their action too late. As a result of this case, Parliament passed the Latent Damage Act of 1986 to remedy the problem created by *Pirelli* – establishing the actual *date* when the damage had occurred.

By 1986 the judges had started to retreat from the decision in *Esso* v. *Mardon*. In *Tai Hing Cotton Mills* v. *Liu Chong Hing Bank* [1986] AC 80, an appeal to the Privy Council, Lord Scarman said that 'there . . . [is no] advantage to the development of the law in searching for a liability in tort where the parties are in a contract'. *D & F Estates* [1988] and *Murphy* [1990] confirmed this position. Collateral Warranties were and are now seen as the answer to the strictures of the House of Lords. Remedies for defective products were to be found in the *contract* between the parties.

Tort in construction

After the decision in *Murphy* it was accepted that the right to sue contractors in tort for negligently constructed work had been ended. The one issue left outstanding was the extent to which construction professionals were liable in tort for negligence in additional to their contractual obligations. Keating (fifth edition.) expressed the view that the law regarding professionals should be re-assessed. It seemed unfair that contractors who negligently constructed buildings could escape liability whilst construction professionals were liable in both tort and contract. *Murphy* is also generally accepted as re-establishing the historical scope of the law of tort. It preceded and receded on a step by step basis. Decisions were made as needed on a practical basis from case to case, on an incremental and pragmatic basis with no sudden surges in liability. *Anns* was described in *Murphy* as such a case. This was because it created a new form of liability for negligently constructed buildings. Some 14 years before its decision in *Anns* the House of Lords in *Hedley Byrne & Co Ltd* v. *Heller & Partners* [1964] AC 465 created a separate form of negligent liability. As a result, the tort of negligence has been moving along two separate lines of liability since then.

The first line effectively starts with *Donoghue* v. *Stevenson* [1932] AC 562. The manufacturers of beer were alleged to have left a decomposed snail in a bottle. Mrs Donoghue who had no contract with the manufacturers (a friend bought the drink), claimed that the beer she had drunk caused her gastro-enteritis. Now the reason why *Donoghue* was important was that a tortious duty of care had up to then, only been imposed between neighbours in a literal sense. Neighbours were the people around you in domestic situations. By a majority, the House of Lords extended the definition of a neighbour into the commercial field. A manufacturer of defective goods could be liable for damage caused to the ultimate consumer of the goods for negligence in the manufacturing process. In other words, the person damaged by the defective goods need not have a contract with the manufacturer. *Anns* and the cases before it, extended that principle to damaged buildings.

The basic principle of *Donoghue* was that liability was limited to physical damage caused by the negligence and did not extend to economic loss. In other words, Mrs Donoghue was entitled to compensation for her physical injuries and for physical damage to her possessions, but she could not recover the economic cost of buying another bottle of beer. This principle was firmly restated by the House of Lords in *Murphy* and is a matter of great importance since the major issue in cases of negligently constructed work, is the recovery of the costs of repairs to the building.

Donoghue was about negligent acts. In 1964, the House of Lords, in *Hedley Byrne*, created a new field of liability for statements made negligently in certain circumstances.

A bank gave a reference for a client to a potential customer of the client. There was no contractual relationship between the bank and the customer's client. The reference turned out to be incorrect and the bank was sued for the resulting losses. The House of Lords breaking new ground, held that where a professional or other person claiming expertise makes a negligent statement knowing that the statement is going to be relied on, the maker of the statement owes a duty of care. This liability differed from that of *Donoghue* in that it also extended to economic loss. After the decision in *Murphy* the question of whether a professional person could still be sued in contract and tort (the dual liability debate) surfaced in several cases. It was finally laid to rest in *Henderson* in 1994. The House of Lords held:

(a) it is well established in law that professionals have liability in contract as well as tort,
(b) in the case of a professional the duty to use reasonable care and skill arises not only in contract, but is also imposed by the law apart from the contract, and is therefore actionable in tort (*Esso* v. *Mardon* followed);
(c) in commercial contracts, where lawyers usually draft the contract, it would not be difficult for the parties to exclude expressly rights and duties arising under the law of tort.

In *RM Turton & Co. Ltd (In Liquidation)* v. *Kerslake & Partners* [2000] NZCA 115, Thomas J said that the principle of *Hedley Byrne* had been formulated in many cases and phrased differently even by the Law Lords in that case. He went on to say at para 75 that 'it has been generally accepted that duty of care will arise under *Hedley Byrne* where the relationship between the parties manifests the following criteria:

(a) the maker of the statement posses a *special skill*;
(b) he or she *voluntarily assumes responsibility* for the statement and it is foreseeable that the recipient will rely on it;
(c) it is *reasonable* for the recipient to *rely* on the statement and he or she does so; and
(d) the recipient *suffers loss* as a result.' (emphasis added).

Lord Keith described *Pirelli* in his speech in *Murphy*, as a case of economic loss suffered under the principle. At p. 466 he said that:

'where the tortious liability arose out of contractual relationship with professional people, the duty extended to take reasonable care not to cause economic loss by the advice given. The plaintiffs built the chimney as they did in reliance on that advice. The case would accordingly fall within the principle of Hedley Byrne v. Heller.'

When does the duty of care arise?

Caparo Industries Ltd v. *Dickman* [1990] AC 695 formulated a single general principle for deciding whether a duty of care existed. This is known as the three-stage test for the imposition of a duty of care in tort (sometimes called the three-fold test). The House of Lords held that the criteria for the imposition of a duty of care was foreseeability of harm, proximity of relationship and the reasonableness of imposing such a duty. Winfield & Jolowicz (2002) considers that the duty of care to fulfils two functions as listed below.

1   In every case the court must decide the question 'Did this defendant owe a duty of care to this claimant?' in order to decide the dispute between the parties.
2   It must also decide the broader issue of 'whether and how far', the law of negligence should operate in situations of a particular type.

This broader issue is usually the preserve of the Court of Appeal and the House of Lords. It arose in 33 reported cases between 1981 and 2000 according to Winfield and Jolowicz (at p. 112). However in *Caparo*, Lord Oliver stressed at p. 633: 'the existence of the nexus [of duty] between careless defendant and the injured plaintiff can rarely give rise to any difficulties'. Winfield & Jolowicz (at p. 113) adds three further warnings concerning the application of the three stage test and these are elaborated below:

1   They are merely steps in a pragmatic inquiry and not a scientific test. Lord Roskill in *Caparo* stressed at p.628 that 'at best they are but labels or phrases descriptive of very difficult factual situations which can exist in particular cases and must be examined in each case before it can be pragmatically determined'. May LJ in *Merrit* v. *Babb* [2001] EWCA Civ 214 described any attempt to accommodate in a single short abstraction every circumstance that can arise, as 'reaching for the moon'.
2   Once a duty has been established in a situation before the court it will be unlikely that it will return to issues such as proximity or fairness. See for example the comments made by HHJ Hicks QC in *London Waste Ltd* v. *Amec Civil Engineering* (1997) 83 BLR 136. He emphasised the point by said that 'the three fold test . . . is for use in grappling with new questions, not justification for altering the answers to old ones'.
3   The *Caparo* formula is not the exclusive approach to the duty of care. Sir Brian Neil in *BCCI* v. *Price Waterhouse* [1998] Lloyd's Rep Bank 85 commented that 'the courts had been following three separate but parallel paths in their approaches to duty'. These were the tripartite approach, the 'assumption of responsibility' test, and the 'incremental approach'.

This duty can also arise where the parties are in a contract. Such duty is also owed to foreseeable third parties under the principle in *Hedley Byrne* discussed further below.

The foreseeability of damage

There must be a relationship of sufficient proximity between the party owing the duty of care and the party to whom it is owed so that prima facie a duty of care will arise between them. Winfield & Jolowicz considers proximity to mean nearness in a spatial sense. The question to be asked is whether the person directly affected by the alleged careless act was directly affected by the failure to take care.

Reasonableness

Is it fair, just and reasonable that the law should impose a duty in the circumstances? This is often a question of policy. *Caparo* is such an example. The House of Lords decided that the auditors owed no duty of care to investors who had lost their investment due to faulty prepared accounts. One reason may be that investors have sophisticated knowledge of investments and would not merely rely on published accounts. *Marc Rich & Co. AC* v. *Bishop Rock Marine Co. Ltd* [1996] AC 211 reflects the approach to fairness in a much more concrete setting as far as the construction professions are concerned. A vessel carrying cargo on a journey from South America to Italy developed a crack in its hull off Puerto Rico. A ship surveyor employed by a marine classification society inspected the vessel and pronounced it seaworthy. Marine societies inspect ships, oil rigs and oil platforms and issue certificates of seaworthiness for insurance purposes. The surveyor decided that the ship was fit to complete its voyage. It sank a few days later with the total loss of its cargo. The liability of the cargo carriers is limited by international conventions governing the carriage of goods by sea. Unable to recover all its losses, the owners of the cargo brought an action to recover their remaining losses against the society, alleging that the surveyor had been negligent. On a preliminary issue (assuming the surveyor was negligent and that the loss of the cargo was reasonably foreseeable), the House of Lords held that though the requirements of proximity were satisfied, there were policy considerations against a decision in favour of the owners. These included (i) the conventions limiting liability, (ii) the possibility of forcing the societies to procure expenditure in maintaining of insurance and (iii) societies refusing to accept high-risk ships. This could lead to them to fulfilling no longer their public duty of ensuring the safety of ships at sea.

Voluntary assumption of responsibility

In *Henderson*, the underwriting members – called 'names' – brought an action against their underwriting agents. The House of Lords held that a duty of care was owed to the members in tort and the duty was not excluded by the existence of a contractual framework regulating their agreement. Thus the members were free to pursue their remedy either in contract or in tort.

Lord Goff (with whom the other Lordships agreed) gave the main opinion. *Hedley Byrne* v. *Heller*, he said at p.766, was the key to the understanding of the principles involved:

> 'We can see that it rests on a relationship between the parties, which may be general or specific to the particular transaction, and which may or may not be contractual in nature. All of their Lordships spoke in terms of one party having assumed or undertaken a responsibility towards the other.' (the reference here is to the speeches in the House of Lords in *Hedley Byrne* v. *Heller*)

Later on he referred to the cases subsequent to *Hedley Byrne*. In many of these cases where negligent misstatement was alleged, the question was frequently whether the claimant fell into the *category of persons* to whom the maker of the statement owed a duty of care. He then referred to the attempts of the courts to restrict the 'category of persons within reasonable bounds'. At p.776 he made the following observations:

1 The 'assumption of responsibility' concept had been criticised as being 'unlikely to be helpful or realistic test in most cases'. In the present case he could see no reason for not invoking the concept.
2 Where liability may arise under a contract or in a situation 'equivalent to a contract', the test applied would be an objective one of whether in the particular circumstances of the case the defendant had assumed liability to the claimant.
3

> 'In addition the concept provides its own explanation why there is no problem in cases of this kind about liability for pure economic loss; for if a person assumes responsibility to another in respect of certain services, there is no reason why he should not be liable in damages for that other in respect of economic loss which flows from the negligent performance of those services. It follows that, once the case is identified as falling within the *Hedley Byrne* v. *Heller* principle, there should be no need to embark upon any further enquiry whether it is 'fair, just and reasonable' to impose liability for economic loss.'

4 Where advice is given on an informal occasion there may be no assumption of responsibility.
5 Similarly the assumption of responsibility may be negated by an appropriately worded disclaimer of responsibility.

In *Merrit* a surveyor was held personally liable for the negligent survey of a property made for valuation purposes. May LJ summarised what he called the repeated attempts to define a comprehensive test to define the circumstances in which a person owes a duty of care to another, not to cause foreseeable economic loss in the giving of advice, or in providing information. He reviewed the authorities and then commented at para. 41:

> 'In such cases especially – but, I think in every case – reliance is an intrinsically necessary ingredient which appears in every formulation of the test. Beyond that, two strands of consideration emerged. These may for convenience be called the *Caparo* strand and the *Henderson* strand. The *Caparo* strand asks whether, in addition to foreseeability, there is sufficient relationship of proximity and whether the imposition of a duty of care is fair, just and reasonable. The *Henderson* strand asks whether the defendant is to be taken to have assumed responsibility to the claimant to guard against the loss for which damages are claimed. The difficulty with the *Caparo* strand is that it is sometimes seen as being unhelpfully vague. The difficulty with the *Henderson* strand is that it was originally often expressed in terms of "voluntary assumption of

responsibility" which tended to import a degree of subjectivity. *Henderson* itself put paid to that and, as Lord Slynn said in *Phelps* "[assumption of responsibility] means simply that the law recognises that there is a duty of care. It is not so much that responsibility is assumed as that it is recognised by the law". Thus, the *Caparo* strand and the *Henderson* strand in reality merge. In my view, it is very often a helpful guide in particular cases to ask whether the defendant is to be taken to have assumed responsibility to the claimant to guard against the loss for which damages is claimed. But I also think that it is reaching for the moon – and not required by authority – to expect to accommodate every circumstance, which may arise within a short abstract formulation. The question in each case is whether the law recognises that there is a duty of care.'

Causation

The claimant must prove that there was actionable damage. On the balance of probability, it must show that there was a connection between the negligent conduct and the resulting damage. That 'but for' the negligence of the defendant the damage would not have happened. The classic example is *Barnett* v. *Chelsea & Kensington Hospital Management Committee* [1969] 1 QB 428. It was proved that the deceased would have died from arsenic poisoning even if the hospital had treated him. Although it was negligent in not treating him, it was not the cause of his death.

Causation was considered in a construction context in *BHP Billiton Petroleum Ltd and ors* v. *Dalmine Spa* [2003] EWCA Civ 170. Rix LJ described it as a difficult concept despite being a matter of common sense. The first step in establishing causation is to eliminate irrelevant causes and that was the purpose of the 'but for' test. (See the reliance on Clerk and Linsell at para. 26). 'In other words, if the damage would have occurred in any event, the defendant's conduct is not a "but for" case.' The appellant, *Dalmine Spa*, provided 12 inch diameter steel pipes for use in the construction of a gas pipeline in the Irish Sea. During inspection and certifying of the pipes after their manufacture, the results were deliberately changed and falsified. After installation the pipes leaked in six failure zones and had to be replaced. In the course of discovery, the appellant's fraudulent misrepresentation emerged. At first instance, the issue was whether the non-compliant pipes caused the pipeline to fail or whether it would have failed anyway even if compliant pipes had been used. Creswell J, giving the judgment, decided in favour of BHP on the balance of probability and did not have to consider this issue further.

On appeal, the appellant conceded in the Court of Appeal that the burden of proving that compliant pipes would have failed in any event was one it could not sustain. Once BHP showed that the damage to the pipeline occurred in the parts containing non-compliant pipes, the burden of proof passed to the appellant. To discharge this, they had to demonstrate that compliant pipes would have failed in any case. They were unable to prove this. Had they been able to do so, under the 'but for' test, they would have had no liability.

Application to construction contracts

In *Henderson*, Lord Goff said that, in the traditional construction contract where there was a chain of contracts:

'It will not be ordinarily open to the building owner to sue the sub-contractor or supplier direct under the *Hedley Byrne* v. *Heller* principles, claiming damages from him on the basis that he has been negligent in relation to the performance of his functions. For there is generally no assumption of responsibility by the sub-contractor or the supplier direct to the building owner, the parties having so structured their relationship that it is inconsistent with any such assumption of responsibility.'

Can the contractor sue the professionals for *inadequately* prepared specifications and drawings under the principles of *Hedley Byrne* v. *Heller*? This was considered in *RM Turton & Co. Ltd (In Liquidation)* v. *Kerslake & Partners* [2000] NZCA 115. The Supreme Court of New Zealand reviewed the current scope of the principle and its application to a chain of contracts in the traditional contract. By a majority it decided that the alleged representation, that the components described in the specification would meet the required output, 'Would cut across and be inconsistent with the overall contractual structure which defines the various relationships of the various parties to this work, and in the circumstances of the case it would not be fair, just and reasonable to impose the claimed duty of care'.

For the contrary position where the *contractor designs and constructs* a building, see *Surrey* v. *Charles Church Development* (1996) 12 Const. LJ 206 per HHJ Hicks at 212. There he held that whist an ordinary builder owes no duty not to cause economic loss, the position is different where the builder carries out the design and the construction. In such a case there was sufficient reliance for the imposition of a duty of care under *Hedley Byrne* v. *Heller*. Such a duty to take reasonable care and skill in contract was co-terminous with the duty in tort.

The liability of professionals

Scruton LJ in *IRC* v. *Maxse* (1919) said that a profession 'in the present use of language involves the idea of an occupation requiring purely intellectual skill, or if any manual skills, as in painting and sculpture or surgery, skills controlled by the intellectual skill of the operator'. *Lanphier* v. *Phipos* (1838) 8 C&P 475 stated that 'every person who enters a learned profession undertakes to bring to the exercise of it a reasonable degree of care and skill'. The category of professional of importance to the construction industry falls into two groups: those who design structures and supervise their construction, such as architects and engineers; and those who give advice or provide information concerning such structures, such as surveyors, valuers and solicitors.

Where the professional is retained, the duty of reasonable care and skill will be owed in contract. At one time, professional negligence was not considered as a liability that arose in tort, but simply a convenient form of words to describe a breach of contract where one of the parties to the contract claimed to practise a particular skill or profession. *Esso* v. *Mardon* (1976) confirmed that a victim of negligent professional advice could choose a tortious basis of action, even though the victim had a contract with the adviser. (See also *Midland Bank Trust Co. Ltd* v. *Hett Stubbs and Kemp* [1978] 3 WLR 167. Duties in tort will also be owed to *foreseeable third parties* who may act on negligent advice given: see *Hedley Byrne* and *Ross* v. *Caunters* [1980] Ch 297.

What is the standard of care owed?

The duty of reasonable care and skill owed in contract by a professional is also the duty of care in negligence. Negligence was defined in *Blyth* v. *Birmingham Waterworks* (1856) 11 Ex 781 as 'An omission to do something which a reasonable man would do', or as 'doing something, which a prudent or reasonable man would not do'.

Whether a person had acted reasonably is judged by an *objective standard*. The standard is not what the defendant thought but what 'the man on the Clapham omnibus' would think of his conduct. That of course applied to jury trials in civil actions. What matters nowadays is the opinion formed by the trial judge of the defendant's conduct.

The test in *Blyth* only refers to the general public. *Blyth* had been refined when applied to professionals in a case of medical negligence called *Bolam* v. *Friern Hospital Management Committee* [1957] 1 WLR 582. A patient was injured undergoing electro-convulsive treatment without any restraint or sedation. Although the experts were divided in their opinion, that was the normal treatment at the time. This is now known as the *Bolam test*. In *Bolam*, McNair J, in giving direction to the jury, said that the test for professional negligence differed from ordinary negligence:

> 'where you get a situation which involves the use of some specialist skill or competence, the test of whether there has been negligence or not is the test of the man on the Clapham omnibus because he hasn't got this special skill. A man need not possess the highest expert skill at the risk of being found negligent. It is a well established law that it is sufficient if he exercised the ordinary skill of an ordinary competent man exercising that particular art'.

Although *Bolam* concerned medical negligence, it is now accepted as the general test for assessing professional negligence. Expressed in *Saif Ali* v. *Sydney Mitchell* [1980] AC 198 by Lord Diplock at p.220:

> 'No matter what profession it may be, the common law does not impose on those who practise it any liability for damage resulting from what in the result turn out to have been errors of judgement, unless the error was such no reasonably well-informed and competent member of that profession could have made.'

Bolam was applied in *Nye Saunders and Partners (a firm)* v. *Alan E Bristow* (1987) BLR 92. There Brown LJ said at 103: 'the duty and standard of care to be expected from [the architect] was accepted as being that which applied to any profession or calling requiring special skill, knowledge or experience'. Professional negligence thus can be seen from the *Bolam test* as being pitched at the level of reasonable competence of the individual being sued for negligence. Such competence is judged by the *standards in use at the time*, without the benefit of hindsight. This is known as the *state of the art defence*.

As far as professional services such as design are concerned, there are numerous dicta from the courts defining the standard of care required. A judgment that explained it particularly well is that of Bingham LJ in the Abbeystead case, *Eckersley, TE and Others* v. *Binnie & Partners and Others* (1988) 18 Con LR 1. He disagreed with his fellow judges and would have allowed Binnie's appeal but, despite that, his judgment sums up the degree of care needed not to be judged negligent. Ward LJ adopted it without qualification in *Michael Hyde and Associates Ltd* v. *JD Williams and Co Ltd* [2000] EWCA Civ 211:

1 The law requires a professional to live up in practice to the standard of the *ordinary* skilled man exercising his special *professional skills*. He does not need to possess the highest possible skills.
2 The law does not impose liability for damage resulting from *errors of judgement* unless the error was such that no reasonably well informed or competent member of that profession could have made it.
3 He should possess the *body of knowledge* which forms part of the professional equipment of the ordinary member of his profession.
4 He does not lag behind other hardworking and intelligent members of his profession in knowledge of new advances, discoveries and developments in his field.
5 He should have awareness as an ordinary competent practitioner of the *deficiencies* in his knowledge and the limitations in his skills.
6 To his professional tasks he need not bring more that the *ordinary competent skills* of his profession; the law does not require him to be a paragon, combining the qualities of polymath and prophet.

Relatively speaking, this standard of care is not particularly high. It applies to the standard of care which is that of the ordinary competent practitioner and then only to obligations of reasonable skill and care, such as the ordinary negligence responsibility (whether arising in contract or in tort) of an architect or engineer carrying out design. Where the contractor undertakes design along with his building obligations, the position at common law is different. A design and build contractor may be held to have impliedly warranted the fitness for purpose of the design.

Variations in the standard of care

A further question that arises is whether an *expert* in his field owes a *higher duty* than that of the non-specialist. There are two contrary views. In *Ashcroft* v. *Mersey Regional Health Authority* (1983) 2 All ER 245, it was said by the court that the more skilled the practitioner was, the higher was the standard of care expected.

This view differs from that expressed by Webster J. in *Wimpey Construction UK*

v. *Poole* (1984) 2 LLR 499 where the matter was discussed at length. This was a case where Wimpey was suing its own insurers and in the process trying to persuade the court that it had negligently carried out the design. In March 1972 Wimpey entered into an agreement with Vosper Thornycroft to design and construct an extension to the fitting-out facilities at Vosper's shipyard at Southampton.

The work consisted of the construction of a new quay wall and the provision of the foundation base for a crane positioned 30 feet behind the face of the quay wall. Construction started in February and dredging in front of the wall was carried out in two sweeps in late June. In July it was discovered that the quay wall had moved and a central section had settled. At the end of July it was confirmed by measurement that the crane base had also settled. Wimpey carried out substantial remedial work and subsequently claimed the costs through their professional indemnity policy. Their insurance company refused to pay out on the basis that the policy covered negligence only, and there had not been any. Wimpey then went to court to argue that they *had been negligent* and thus were entitled to claim under their policy.

Wimpey argued that the *Bolam test* should be *qualified* in three ways:

1 The *Bolam* test was not applicable where the client obtains and pays for someone with especially high skills. The judge rejected this since he had no way of knowing whether Wimpey as a company had especially high skills. In any case he was bound by the *Bolam test* as approved without qualification in *Whitehouse* v. *Jordan* [1981] 1 WLR 246 by the House of Lords.
2 The judge accepted the second qualification, which was that the duty of a professional man had to be judged in the light of his *actual* knowledge. This could not be answered by reference to a lesser degree of knowledge than he had simply on the grounds that an ordinary competent practitioner would only have had that lesser knowledge. The judge accepted this, saying that it flowed naturally from Lord Aitkin's test in *Donoghue* v. *Stevenson*. However, even on that basis Wimpey had not been negligent.
3 The third qualification was that the test should be applied to Wimpey's most experienced designer or to Wimpey as a company, assuming that it employs the most experienced of designers and at least one designer of exceptionally high qualifications. The judge decided that was not necessary to decide this point since even on this test Wimpey was not negligent.

Wimpey was thus unable to prove on the balance of probability that they had been negligent and were unable to recover the cost of the repairs under their insurance policy.

Expert witnesses

As a general rule the witness called to give evidence is not permitted to express his own opinions. He is not permitted to speculate on the course of events or draw inferences from the facts. It has, however, long been recognised that there are matters where specialist skill, knowledge or expertise may be required to draw inferences from the evidence given by witnesses. As early as 1553 Saunders J said in *Buckley* v. *Rice-Thomas* (1554) 1 Plowd 118 that 'if matters arise in our law which concern other sciences or faculties, we commonly apply for aid of that science or faculty which it concerns'.

In *National Justice Compania SA* v. *Prudential Assurance Co Ltd* [1993] 11-CLD-09-17, Creswell J commented that: 'A misunderstanding on the part of some expert witnesses has taken place concerning their duties and responsibilities, which has contributed to the length of the trial.' He then gave an outline of the duties and responsibilities of expert witnesses in a civil trial:

1 Expert evidence presented to the court should be, and should be seen to be, the independent product of the expert uninfluenced as to form or content to the exigencies of litigation; see *Whitehouse* v. *Jordan* [1991] 1 WLR 246, 256 per Lord Wilberforce.
2 Independent assurance should be provided to the court by way of objective unbiased opinion regarding matters within the expertise of the expert witness: see *Polivitte Ltd* v. *Commercial Assurance Co plc* [1987] 1 LLR 379, 383 per Garland J, and *Re J* [1990] FCR 193 per Cazalet J. An expert witness in the High court should never assume the role of advocate.
3 Facts or assumptions upon which opinion was based should be stated together with material facts which could detract from the concluded opinion.
4 An expert should be clear when a question fell outside his expertise.
5 If the opinion was not properly researched because it was considered that insufficient data was available then that had to be stated, together with an indication that the opinion was provisional (see *Re J*). If the witness could not assert that the report contained the truth, the whole truth and nothing but the truth, that qualification should be stated on the report: see *Derby & Co and others* v. *Weldon and Others* (no 9) The Times, 9 November 1990, per Lord Justice Staughton.
6 If, after exchange of reports, an expert witness changed his mind on a material matter then the change of view should be communicated to the other side through legal representatives without delay and, when appropriate, to the court.
7 Photographs, plans, survey reports and other documents referred to in expert evidence had to be provided to the other side at the same time as the exchange of reports.

The application of the Bolam test requires the assistance of expert witnesses, since only specialists can assist the court in deciding whether the professional fell short of the standard required of the average competent professional. A useful and informative example of the approach of a judge to assessing the value of the evidence of the expert witnesses is contained in *Wimpey Construction UK* v. *Poole* (1984). The experts were amongst the most prominent practitioners in the field of geotechnical engineering, a virtual 'who's who' of the day. The judge found that most of their evidence was of little assistance in deciding the standard required of the average competent designer of structures in soil, indicating perhaps that getting the 'top man' may not always help your case?

In *Michael Hyde and Associates Ltd* v. *JD Williams & Co. Ltd* [2000] EWCA Civ 211, the Court of Appeal considered *the approach* to be taken to the evidence of experts. It decided that the Bolam type test was appropriate where the negligence lies in the *conscious choice* of available courses made by a trained professional. Its use can be inappropriate where it is in the 'oversight where the neglect is said to lie' (Smedley LJ at p.14). Where there are two bodies of opinion having opposing views, a man is not negligent merely because he has chosen one above the other.

There are a number of qualifications to the Bolam test:

1 Where the professional opinion is not capable of withstanding logical analysis, a judge is entitled to hold the opinion is not reasonable or logical. It would be rare for a judge to reach such an opinion (per Ward LJ at p.9).
2 Where the judge considers that the evidence did not constitute a responsible body of professional opinion of architects in the case *Nye Saunders* above.
3 Where the judge considers *no specialist skill* is required to decide whether the defendant has fallen from the standard required.

Michael Hyde can be considered to be such a case. The central question was whether the architect should have relied upon the assurances given by British Gas. These assurances contained a disclaimer given by them in their quotation. The client argued that it should have been advised that further inquiry was necessary before it went ahead with the installation of the modern heating system in its distribution centre. The client was a mail order company and stored vast amounts of clothing in its warehouses. It purchased a five-storey derelict cotton mill and engaged Michael Hyde and Associates Ltd (MHA) to provide 'all architectural, clerk of works, survey, quantity surveying and structural engineering services' necessary for the project.

British Gas suggested a direct fired system which was cheaper and provided better ventilation in the summer. Their quotation for the work carried a *disclaimer* against liability for the discoloration of products caused by the system. After installation phenolic yellowing occurred in the textiles stored in the warehouse. The direct cause was the use of the direct-fired system. As a by-product it produced oxides of nitrogen, which combined with yellowing precursors present in plastic packing, cardboard and paper. The textiles then absorbed these yellowing products leading to the discoloration.

The client alleged that MHA ought to have been aware of the discoloration and warned them of it. After a hearing of five days a High Court judgment was entered against MHA in the sum of £365,325. They appealed against the order.

The Court of Appeal upheld the appeal, finding that the claimant had not established that further investigation would have in fact have uncovered an unacceptable risk of discoloration. Ward LJ adopted Oliver J's formulation in *Midland Bank Trust Co. Ltd* v. *Hett Stubbs & Kemp* (1978) Ch 384 at 402:

'Clearly, if there is some practice in a profession, some accepted standard of conduct laid which is laid by a professional institute or sanctioned by common usage, evidence of that can and ought to be received. But evidence which really amounts to no more than an expression of opinion by a particular practitioner of what he thinks he would have done had he been placed, hypothetically and without the benefit of hindsight, in that position of the defendant, is of little assistance to the court.'

Sedley LJ suggested that the Bolam test was not applicable to all allegations of professional negligence. It was, he said at p.14:

'Typically inappropriate where it is an *oversight* where the neglect is said to lie. . . . In the present case the court is asked to say that the expert evidence called by the defendant concludes the case in his favour, since it establishes the existence of a respectable school of practice which would have done as he did. That the judge preferred an opposing school is, it is argued nothing to the point. Although my reason for rejecting this argument is not quite that of the judge – who based himself on whether the court needed expert help at all – my conclusion is the same as his: the defendant overlooked a risk which he ought to have brought to the claimant's attention and later proved real. This requires no inquiry into competing schools of professional practice: it requires the court to decide, with what ever help the expert evidence affords, whether a competent architect would have overlooked the implications of the disclaimer.'

Experts under the new Civil Procedure Rules

The new rules introduced in April 1999 allowed the court to impose a *single expert* on the parties. The fear was expressed that this would be done in cases where this was quite unsuitable. Since the rules came into force the practice in the TCC has been for parties to display a greater willingness to propose the appointment of a single expert.

HHJ Dyson QC suggested (Const L J 1999) that the following type of case might be suitable for a single expert:

(a) where the sums at stake in litigation are small in relation to the costs likely to be incurred;
(b) where the expertise consist of personal judgement or is derived from personal experience (such as valuation evidence);
(c) where the evidence the court needs to have explained is relatively uncontroversial;
(d) where the issue is relatively peripheral to the case.

Other examples would be issues of the quantification of loss (e.g., variations, loss and expense claims) and dilapidation cases.

Immunity of expert witnesses

In *Stanton* v. *Brian Callaghan* [1999] 2 WLR 745, [1999] BLR 172, the Court of Appeal considered in detail the question of the immunity of expert witnesses. The claimant's expert agreed at a meeting of the experts that a less expensive remedial scheme was appropriate. The claimant alleged that the expert acted negligently in so doing. The appeal was struck out in the Court of Appeal on the grounds of *public policy*. It required that experts should be able to reach agreements at meetings between experts without the fear that they would be sued in negligence. In doing so they were immune from such claims.

Fitness for purpose

If the obligation of the designer is only to use reasonable care and skill in carrying out the design, does this mean that the design should also be fit for its purpose? What is the difference between the two obligations? Fitness for purpose is the greater obligation and the importance of the distinction is explained below:

1 A product or design may not be fit for its purpose without any allegation of fault. Where a particular purpose is made known by the employer, negligence need not be proven. All that has to be shown is that the particular purpose specified has not been achieved.
2 As far as reasonable skill and care is concerned, it has to be proved that the design was negligent and that the designer did not use reasonable care and skill.

This issue arose in *Greaves (Contractors) Ltd* v. *Baynham Meikle & Partners* [1975] 3 All ER 99. The contractors by a contract agreed to design and construct a warehouse and office for a company which intended to use the warehouse for storage of barrels of oil. The contractors appointed structural engineers to design the structure of the warehouse and, in the course of giving them instructions, told them that the first and upper floors of the warehouse would have to bear the weight of fork-lift trucks carrying heavy barrels of oil. The warehouse was built using the design provided and brought into use. Later on, the floors began to crack and the premises became too dangerous to use. It was found at trial that the cracks were the result of resonance of two different frequencies, one caused by the movement of the floor and the other by the vibrations of the forklift trucks.

The Court of Appeal held that there was an *implied term* in the designers' contract that they should design a warehouse fit for the purpose for which they were required, namely reasonably fit for fork-lift trucks carrying barrels of oil. As the structural engineers had not produced such a design, they were liable to the contractor for the cost of remedial works required by the building owner.

In the alternative, if there was no such term to be implied, the structural engineers were in breach of their duty to use reasonable care and skill because they knew (or ought to have known) that the purpose of the floors was to carry fork-lift trucks laden with barrels of oil. Lord Denning in *Greaves* found the implied term had been implied in fact (i.e., if the parties had thought about it, they would have agreed it). Therefore no general assumption about terms being implied into a designer's contract, that the design will be fit for the purpose, can be drawn from this case. However, he went on to observe:

'What is the position when an architect or engineer is employed to design a house or a bridge? Is he under an implied warranty that if the work is carried out to his design, it will be reasonably fit for its purpose or is he only under a duty to use reasonable care and skill? This question may be required to be answered some day as a matter of law. But, in the present case I do not think we need to answer it. For the evidence of both parties were of one mind in the matter. Their common intention was that the engineer should design a warehouse, which would be fit for the purpose required. That common intention gives rise to a term implied in fact.'

By way of emphasis, Brown LJ said that the decision in the case *laid down no general principle* as to the obligation and liability of professional men; it was a case that depended on the special facts and circumstances that had arisen.

The implications of *Greaves* were considered in *George Hawkins* v. *Chrysler (UK) Ltd and Burne Associates* (1986) 38 BLR 36 by the Court of Appeal. The plaintiff was an employee of Chrysler who slipped on water in a shower room after using a shower. He sued Chrysler, his employer and Burne Associates (who had designed, specified and supervised the installation of the showers) for damages for his injuries. Chrysler also claimed against Burne, but settled with the employee. They carried on the case against Burne Associates. In their action they claimed:

(a) it was an implied term of the contract that Burne would use reasonable care and skill in selecting the material to be used for the floor of the showers; and
(b) there was an implied warranty that the material used for the floor would be fit for use in a wet shower room.

The court *declined to imply terms* into the contract either as a matter of fact or of law. It decided that there was nothing in this case that gave rise to any inference that a higher duty than reasonable care and skill was required to be exercised by the designers. It stated that while there may be anomalies between the position of contractor and sub-contractor in this respect (i.e., obligations of fitness for purpose), as opposed to professional people: 'in the absence of special circumstances, it was not open to the court to extend the responsibilities of a professional man beyond the duty to exercise reasonable care and skill in accordance with the usual standards of his profession'.

The duty to warn

Novel and risky design

What is the position if a design has not been carried out before and it fails when constructed? In *Independent Broadcasting Authority* v. *EMI Electronics and BICC Construction Ltd* (1980) 14 BLR 1, it was accepted that the design of a cylindrical mast, which collapsed after being built, was 'both at and beyond the frontier of professional knowledge at that time'. The judges in the House of Lords did not regard the fact that there was no precedent for a design of such a tall cylindrical mast as being any reason for excusing the designers when the mast collapsed. It took the view that the designers needed to take added precaution in order to discharge their duty of reasonable skill and care in the circumstances of novel design. This included warning the employer of the risky nature of the proposed design.

Failure to use reasonable skill and care

In *QV Limited* v. *Frederick F Smith & others* [1998] OLL 1403, OR, QV employed Mr Smith, a chartered builder and project manager, and DF Green & Sons Ltd to construct a cold store. Mr Smith specified cladding by Eternit UK Limited and a spray foam insulation system to be used in combination with the cladding. After completion in 1992, the roof cladding cracked and rain leaked into the store. The cracking was due to thermal movement in the cladding being restrained by the rigid spray foam insulation.

QV sued Mr Smith in contract and tort, alleging that he failed to provide the basic design of the building with reasonable care and skill. Greens and Eternit were joined as third parties for constructing and supplying products not fit for their intended purpose and for not being of merchantable quality.

Held: Mr Smith owed a duty of reasonable care and skill in the basic design of the building. He was in breach of that duty. A reasonably competent designer would not have specified spray foam insulation to be used in combination with Eternit cladding.

This case is of interest in that Mr Smith was a chartered builder and project manager. An architect described by the judge 'as a qualified architect with as he said, no particular speciality' gave expert evidence for the client. The judge also made the point that 'at one stage I was under a misapprehension that Mr Smith was a qualified architect'. He went on to hold that the test to be applied to him was that of an ordinary competent designer in designing the cold store 'who had professed to have the necessary skill to carry out the required design'.

Normally, expert evidence has to be given by a person in the same profession. In *Samson & Another* v. *Metcalfe Hambleton & Co.* [1998] 15-CLD-02-15, the Court of Appeal upheld a chartered surveyor's appeal against a finding of negligence. In preparing a report on the structural design of a house he had failed to notice a significant crack in a wall. At first instance the judge found him negligent in not noticing it and failing to advise the purchaser to investigate the matter further. In reaching this decision the judge rejecting the evidence of the surveyor's expert, another surveyor. He preferred the evidence of the purchaser's expert, a structural engineer. Their failure to call a chartered surveyor to give evidence of the standard and skill expected of a surveyor was fatal to their claim. Their expert was not qualified to give evidence of the practice accepted as proper by a responsible body of chartered surveyors.

Warning: design and competence of other professionals

In *Tesco Stores* v. *The Norman Hill Partnership* (1997) CILL 1301, a fire destroyed a shopping development at Grove Park, Maidstone. The Norman Hill Partnership 'NHP' provided the developer Investments with the design and fit-out drawings for the shell of a supermarket. NHP under a separate contract also provided Tesco with drawings for the fit-out work, attended the site and made periodic inspections of the fit-out work. The fire spread rapidly and it was accepted that there were defects in the construction, which contributed to the rapid spread of the fire. Both Investments and Tesco claimed that NHP should have warned them that they had spotted faults in the construction of the shell whilst inspecting the fit-out work. It was held that NHP owed a duty of care to both parties in respect of defects that they had become aware of in the course of their contractual duties. This did not, however, impose a positive obligation to actually inspect for defects.

The extent of the duty to warn in relation to fellow professionals was considered in *Chesham Properties* v. *Bucknell Austen Project Management Services and Others* (1997) 82 BLR 92 by HHJ Hicks QC. Chesham were property developers of the Royal Court House in Cadogan Place, London. The defendants were their professional advisers engaged in the development of the scheme. Out of the performance of the work they made several of allegations of professional negligence against their project managers and a number of their professional advisers. They claimed that the project managers and architects wrongly granted extensions of time and awards of loss and expense. In addition they claimed that the professional team owed a duty to the client to warn of actual or potential deficiencies in the performance of the other defendants. In giving judgment the court decided the following:

1 The project manager was under a duty to advise and/or inform the claimants of actual or potential deficiencies in the performance of the others. This arose out of their contractual obligation to 'implement all monitoring procedures including the performance of consultants'. Monitoring, in such a context could not sensibly be confined to passive observation only; it must include reporting to the principal on the performance being monitored by reference to the standards that should be achieved.

2 The architects were obliged to advise of actual or potential deficiencies in the performance of the structural engineers and quantity surveyors in contract and at common law. They were not obliged to advise and/or warn of their own actual or potential deficiencies. Note the explanation for the existence of the duty arose out of the requirement of the RIBA booklet which described at para. 2.38: 'Provide management . . . appoint and co-ordinate consultants etc; monitor time, cost and agreed targets' and which had been incorporated into their contract. Furthermore, the paragraph also extended to the structural engineer and quantity surveyor, who in the circumstances must also qualify as "consultants". The obligation to "co-ordinate" and to "monitor" must include a duty to report on their performance.

3 The structural engineer did not owe any duty in contract or a common law to advise/warn the plaintiff of actual or potential deficiencies on the part of the architects or quantity surveyors.

4 In the absence of any express provision no term beyond the normal duties of a quantity surveyor would be implied. These were limited to exercising reasonable care and skill in:
    (a) preparing bills of quantities,
    (b) carrying out the function of quantity surveyor under the building contract,

(c) ascertaining the value of contractor's work as executed and preparing monthly accounts and an estimate of the contractor's final account.

Design liability in design and build contracts

Amongst the areas of conflict that the Latham Report identified was the following concerns. In Item 3.10 it was the lack of co-ordination between design and construction and in Item 4.18 the lack of distinction between the duty of reasonable care and skill and the requirement of fitness for purpose (the so called 'fuzzy edge disease'). The traditional method of contracting draws a clear distinction between the two. Design work is firmly in the hands of the employer's professional designer and construction work firmly in the hands of the contractor. This distinction is emphasised in the JCT 98 standard form of contract, which provides that design is not the province of the contractor. The ICE 7th by contrast do provide that certain design responsibilities may be allocated to the contractor, but stresses that design liability is limited to the exercise of reasonable care and skill.

Quite often in civil engineering works the contractor has to provide substantial temporary works. In doing the design of this work the contractor's liability will be limited to that of reasonable skill and care. Given these distinctions, the employer when faced with a defective building or structure may have to bring claims against both the designer and the contractor. The reason for this is quite straightforward. Many defects do not conveniently fall into either the design or the workmanship category and quite often overlap between the two.

The conflict between workmanship and design

John Mowlem, discussed earlier in Chapter 6, demonstrates this. There, in a contract carried out under the then standard form of contract JCT 63, design parameters were included in the bills of quantities. Faults developed in the work being constructed, resulting in expenditure in remedying the work, and subsequent delay in completion and occupation. In an action against the structural engineers and the contractor two issues had to be decided. Whether (i) the defect was the result of inadequate design and/or faulty workmanship and (ii) whether the bills of quantities imposed a performance specification on the contractor. The judge held that faults in the design resulted from a lack of reasonable care and skill by the engineers in carrying out the design. Since design was not the responsibility of the contractor, no such obligation could be placed on it using the bills of quantities included in a traditional JCT contract.

Design and build contracts

One of the answers suggested to the problem of the interaction between workmanship and design, is the 'design and build contract', or package deal or turnkey as the project may be described in the process industries. Initially, the employer provides the contractor with a document called 'the employers requirements'. This will include the preliminary design that has been carried out by the employer's design team. The contractor's tender will include a document containing 'the contractor's proposal'. (For further details of the process consult the NJJC

Code of procedure for Design and Build). The NJJC supports the transfer of the design team to the successful contractor's employ so that the design can be 'sympathetically developed'. Novation is the legal description of this process which requires the agreement of all three parties. The NJJC also recognises that the employer's agreement with its design team may overlap with the agreement between the contractor and the design team. It suggests that there may be some incomparability and some conflict of interest. What it does not address is the fact that this incomparability is not simply a matter of ensuring that there are no gaps or overlap. Its approach does seem to suggest that there is in practice, no practical difference between a design carried out by (i) an independent design professional and (ii) a contracting company. It is suggested that this is not entirely correct. A designer may not approach a project from cost perspective as a matter of course. The actual design may considered to be a far more important starting point. This is not to say that cost is an irrelevant consideration. The designer may well, from a personal and professional viewpoint, seek to achieve the best possible design when constrained by cost. Whereas, a contractor's approach to design must in the first place be from commercial perspective and be driven by the need to make profit from its activities. It will therefore design to a price because that is what it is in business to do.

The contractor in this type of contract not only takes on the job of constructing the works but also of designing it. This fundamentally alters the contractor's obligation in respect of design. This is because there are a number of terms implied into building contracts, whether they are design and build or a traditional building one. The three important terms, relevant to design, that are usually implied into building contracts are that the contractor will:

(a) carry out the work in a good and workmanlike manner
(b) use materials of good quality and
(c) ensure that the materials and the work will be reasonably fit for their purpose.

Between two contracting parties, the implication of the term that the work will be carried out in a good and workmanlike manner will cause few difficulties. Where there is a chain of contracts, the court may refuse to imply such a term into a warranty given to the employer by the sub-contractor: see the case of *Greater Nottingham Co-operative* discussed earlier in chapter 6. Similarly, the warranty to use materials of good quality will also be implied between contracting parties only. An obligation of fitness for purpose is readily excluded in traditional building contracts: see *Cable (1956) Ltd* v. *Hutcherson Brothers Pty Ltd* (1969) 43 ALJR 321. There, it was held that the obligation on the contractor was to carry out the work described in the drawings. The position is different in contracts for design and build, where the employer does rely on the contractor in the selection of materials. Whether the warranty as to fitness for purpose extends to the design in design and build contracts is now settled. In *Greaves* for example Lord Denning observed that 'It was . . . duty of contractors to see that the finished work was reasonably fit for the purpose for which they were required. It was not merely an obligation to use reasonable care, the contractors were obliged to ensure that the finished work was reasonably fit for its purpose.'

That statement was made in the context of a claim against an engineer not the contractor. *Viking Grain Storage Ltd* v. *T H White Installations Ltd* (1985) 33 BLR 103 however, settled the issue. An obligation of fitness for purpose will be imposed at common law. The parties in that case had been unable to agree the express terms of their contract. A preliminary issue was what *terms* were to be implied into their contract. The employer argued that:

(a) the design and build contractor would use materials of good quality and reasonably fit for their purpose; and
(b) that the completed work would be fit for their purpose.

The works in this case, were a grain drying and storage installation. The Official Referee had no difficulty in holding that there was reliance by the employer on the skill and judgement of the contractors. It followed that those terms were implied into the contract. It should be noted that in *Viking Grain* the parties throughout the construction of the works were involved in negotiating the express terms of their contract. It was not in place when the collapse occurred, and thus the court had to consider what terms were to be implied into their agreement. The contractor may of course expressly assume liability for design, especially under package deal, turn-key or design and build contracts. In such a contract the contractor therefore assumes both design responsibility and liability for workmanship and materials. In the absence of a contractual condition such as found in the standard forms of design and build contracts, the contractor promises that the work shall be reasonably fit for its purpose. The parties may also expressly agree that the contractor shall only have the design liability of a competent professional.

The problem for the design and build contractor is to what extent the liability of professionals employed by it differs from its own. If it does differ, then there is a mismatch of liabilities owed by the contractor to the employer, and that owed by professional specialist to the contractor. Where the JCT Standard Form of Building Contract with contractor's design, 1998 is used, the design liability of the contractor is limited. Condition 22.5.1 states that it is the same as that of an architect or other appropriate professional designer. The position is similar where the JCT 98 with contractor's designed supplement is used. The ICE contracts also limit the design liability of the contractor. Normally, a fitness for purpose obligation will not be implied into a professional's contract (see for example the case of *George Hawkins* above). Even where there is no implied term that the design should be fit for its purpose, the contractor, in order to discharge the obligation to use reasonable care and skill, has still to carry out the design in such a way that it is fit for any purpose made known to it by the employer.

Consultants' agreements with design and build contractors

With the increase in design and build contracts it is becoming quite commonplace for the architect or engineer to be engaged by the contractor or by a specialist sub-contractor. The problems that can arise within this complex relationship have surfaced in a number of cases. How is the mis-match mentioned earlier demonstrated and what is the role of expert evidence in these types of disputed liability? Not only has the relationship between design and workmanship been considered in a number of cases, but also the limit of other duties. How does the role of the professional differ when providing pre-contract advice, working drawings, and supervisory services to the contractor in such a role?

PSC Freyssinet v. *Byrne Brothers (Formwork) Limited* (1997) 15-CLD-O9-24 illustrate the difficulty of assigning design liability. The court had to decide where the liability for a faulty design lay. The main contractor was employed under a design and build contract. It sub-contracted the design and construction of two car park superstructures at Lakeside, Thurrock to a specialist sub-contractor Byrne Brothers (Byrne). They in turn appointed consulting engineer Ridd Wood Partnership (RWP) to design and co-ordinate the construction of the car parks. The design and fixing of the ducts, anchors, stands and associated reinforcement for it, and post tension and grout tendons within the post tensioned beams was sub-contracted to PSC Freyssinet (PSC). The construction began in May 1993, and sometime later cracking was discovered in the beams and columns. After the work was remedied PSC in December 1993, claimed payment from Byrne for sums due under their contract including the costs of remedial work to the beams and columns. Byrne defence was to deny that PSC was entitled to any money. The breaches of contract due to the cracking in the beams and the columns had cost them an additional £500,000s to correct. They counter-claimed for these sums. HHJ Wilcox QC, Official Referee, tried a number of these preliminary issues in December 1996 including the allocation of liability for the defective work.

Byrne had, after prolonged negotiation, formally appointmented RWP as their design consultants for the project. Their brief included 'design, development and detailed structural engineering' as well as 'co-ordination and checking of the post tension spine beam system'. They however, made it clear that while they were in favour of appointing PSC, they themselves had no knowledge or expertise in pre-stressing. In accepting the role of lead designers and structural engineers, they accepted the obligation to co-ordinate all technical matters in accordance with the then British Standard 8110. In this role they had to examine the methods of construction and check its buildability. Note that RWP had no one in-house who could check the post-tension system and neither did they employ an outside specialist consulting engineer with the expertise in this type of work.

As far as PSC was concerned, they were appointed as specialist sub-contractors 'to design, supply and fix ducts, anchors, strand and associated reinforcement tensioning and groutings'. They were appointed as specialist sub-contractors with limited design responsibilities and did not receive full details of the whole structure or a detailed design of the columns. Their role and expertise lay mainly in (i) the manufacture or design of pre-stressed anchorage units (ii) fitting and installation of these in the beams and (iii) operating the jacking process to tension them.

To carry out the post tension work on the beams constructed by Byrne, PSC specified a minimum concrete strength (a C40 mix) but left to them the precise mix of cement and type of aggregate. They also gave further advice on the actual strength ($33Nm^3$) required before tensioning could take place, the striking of the

form work and the need for back propping of the beams during construction.
The case for was based on the following points:

1 PSC had to design the whole beam.
2 PSC assumed duties that went beyond their contractual obligations.
3 A term should be implied into the contract with PSC that the design of the
   works should be reasonably fit for their intended purpose.

The judge made a number of findings as follows:

1 PSC did not design the beams. They only considered the beams in relation to
   a sub- frame and not the whole structure.
2 Where they performed duties beyond their contractual obligation, they owed a
   duty in tort concurrent with their contractual duty to exercise reasonable care
   and skill appropriate to a specialist post-tensioning engineer.
3 There was no implied term that PSC should have designed the whole of the
   beam to ensure it was reasonably fit for its purpose.

As far as their case against RWP was concerned, Byrne argued that RWP (and
PSC) had not designed the beams with reasonable care and skill because PSC, as
designers of the beam, and RWP, as checkers of the design or alternatively as
designers of the beam:

(a) failed to consider adequately the effects of early thermal movements;
(b) failed to consider their method of construction in relation to the construction
     sequence in the design of the works.

Neither RWP nor PSC took account of early thermal movement in their calculations or design. However the difficulty for Byrne was that in order to prove their
case, they relied on an expert to show that a reasonably competent designer/engi-
neer exercising proper skill and care should have been put on notice as to early
thermal movement and its effects. There was plenty of evidence in the literature
and research papers. Unfortunately for Byrne, he failed to identify thermal movement as the cause of the cracking in his first report (this destroyed his credibility
in the eyes of the judge). IHHJ Wilcox observed that 'there were aspects of his
approach that troubled me . . . He was apt to condemn those who disagreed with
his view when overlooking a tenable alternative explanation to his. In detail there
was a want of care in the preparation of his report.'
    By contrast, RWP's expert argued that there was nothing in the literature that
early thermal movement was a phenomenon that a competent engineer should
have taken into account. All the experts agreed that BS8110, giving recommendations for the structural use of concrete in buildings was the standard that
should have been used. BS8110 refers expressly to elastic deformation, shrinkage
and creep but not to early thermal movement. Giving judgement he held further
that:

1 In deciding whether PWP/PSC were negligent in the design of the beams it was necessary to determine whether they acted in accordance with practice accepted as proper by a reasonable body of like professionals skilled in that particular role (note that this the application of the *Bolam* state of the art defence mentioned above).

2 Neither party was negligent in not taking early thermal movement into account in the design. A conscientious consultant and competent consulting engineer would not have taken that into account in the spring of 1993.

Design versus workmanship

The case of *Cliffe (Holdings) Ltd* v. *Parkman Buck Ltd and Edmund Thompson Wildrinton and others* (1997) 14-CLD-07-04 involved the employment of architects and engineers as sub-contractors and sub sub-contractors. The developer employed the contractor under a JCT Design and Build contract. The development consisted of four blocks containing 28 self-contained units of varying sizes together with forecourts and service roads in Rochester, Kent for an all-in price of £1.15 million. The contractor sub-contacted its engineering and architectural requirements to consultants Parkman Buck Ltd. (PB). As they employed no architects, they in turn sub-contracted architectural services to E.G. Joyce (Joyce). They did not employ any architects themselves. Their fee was to be £5000 payable out of the £28,000 due to the consultants.

The developer withheld £220,000 from the contractor, complaining of a number of defects. Arbitration resulted with the contractor seeking to recover this sum plus interest. Later they compromised the arbitration and paid the developer its costs in defending the claim in arbitration. They then claimed from the engineer and architect, the sum withheld by the developers plus all the cost of the arbitration. They alleged breaches of contract and negligence in a number of areas with defective work. These included (i) inaccurate drawings (ii) inappropriate advice and (iii) preparing inaccurate specifications. Since the engineers had sub-contracted work to architects, the judge had first to decide a number of preliminary issues.

First was whether the engineers were liable in contract for any failure to exercise reasonable care and skill by the architects (their sub-contractor). It was held that a contractor remains liable for the work not properly performed by its sub-contractor. In failing to appoint the architect under the ACE agreement, the engineer was liable for the work sub-contracted. This is because clause 5.2 of the agreement allows an engineer with the agreement of the employer to delegate work to another specialist. The second issue was whether the architect, as a sub-contractor, owed a duty of care in tort to the main contractor. Since the sub-contractor was hired for its professional expertise, the judge held there was no reason why principles of *Hedley Byrne* should not apply.

The question to be answered was whether the architects had assumed responsibility for giving advice to the contractor in circumstances where they knew it was likely to be relied on. If it was, was it reasonably foreseeable that any advice if relied on or carelessly given might result in economic loss? It was also held that the position of a professional differs from that of an ordinary trade sub-contractor. Note here that an ordinary trade contractor does not usually provide advice, and where advice is given, it would not usually be accepted without further checks being made on its accuracy. The architects were found to owe a duty of contractual duty of reasonable care and skill to the engineers concurrent with a care in tort to the contractors. The application of these principles to the facts produced the following results:

1 An access ramp for the disabled was constructed with a slope of one in six, using the builder's measurements. Whereas the drawing produced by the architects to obtain Building Regulation approval, was one in twelve. This resulted in remedial work being necessary when the ramp was being constructed. The contractor claimed against both engineer and architect for the costs involved. In dismissing the claim the judge made a number of observations on the nature of the contractor's duties: (i) he should have checked the drawing for the slope and layout before starting work (ii) that the engineers and architects were entitled to expect that the contractor was a reasonably competent one and (iii) that he would organise his work properly, plan it and carry it out 'regularly and diligently'.

2 Glazed double doors suffered serious defects. These had been designed and supplied by a specialist manufacturer and installed by the contractor. The architect had written to the contractor commenting on the suitability of the doors, but had stressed that further advice should be sought from the supplier. The contractor's case was that the doors were unsuitable for use on an exposed site. It claimed that the architect, and hence the engineer, had a duty to investigate the propriety elements involved. Where there was a doubt about its suitability, they had to ensure that the contractor was aware of the practical consequence of such an investigation. It was held however, that there was no such duty to be implied.

3 In each of the 28 units a 50 mm screed of sand and cement was laid on the first floor. The architect's brief was to develop the contractor's specification in order to obtain building regulation approval. Unfortunately, the contractor's specification differed from the developer's specifications and the architects failed to pick-up this discrepancy. The contractor preferred insitu reinforced concrete structural topping as screed since it could employ its own operatives to carry this out. Various complaints were made about the quality of the work. Floors were out of tolerance; there were hollow areas and cracking. It was held that the engineers should have spotted the diverge between the contractor's specification and the architect's drawings. Since no permission had been sought to change the design, the engineers were held to be liable for the omission. However the departure from the specified design had led to no damage. The load bearing capacity of the floors was unaffected and the cost of the screed was less that of that of one with structural topping floors. It was also held that it was in fact poor workmanship that had caused the defects rather than a departure from the specification.

HHJ Wilcox QC, OR made some observations on the nature of a contractor's duties in a design and build contract. In such a contract '. . . the contractor may supplant the architects in some of their traditional roles as exemplified under a full RIBA agreement. In particular, involvement in the choice and the co-ordination of the specialist systems and sub-contractors, the integration of those systems, the approval of all drawings and site supervision. The result may achieve economies and render the tendering contractor more economic. It does however put a premium upon the strength of the organisation and experience of the contractor in the enhanced role it plays in such a contract. Merely by employing an architect in a restricted role, such in this case, can it expect the architect to compensate for the shortcomings of the other sub-contractors?' These views were echoed in two other recent Court of Appeal decisions.

In *J. Jarvis & Sons* the design and build contractor started work on site before planning permission was granted. The architect had been involved in the planning application on behalf of the employer. Its contract was novated to the winning tenderer. The contractor sued the architect in contract and tort for failing to advise it that planning permission had not been granted. The Court of Appeal held that where the client has relevant experience in the relevant area (and Jarvis was a very experienced contractor with particular expertise in design and build), the duty was to only give advice if it was sought. Since they had started work when they were well aware that no concluded agreement had been reached on their proposals, any advice the architects gave would have been superfluous (see Ch.2 for further discussion of the case).

In *Bellefield Computer Services and Others* v. *E Turner & Sons Ltd and Others* [2002] EWCA Civ 1823 the issue was to what extent the contractor could rely on its professional advisers. The contract was a traditional one, although the contractor effectively acted as a design and build contractor by engaging a firm of architects to assist it. The court concluded that the function of the architects was limited to producing drawings and/or instructions necessary to demonstrate the final result required by the contractor and its sub-contractors. As far as the undertaking to the local authority to provide a sufficiently safe design the court dismissed any suggestion that the architects had enlarged their duty. At para. 70 Ward LJ said:

> 'again one comes back to the position that Watkins were not supervising or inspecting architects; they were working with an experienced contractor who had, so far as the evidence showed, given Watkins no reason to think that their design intentions were not being followed on site.' (see Chapter 14 for further discussion of the case)

If then, there is no duty on the designer to check whether its design intentions are being followed during construction, to what extent can the contractor depend on the design provided by the employer's professional advisers? The case of *Co-operative Insurance Society Ltd* v. *Henry Boot Scotland Ltd and ors* [2002] EWHC 1270 (TCC) examined this issue. A contract to demolish, design and reconstruct a building in Glasgow was let under JCT 80 conditions of contract with contractor's design supplement 1981 edn., (revised July 1994). Clause 2.2.1.4 places on the contractor the 'like liability to the Employer, whether under statute or otherwise, as would an architect or, as the case may be, other appropriate professional designer holding himself out to be competent to take on work of such design . . .' Note for example that s. 13 of the Supply of Goods and Services Act 1982 implies a duty of reasonable care and skill into a contract for services. At common law there is also a duty of reasonable care and skill in a designer's contract.

The scope of work contained in the design supplement was the design and construction of the basement walls. During the excavation of the works, water and soil flooded the sub-basement excavations. Consultants employed by the employer had carried out the design. The contractor argued that its obligation was only to prepare working drawings from the design provided and not to check its feasibility. HHJ Seymour QC said at para. 68 that in his judgment clause 2.1.2 of the conditions required the contractor to develop the conceptual design provided by the engineers 'into a design capable of being constructed'. In order to complete the design the contractor was required to:

1 examine the design at the point at which it was taken over,
2 assess the assumption on which it was based,
3 form an option on whether those assumptions were valid.

On that basis someone who undertakes an obligation to complete a design begun by someone else will, before the process of completion begins, examine whether it had been prepared with reasonable care and skill. Completion requires a need to understand what has been done and judging its sufficiency. It should be noted that these comments relate to the particular wording of the contract conditions. However, they do contain a general warning to contractors accepting a pre-existing design element that they are required to complete. They might be better off limiting their liability for the suitability of that design by a suitably worded clause. One that confines responsibility to the design described in the contract, and limiting their liability for the designs prepared by third parities.

An additional argument made by the contractor in the case, was whether the site investigation report had been (i) incorporated into the contract and (ii) whether the water levels recorded in it amounted to misrepresentation. As far as incorporation was concerned this was disposed of when the court accepted that it was not mentioned in the list of 'contract documents'. Mere reference to it in a drawing was insufficient to override the 'deeming' requirement of clause 2.2.2.4. This clause requires the contractor to satisfy himself when submitting his tender that he has (i) examined the site and its surroundings and (ii) the form and nature of the site including the ground and sub-soil. Since the levels recorded in the site investigation report varied across the site, there was held to be no misrepresentation. The fact that the report contained reservations about the levels and the nature of the site was to put a contractor on notice about the risk contained in the site.

Further problems with responsibility for a pre-contract design can arise where as a condition of the contract, the contract with the designer is novated to the successful design and build contractor. Can the contractor sue the designer for pre-novation defects in the design work carried out for the employer? This matter was addressed in *Blyth and Blyth Ltd* v. *Carillion Construction Ltd* [2001] ScotCS 90. A deed of novation was entered into and the terms of the deed had to be construed by the court. The contractor failed in its claim against the designers based on breaches of contract alleged to have occurred prior to the deed of novation. Part of the claim was for inaccuracies and deficiencies in the pre-tender information. Acting on the accuracy of this information the contractor submitted a lower tender than it would have done had the designer carried out its duties with reasonable care and skill. This failure to obtain compensation for pre-tender breaches does to some extent reflect the position in the traditional contract. See *Pacific*, which confirmed that the contractor could not sue the engineer in tort for providing inaccurate information at the tender stage. The difference between the two cases being that the claim in *Blyth* was in contract not tort. See also *Aldi Stores* where a similar claim for inaccurate pre-contract advice was alleged. The case deals primarily with limitation periods, which can provide an additional difficulty for the design and build contractor. The limitation period in a collateral warranty agreed with third parties was different from the limitation in the contract with its structural engineer (discussed further in Chapter 14).

Summary

1 Professionals owe a duty of reasonable care and skill in contract and in tort: see *Henderson*.
2 The duty in tort can be excluded in their contract by an appropriately drafted disclaimer of liability.
3 The duty in tort arises from the reliance principle of *Hedley Byrne*. Where a professional makes an inaccurate statement that causes loss, this can extend to economic loss. The application of the principle requires the following criteria: a special skill, voluntarily assumed responsibility, the foreseeability of loss and reasonableness of the reliance.
4 The advantage of suing in tort is that the claimant can take advantage of the longer limitation period.
5 The test for professional negligence is *Bolam*. It requires the use of expert witnesses to assist the court in deciding whether the professional has fallen below the standard of reasonable care and skill required by that profession.
6 The duty is that of the ordinary competent professional in that particular field.
7 A professional does not warrant that the design will be fit for its intended purpose.
8 A contractor who carries out a design warrants at common law that it will be fit for its intended purpose. This duty is, however, limited by the standard forms of contract to the exercise of reasonable care and skill.
9 Where the contractor employs a professional there may well be a mismatch of liabilities.
10 A professional employed by a contractor need only give advice if asked and can assume that the contractor is competent and that it will carry out the work regularly and diligently when undertaking a design and build contract.

11 When entering into a design and build contract, the contractor may be expressly accepting liability for the accuracy of the pre-contract design.
12 The dichotomy between design and workmanship is not resolved by placing design liability onto the contractor. It merely shifts the resolving of this conflict further down the chain of contracts formed.

14   Liabilities Post-completion

Background

Commercial buildings are rarely constructed for the sole use of the employer under the construction contract. The employer may be a building developer, a public body, a speculative builder or a housing cooperative, for instance. For this reason allocating responsibility for latent defects that occur post-completion is of major importance when selling or leasing the completed building or development.

Defective work during construction

There are a number of ways in which the standard forms of contract deal with defective work occurring during construction. At their simplest these provisions allow either (i) for the removal, repair or replacement of defective work or (ii) for the keeping of the sub-standard work and a subsequent reduction in the contractor's price. A failure to remove defective material is further subject to powers given to the employer to bring the employment of the contractor to an end.

Practical completion

The issue of a certificate of practical completion triggers the defects liability period. At the end of the period a list of defects is produced that the contractor must rectify in a reasonable time.

The final certificate

With the JCT 80, the issue of the final certificate was held in *Crown Estate* to be 'conclusive evidence that all work has been properly carried out'. It acts as an *evidential bar* to the allegation that the work has not been properly carried out. This applied to both patent and latent defects. To prevent this result the solution is to use amendment 15. This is now incorporated in JCT 98. It provides that the final certificate is only conclusive of matters left to the reasonable satisfaction of the architect.

Other contracts

Outside the final certificate regime of the JCT contracts, the position is different. Whether or not the final certificate provides a *bar* and is conclusive evidence depends on the wording of the contract. Compare this position with the ICE contract, which does not have a final certificate. The issue of a certificate of substantial completion does not act as an evidential bar to allegations of defective work.

The effect of practical completion

For all practical purposes, the limitation period begins. In a simple contract the period is six years from the date of the breach. For contracts under seal the period is 12 years. The Limitation Act 1980 provides a *defence*. It enables the defendant to defend himself by raising a defence that the action is statute-barred.

How long should a building last?

At the heart of the problem of post-completion liabilities in England is the resolution of this question. What if, after completion, the building should fail in some way? Who should be responsible for the costs of making good those defects?

There are a number of problems:

1 The *caveat emptor* rule – let *the purchaser beware*. For the sale of goods, this rule has been abolished. A purchaser of a defective commercial building is subject to the rule and normally has *no rights* against the seller. Such a right does exist for dwelling houses under the Defective Premises Act 1972. Section 1 (1) provides that a person taking on work for, or in connection with, the provision of a dwelling owes a duty to the person ordering the work and to any person acquiring an interest in the dwelling. The duty is to see that the work is done in a workmanlike or 'professional manner', with proper materials and that the house should be fit for human habitation.
2 The *privity of contract* rule. It is a basic principle of English law that only parties to a contract can have rights and responsibilities under it. Third parties therefore are unable to sue the contracting parties responsible for construction of the defectively constructed building. This rule is now subject to the Contracts (Rights of Third Parties) Act 1999. The act allows parties to opt out of the rule should they choose to do so.
3 The history of post-completion liability is the story of attempts to evade the effects of the doctrine by suing in negligence instead. The added advantage was to side-step the limitations periods in contract.

The background

The tort of negligence is concerned with breach of duty to take care. To succeed in an action in the courts for negligence, a claimant (the person who suffered from the failure of the duty to take care) must prove:

(a) the defendant (person being sued) owed the plaintiff a duty of care; and
(b) the defendant was in breach of that duty; and
(c) the plaintiff has suffered loss as a result of that damage.

The legal duty of care referred to above arises independently of there being a contract. The modern law of negligence began in 1932 in *Donoghue* v. *Stevenson* [1932] AC 562, commonly known as the snail in the ginger beer bottle case. In this case the House of Lords was asked whether Mrs Donoghue (the plaintiff) could bring an action against the manufacturer of poisoned ginger beer when the product had been purchased by someone else. They held that a manufacturer who sold products which were likely to reach the consumer in the same state they left, without any possibility of any intermediate examination, owed a duty to the consumer to take reasonable care to prevent injury.

In this case Lord Aitkin formulated the neighbourhood principle so as to test whether a duty of care existed: 'You must take reasonable care to avoid acts or omissions which you can reasonably foresee would be likely to injure your neighbour.' Who then are these neighbours? 'People so closely and directly affected by my acts that I ought reasonably to have them in contemplation as being affected.'

The test created by *Donoghue* v. *Stevenson* allowed judges to assess whether a duty of care was owed in *new* and *novel* situations. This test has been refined in many cases since then. In deciding whether a duty of care was owed, the court will consider the following factors which are discussed further in Chapter 13:

1 Was the harm caused reasonably foreseeable?
2 Was there proximity of relationship between the parties?
3 Would it be just and reasonable to impose a duty?

Throughout the 1960s and into the 1970s there was a rapid expansion of negligence, particularly in construction cases. In 1964 the principle of *Donoghue* was extended to *statements* which were given negligently: see *Hedley Byrne*. This is now one of the most important cases in English law. See Chapter 13 for a fuller account of the principle in relation to the liabilities of construction professionals and contractors.

The next major extension of Lord Aitkin's test in *Donoghue* v. *Stevenson* was in *Dutton* v. *Bognor Regis Urban District Council* [1972] 1 All ER 462. A house was built on a rubbish tip and Mrs Sally Dutton was the second owner of the house. The walls and ceilings developed cracks, the staircase slipped and the doors and windows would not close. She sued the builder (with whom she settled before the hearing) and the local authority.

The Court of Appeal found in her favour and held the following:

1 The local authority, through its building inspector, owed a duty of care to Mrs Dutton to ensure that the inspection of the foundations of the house was properly carried out and that the foundations were adequate.
2 The local authority were liable to Mrs Dutton for the damage caused by the breach of duty of their building inspector in failing to carry out a proper inspection of the foundations.

Lord Denning MR said: 'I should have thought that the inspector ought to have subsequent purchasers in mind when he was inspecting the foundations. He ought to have realised that, if he was negligent, they might suffer damage.'

Two other important points arise from this case:

1 In order to hold the local authority liable, the court had to hold that the builder too was liable in negligence. It had always been settled law before this that the builder owed no liability in negligence to future owners for negligently constructing a building. See *Bottomly* v. *Bannister* (1932) and *Otto* v. *Bromley* (1936).

2 The court of Appeal had in fact extended *Donoghue*. That case held that a manufacturer was liable if the product caused injury to persons and their property. The ginger beer bottle was the product. Here the court was saying that you could sue for a *product* which was defective (i.e., the damaged house in *Dutton*).

In 1978 *Anns* v. *London Borough of Merton* [1978] AC 728 came before the House of Lords. It was alleged that in 1962 a builder had erected a block of maisonettes with defective foundations. A house on the site containing a large cellar had been demolished and not properly filled and subsequently foundations did not go down to the depths of the approved plans. The plaintiffs had taken long leases on some of the maisonettes. In February 1972 writs were issued against the builder and the local authority. The writ claimed that the builder had negligently constructed the building. The writ against the local authority claimed that it had negligently carried out its powers of inspection under the by-laws in that it had allowed the contractors to build foundations in breach of them, resulting in underpinning and super-structure repairs.

The case was taken to the House of Lords on what is called a preliminary issue. Instead of having a full trial, the Lords were asked to state what the law was, given certain facts. The local authority here argued that Anns could not bring an action because they were outside the limitation period. In any event the Lords agreed that action was not barred by the limitation period. In the course of this action the House of Lords also agreed to hear argument on the liability of the local authority under the by-laws.

The decision in *Anns*

It attempted to lay down a principle by which a court could decide whether a duty of care in tort would be owed in a new or novel situation not covered by existing case law. This became known as the two-stage test:

1 The plaintiff must first establish that the defendant ought reasonably to have foreseen that the defendant's conduct might cause the plaintiff some loss or damage.
2 If loss or damage was caused, it was up to the defendant to show that for some reason of policy he was immune from liability.

More specifically it was held that the local authority in carrying out its building inspection functions under the by-laws owed a duty to *future* owners or occupiers to ensure that the property would not become a danger to the health and safety of the occupiers. If they failed in this duty, then the owner and occupier would be able to recover damages based on the cost of restoring the property to a state in which it was *no longer dangerous*. The House of Lords laid great stress on the need for a present or imminent danger to the health and safety of the occupier. In order to arrive at this decision the House of Lords had to define the liability of the builder in such a situation. It was felt that it would be unjust to saddle the local authority with responsibility while the builder escaped liability. It therefore held that the builder owed a duty of care to subsequent owners and occupiers and that, in addition, he would be liable for breach of statutory duty if the building did not comply with the *by-laws*.

Duncan-Wallace, prescient as ever, writing before the case went to the House of Lords, said: 'Should the House so treat it, therefore, *Anns* v. *Walcroft Property Co Ltd.* . . . may live to rank as perhaps the most important decision in the law of tortious negligence since *Donoghue* v. *Stevenson*' (1986, p.21).

The effect of *Anns*

There were three principal effects of this case:

1 It increased litigation since an owner/occupier of a defective building could sue not only the builder and local authority but also all the other parties involved in the construction.
2 It opened up the option of suing both in contract and in tort.
3 It made available the *longer* limitation periods in tort to owners of defective premises.

The development and extension of the law of tort probably reached its climax in the House of Lords in *Junior Books* v. *Veitchi Co. Ltd* [1982] 3 WLR 477. This was an appeal from Scotland. Junior Books entered into a contract with a company to construct a factory. Specialist flooring sub-contractors Veitchi were nominated to carry out the flooring work. Two years after completion, the floor developed cracks in the surface, and the owners were faced with the prospect of continual maintenance costs to keep the floor usable. The owners brought an action against Veitchi, claiming the floor was defective because it had been laid negligently. As damages, they claimed (i) the cost of repairing the floor, (ii) the economic loss of moving machinery, closing the factory and the payment of wages and overheads and (iii) loss of profit during the period of replacement of the floor. Veitchi in return claimed that they had no cause of action in tort because the defective floor was not a danger to the health and safety of any person or likely to cause any danger to any other property of the owners.

Veitchi was found liable and appealed to the House of Lords against the judgment. The Court of Sessions accepted the owners' argument that Veitchi owed them a duty of care in negligence and had breached that duty.

The House of Lords, by a majority, dismissed the appeal for the following reasons:

1 Where the relationship between the producer of the faulty work and owner was sufficiently close, a duty of care was owed to prevent harm being done by faulty work and this extended to the cost of repairs.
2 Veitchi had been nominated as a specialist, so the relationship was so close as almost to amount to a contract.
3 Veitchi knew that the owners relied on their skill and experience to lay a proper floor.

4 The damage was a direct result of their negligence in laying a defective floor.
5 They were liable for all costs of relaying the floor including loss of profits.

This decision led to enormous difficulties for the construction industry. What after all is so special or unusual about an employer engaging a contractor who in turn engages sub-contractors? And what is so special about the process of nomination?

By 1983 architects, engineers, contractors and sub-contractors were at risk from claims in negligence from a fairly wide range of potential plaintiffs. These included persons with whom they were in contract, such as developers, but also subsequent owners, including tenants and sub-tenants. The criticism of this state of affairs started to mount, particularly from the construction professions.

It must be realised that English law has never allowed recovery of damages in tort without physical damages. However, after *Anns* it was possible to sue in negligence for *building defects only*. *Anns* confirmed that it was possible to sue not only the contractor but also the local authority who had (negligently) approved the work. *Anns* was not really about defective work; it was about *when limitation* periods start. Purchasers and tenants needed to sue to recover the costs of remedying defects in their building but had no contractual rights. Contractual parties whose limitation periods in contract had expired and employers who wanted to sue sub-contractors where the main contractor was insolvent, also needed the remedy provided by the law of tort.

The advantages of suing in tort

Closely allied to the issue of liability for post-completion defects is the English law of *limitation of actions* and the doctrine of *privity of contract*. Thus an review of the earlier cases on liability in tort were concerned with the *start* of the limitation period rather than the substantive issue of whether liability existed at all. Limitation is the name given by lawyers to the rules that restrict the period within which court actions must be started. Failure to start an action within time limits does not in theory stop an action being brought. What it does is to allow the party being sued to plead the defence that the claim is barred by statute.

In the Law Reform Committee's *Final Report on Limitations of Actions*, published in 1977, the important reasons for the existence of such rules were summarized:

(a) to protect defendants against stale claims;
(b) to encourage plaintiffs to institute proceedings without unreasonable delay and thus enable actions to be tried at the time then the recollections of witnesses were still clear;
(c) to enable a person to feel confident, after the lapse of a certain period of time, that an incident which might have led to a claim against him is finally closed.

The current law is substantially governed by the Limitation Act 1980. The most important sections dealing with construction contracts are:

Section 5:   in contract an action on a simple contract must be brought within *six years* of the breach of contract.

Section 8:   in a contract under seal the period is *12 years*.

Section 2:   in tort the period is *6 years* from the date the cause of action accrued.

Section 32: there is also a provision postponing the limitation period where there has been fraud, concealment or mistake.

The problem with limitation

The difficulty with these rules is identifying when actionable damage occurs in a building or structure. In contract this is not a particular difficulty since in the construction industry time starts to run on completion of the project, so the issue of a certificate of practical or substantial completion normally signals the starting date. No breaches of contract are likely to occur after that date. In the law of tort, however, the law of limitation went through a long period of change and uncertainty. This was accelerated by the growth of claims for defects in tort that started in the 1970s. As a result of the cases such as *Dutton* and *Anns*, which held that the builder and the local authority could be held liable to successive owners of property for negligent constructing and inspecting of buildings, the problem of *when actionable damage occurred* assumed greater importance.

In *Spartam-Souter* v. *Town and Country Developments (Essex) Ltd* (1976) 3 BLR 72, [1976] QB 858, the Court of Appeal held that the six-year period in tort does not 'begin to run until the plaintiff discovers, or ought with reasonable diligence to have discovered the damage'. This decision was overturned by the decision of the House of Lords in *Pirelli General Cable Works Ltd* v. *Oscar Faber and Partners* (1982) 21 BLR 99 in 1983. In *Pirelli* the plaintiffs had a 160 foot high chimney built at their factory. The exterior was made of pre-cast concrete and the interior clad in a refractory material to protect the outside from the high temperatures inside the chimney. The work was completed in July 1969. Cracks developed in the refractory material, which then spread to the pre-cast concrete casing. These defects resulted in the chimney being partially demolished and rebuilt.

The chimney had been designed and constructed by a nominated sub-contractor who at the trial was in liquidation. At trial it was agreed that damage had started to *occur* no later than April 1970. The plaintiffs discovered the damage in 1977 and issued a writ in 1978.

At trial it was accepted that although the nominated sub-contractor had undertaken the design, the defendant (as consulting engineers) had accepted responsibility for it. The limitation period having expired in contract, the defendants brought their action in tort for carrying out negligent design.

The defendants argued that the decision in *Spartam-Souter* was wrong. It their view there were three possible dates when the cause of actions occurred:

(a) when the plaintiffs first decided to act on their advice and install the chimney; or
(b) the date on which the chimney was completed; or
(c) the date on which the cracks first occurred.

All the resulting dates were six years *before* the writ was issued. As a result the plaintiffs were out of time for their action.

The House of Lords found itself unable to agree with the decision of the Court of Appeal in *Spartam-Souter*. It held therefore that: 'The plaintiff's *cause of action* will not accrue until *damage occurs*, which will commonly consists of cracks coming into existence as a result of the defects even though the cracks may be undiscovered or undiscoverable.'

The plaintiffs lost their case because it was outside the limitation period. The House conceded that the position was unsatisfactory since the time could start to run without the plaintiff was being aware of it. They recommended statuary intervention in amending the law. Parliament responded by passing the Latent Damage Act 1986.

The problems created by Pirelli

On the surface the limitation period in tort is the same as in contract (i.e., six years). Prior to *Pirelli* time started to run when the damage was *discovered*. Pirelli created a problem in that it was difficult to identify when *time* started to run. In addition time could have started to run long before the damage was discovered. This resulted, for example, in the plaintiff in *Pirelli* itself being out of time when they discovered the damage to their chimney. What is even more important to realise is that the limitation period in tort could be significantly *longer* than in contract. Using the principle established by *Anns*, third parties could sue for damage discovered (thus effectively evading the privity rules in contract). For these parties, establishing *when damage first occurred* was central to their case. This required *expert evidence*. In practice, establishing the actual date of damage in a building can be quite difficult.

A good example of the problems created is *Kensington Area Health Authority* v. *Adams Holden & Partners and Ano* (1984). The judge had to consider movement in the artificial stone mullions which were part of the extensions to Westminster Hospital. He took the view that damage occurred when movement started in the mullions because they were no longer safe and not keeping the building waterproof. He went on to find that if the movement was to constitute damage it had to be more than slight movement, and he was able from the evidence to establish a date when substantial movement first occurred. See also the difficulty in establishing the date of damage illustrated by *London Borough of Bromley* v. *Rush* v. *Tomkins* (1985) 34 BLR 94.

The Latent Damage Act 1986

The Act was passed to correct what was felt to be the injustice created by *Pirelli*. Most commentaries on the act agree it applies only to negligent actions even though the act does not define negligence anywhere in it. The scheme of the act is that negligence claims become statute-barred through one of the following:

1 Six years from the date on which the cause of action accrued, the date to be established according to the old rules in *Pirelli*
2 Three years from the date when the plaintiff knew he had an action: the 'starting date' or whichever of these happened first.

3 Subject to a *long-stop* of *15 years* from the date of the negligence complained of, that ended all claims.
4 The Act confirmed that *third parties* had the *right* to sue under it.

Successive owners

The act confers rights of successive owners. *Perry* v. *Tendring District Council* (1984) 30 BLR 118 had held that the successive owner of property already latently damaged on being sold had no right of action. This was because he had no interest in the property when the damage occurred. Section 3 of the act cures this gap in the law; the effect of the section is that a person who acquires an interest in the property after the damage occurs but before knowledge of it has arisen is given fresh cause of action. The new owner is put in the same position as the person who owned the property at the time the damage occurred. The extension runs for 3 years from discovery of the damage, subject to the long-stop.

The retrenchment

By 1985 a retrenchment in the law of tort had started. In *Peabody Donation Fund Governors* v. *Sir Lindsay Parkinson* [1985] AC 210, it was held that a local authority owed no duty of care to a developer to prevent pure economic loss. The two-stage test in *Anns* was also criticised. It was said by the court that it was not a universal test for negligence. *Anns* received widespread criticism in successive cases before the case of *D & F Estates* finally arrived before the House of Lords in 1988. In addition, the decision in *Junior Books* had been restricted by the courts to the particular facts of that case.

The problem with the Act

The basis of the Act was the right to sue for defects created by *Anns*. This right was abolished in 1990 by a combination of two cases. These were the decisions of the House of Lords in *D & F Estates* v. *Church Commissioners* [1988] also in *Murphy* v. *Brentwood District Council* [1990].

The liability of the builder

D & F Estates v. *Church Commissioners* [1989] AC 177 concerned defective plastering carried out by sub-contractors. Wates Ltd built a block of flats in the early 1960s on land owned by the Church Commissioners. The plastering work was sub-contracted to a firm called Hitchens (described by the Court of Appeal as not worth suing). When the flats were complete the Commissioners let one of them to the plaintiff company (wholly owned by a Mr and Mrs Tilman) who went to live in the flat. In 1980, while they were away on holiday and the flat had been decorated, some plaster fell from the ceiling. The decorators then discovered that much of the plaster on both walls and ceilings was loose. All loose plaster was hacked off and replaced at a cost of £10,000. Within a few months the plaintiffs had started a legal action to recover their loss. Before the case came to trial, further investigations revealed more defective plaster in both walls and ceiling. The estimated cost of replacing this was put at £50,000. This figure, together with a figure representing rent while repairs were carried out, was added to the claim.

It was a difficult claim to frame in the law of tort because a contractor has no liability for the torts of his independent contractors. The plaintiff therefore argued that Wates owed them a *duty of care* in the selection of their sub-contractors. The plaintiffs succeeded in full before an official referee; the Court of Appeal confirmed the £10,000, but set aside the rest of the award. On further appeal, the House of Lords unanimously held that the plaintiffs were only entitled to £50 for cleaning the carpets. This was damage to other property. The rest of the claim was *pure economic* loss.

It ruled that Wates Ltd could not be held responsible for the default of its sub-contractors; and, in any case, the tort of negligence did not permit recovery of the losses the plaintiffs had suffered.

Basis of the decision in D & F Estates

There were two main speeches given by Lord Bridge and Lord Oliver. Both were at pains to define the nature of losses for which a negligent builder could be liable in tort. Lord Bridge regarded the builder as the same as a manufacturer of products. His starting point was *Donoghue* v. *Stevenson*. He concluded that a manufacturer or builder would only be liable in tort for negligence where his defective product or building caused either personal injury or damage to property other than the defective property itself. Damage to the product itself must be pure financial loss (since its only effect would be to reduce the product's value). In his view such an action could only be brought in contract, not tort. He saw no reason to depart from his view purely because a property was dangerous. If the owner incurred expense in making it safe before it could harm persons or other property, his loss was purely a financial one. This finding, it must be noted, was the exact opposite of the decision in *Anns*.

Lord Oliver was rather more concerned with *Anns*. From his speech it appears that he would have liked to overrule it and all the defective building cases based on it. He pointed out that 'a builder of a house or other structure is liable at common law only where actual damage either to person or property results from carelessness on his part in the course of construction'. *Anns*, he argued, 'could not be treated as authority for anything other than making a builder liable for repairing a dangerous defect'.

Both judges floated the idea of complex structures in which it could be said that part of a building could cause damage to other parts. In such a case the application of *Donoghue* v. *Stevenson* might result in the plaintiff recovering all his costs of repair.

Murphy v. *Brentwood District Council* [1991] AC 398 came before the House of Lords in July 1990. The raft foundation design for the plaintiff's house had been submitted to council for approval under s. 64 of the Public Health Act 1964. The council's building control staff were not qualified to judge the suitability of the design, so they sought the advice of independent consulting engineers, who advised that the design was suitable.

The council, acting on this advice, approved the design in 1969. The plaintiff bought the house from the developers in 1970. In 1981 it was found to have inadequate foundation overfill, and had suffered damage from differential settlement. The raft, some walls and service pipes had cracked. The plaintiff sold the house at a diminished value rather than repairing it. Both the judges at first instance and Court of Appeal held that the plaintiff had a good cause of action by virtue of the decision in *Anns* v. *London Borough of Merton*. The Council appealed to the House of Lords on the basis that *Anns* had been wrongly decided. A seven-man House of Lords assembled to hear the appeal. Lord Keith summarised the duty imposed on local authorities imposed by *Anns*.

It was a duty to take reasonable care to avoid putting a future owner of a house in a position in which he was threatened, by reason of a defect in the house, with physical injury. In order to avoid risk to personal health he would be obliged to spend money on rectifying the defect in order to continue living there.

The decision in Murphy

Murphy was an important case in that several strands of English law came together. First, the Court assembled to hear the appeal consisted of seven judges, second, the House of Lords departed from its own judgement and third, important decisions were made about the nature of the law of tort. A summary of the decision is given below:

1. The House of Lords departed from its own decision in *Anns*, the effect being that the decision no longer represented the law.
2. *Dutton* and all the cases decided on the basis of *Anns* were overruled.
3. The decision in *Donoghue* v. *Stevenson* was about defects that were latent defects (i.e., a defect that could not be detected by mere examination). A builder of premises was under a duty to take reasonable care to avoid injury through defects in the premises to persons or damage to property other than the premises itself.
4. A builder is under not liability in tort under *Donoghue* for dangerous defects if the defect was discovered before it caused any personal injury or damage to other property.
5. The Council did not argue that it owed no duty to persons who might suffer injury through a failure to take reasonable care to secure compliance with the by-laws. The House of Lords in any case reserved its opinion until such a case should arise.
6. The complex structure theory floated in *D & F Estates* was unrealistic as far as it concerned a building erected and equipped by one contractor.
7. *Anns* was about economic loss, not dangerous defects. The only right to recover economic loss not following from physical injury lay in negligent misstatement as in *Hedley Byrne*.
8. Liability in *Anns* was based on a present or imminent danger to health and safety. The cost of averting such danger was pure economic loss.
9. *Anns* did not proceed on any basis of principle and constituted a remarkable example of judges making legislation.

10 The two-stage test did not represent the law; it was better that the law of negligence should advance by increments than to open up the area by means of a test with such wide application.

11 The decision in *Junior Books* was based on the reliance principle of *Hedley Byrne*. Other cases such as *Pirelli* also came into the category.

The decision in *Bates*

The Department of the Environment v. *Thomas Bates and Son Limited* [1991] AC 499 was decided on the same day as *Murphy*. As far as the construction industry was concerned this case is probably of *greater significance* than *Murphy*. The owners of a building site granted a lease of it to lessees who contracted with Bates to build a complex including a two-storey building with a flat roof and an 11-storey tower block. In September 1971 the lessees granted an underlease to the Department of the Environment (DOE) including the tower block. During remedial work to the flat roof it was discovered that some of the concrete used was soft. Subsequently it was discovered that the concrete in the pillars was soft as well. Examination showed that the mix contained an excess of concrete and a deficiency of cement. The DOE's experts were of the opinion that the pillars were insufficiently strong to carry the design load and they were strengthened. The DOE issued a writ on October 1982 claiming, amongst other things, the cost of strengthening the pillars and cost of alternative accommodation. They pleaded negligence, as they had no contract with Bates. At trial the judge said the pillars posed no imminent danger to health or safety of either of employees or the public. The DOE appealed to the Court of Appeal, where they lost and then to the House of Lords in July 1990.

The House of Lords held that the tower block had not been unsafe by reason of the defective construction, but had merely suffered from a defect in quality. This made the plaintiff's lease less valuable since the building could not be used to its design capacity unless it was repaired. The loss suffered was pure economic loss and not recoverable in tort.

In summary these cases decided the following:

1 *D & F Estates* v. *Church Commissioners* – the building contractor is only liable in tort for:
   (a) injury to persons,
   (b) damage to other property caused by the damaged building,
   (c) claims for the *costs of defects* were claims for *economic loss* and could not be recovered in tort.

2 *Murphy* v. *Brentwood District Council* –
   (a) The local authority could not be sued for building defects because that was economic loss.
   (b) *Anns* and the cases based on it were no longer good law. The House refused to consider the position of the local authorities for claims for injury to persons. The law should therefore be clarified when such a case happens.

Before dealing with the response to these decisions by the construction industry it is useful to reflect that this was a return to the orthodox position. Humphrey Lloyd QC, then practising at the bar, described the decision in *Murphy* as the removal of the *last remedy for shoddy work*.

In *Donoghue* v. *Stevenson* Lord Buckmaster delivered a perceptive dissenting judgment when he pointed out at p.577 that:

> 'The principle contended for must be this: that the manufacturer, or indeed the repairer, of any article, apart entirely from contract, owes a duty to any person by whom the article is lawfully used to see that it has been carefully constructed. All rights in contract must be excluded from consideration of this principle; such contractual rights as may exist in successive steps from the original manufacturer down to the ultimate purchaser are *ex hypothesi* immaterial. Nor can the doctrine be confined to cases where inspection is difficult or impossible to introduce. This conception is simply to misapply to tort [a] doctrine applicable to sale and purchase. The principle of tort lies completely outside the region where such considerations apply, and where the duty, if it exists, must extend to every person who, in lawful circumstances, uses the article made. There can be no special duty attaching to the manufacture of food apart from that implied by contract or imposed by statute. If such a duty exists it seems to me it must cover the construction of every article, and I cannot see any reason why it should not apply to the construction of a house. If one step, why not fifty? Yet if a house be, as it sometimes is, negligently built, and in consequence of that negligence the ceiling falls and injures the occupier or anyone else, no action against the builder exist according to English law, although I believe a right did exists according to the law of Babylon.'

In *Young & Marten* [1969], Lord Pearce had to consider whether the manufacturer of defective tiles owed a duty to employer in the absence of a contract. Having considered the possibilities in tort, he concluded that it was *not possible* to extend the principle in *Donoghue* further than it had gone until then.

It must also be remembered that English law had always required the presence of physical damage in order to award economic loss. A clear illustration of the principle is given in *Star Village Tavern* v. *Nield* (1979) 71 DLR 3d 439. The defendant negligently collided with a bridge across a river, resulting in it being closed for repairs for a month. The plaintiff was the owner of a public house on the far side of the river. Due to the closure customers had to travel 15 miles to use it, instead of under two miles. The plaintiff sued to recover his loss of profit but failed since he only suffered economic loss. The same principle was applied in *Spartan Steel* v. *Martin & Co* [1972] 3 All ER 557 CA. The contractor negligently cut the power cable supplying electricity to the plaintiff's factory where steel alloys were being manufactured. A 'melt' in progress at the time of the power cut had to have oxygen added to stop it solidifying. This reduced the value of the product and represented a further loss of profit had the melt been completed. The plaintiff claimed the value of the damaged melt and profit they would have made had further melts been processed during the time the power was cut.

Held: Allowing the appeal, the plaintiff could only recover for physical damage to the melt in progress plus the loss of profit. They could not recover the profit they would have made during the time the power was off.

Damage to other property

After *D & F Estates* and *Murphy* the extent of the limits on suing in tort arose in a number of cases. If the contractor was not liable to third parties for defective construction, what was the liability of the architect for defective design? Another arose within the context of latent defects causing damage to other property. Completed buildings rarely collapse suddenly, and if routine maintenance is carried out, the damage is usually discovered before the defect becomes dangerous. Now the essence of *Donoghue* v. *Stevenson* was that the consumer was given no opportunity to examine the product before it was consumed, the bottle in that case being opaque. What happens then, when a defect was examined before it became dangerous: is the defect then patent or latent?

Both issues arose in *Baxall Securities Ltd and Norbain SDC Ltd* v. *Sheard Walshaw Parts and others* [2002] EWCA Civ 09.

The claimants (Baxall) jointly occupied an industrial unit in Manchester. One was the tenant, manufacturing security systems; the other distributed the systems as an associated company and occupied the building under a licence. Under the lease the tenant was responsible for patent defects and the landlord for latent defects arising out of the construction of the work.

The building itself was built for light industrial use. It consisted of a concrete floor with walls and roof supported by a steel frame. The walls had brick facing up to mid-height with metal cladding above. The roof was also clad in metal. Its design consisted of twin pitches running lengthwise separated at the inner eaves by a valley gutter. There was no seal between the lip of the gutter and the metal roof because it was impractical to provide one. The design allowed storm water to drain into the valley gutter and into the perimeter gutters. In effect the building was a large shed with a small office for the occupants along one side of the building.

Before the claimants went into occupation they appointed building surveyors to inspect and report on the condition of the property. After the claimants went into occupation, flooding occurred through the roof on two separate occasions. It was caused by the drainage system being unable to cope with heavy rain. Damage was caused to their property and they brought an action against the architects, engineers, contractor and two specialist sub-contractors. Having no contract with those parties the action was brought in tort for negligence.

The experts agreed that the valley gutter had a fundamental flaw in not making provision for overflows. Wedge-shaped cuts known as weir overflows were impossible because the ends of the gutters were set against the metal girders forming part of the steel frame. A system that could have been used was upstanding pipes at intervals along the gutter. Technically it was possible to provide a system to dispose of the storm water. The issue was whether the defendants were responsible for the defect and whether a duty was owed to later occupiers in tort for the costs of the damage caused by the floods.

By the time the matter came to trial all the other defendants had dropped out and the action continued against the architects, Sheard Walshaw partnerships (SWP). HHJ Bowsher QC decided that in the absence of authority he had to decide the matter from first principles. At para. 96 he observed:

'In *Murphy* v. *Brentwood DC* the House of Lords took the 50th step feared by Lord Buckmaster and declared that a duty rested on builders toward subsequent occupiers with whom they did not have a contract. Should I now take the 51st and find the architects owe a duty in similar circumstances?'

He then concluded that the architects owed a duty of care to future occupiers in respect of latent defects that they could not have discovered by inspection. On the facts he held that the first flood resulted from blockages in the drainage system and could have been discovered by inspection. The second flood resulted from a defects in design, which the claimant could not have discovered by inspection. He awarded agreed damages of £612,153.02 with interest. The architects appealed on the basis that the defect was patent not latent, and that the judge was wrong to conclude that there was a shortfall in the design of the gutter. In allowing their appeal Ward LJ gave the only speech (with which Hale and Brooke LJJ concurred). Giving judgment, he made a number of observations.

1 The principle of *Donoghue* v. *Stevenson* only applied where the defect is hidden.
2 At para. 46 he said:

'The concept of a latent defect is not a difficult one. It means a concealed flaw. What is a flaw? It is the actual defect in the workmanship or design, not the danger presented by the defect. (A good example of the distinction was contained in *Nitrigin Eireann Teoranta* v. *Inco Alloys* [1992] 1 WLR 498.) In my judgement, it must be a defect that would not have been discovered following the nature of the inspection that the defendant might reasonably anticipate the article would be subjected to.'

3 At para. 53 he summarised the position where in the normal course of events a surveyor is engaged to survey a building for the prospective purchaser. If with due diligence the surveyor could have discovered the defect, 'that defect is patent whether or not a surveyor is in fact engaged, if engaged, whether or not the surveyor performs his task competently'.

The Court of Appeal held that the *opportunity to inspect* broke the chain of causation. Although the architect owed a duty to future owners, on the facts of this case they were not liable for the damage.

The same issues arose in *Bellefield Computer Services and ors* v. *E Turner & Sons Ltd and Others* [2002] EWCA Civ 1823. A fire broke out in storage area of a Unigate dairy. It spread quickly and caused extensive damage to the both buildings and equipment. The works had been constructed under a JCT standard contract, 1975 revision, between Turners and Unigate Western. Although the contract was a traditional kind, Turners effectively acted as a design and build contractor by the following actions.

1 Taking responsibility for work of design required for the factory.
2 Engaging as sub-contractors the architects, H. D. Watkins and Associates (HDW) for architectural and design services. In the High Court the judge found their role was what he called 'responsive' in nature. They produced drawings as and when required by Turners.
3 Employing specialist sub-contractors to design and install compartment walls and fire lining to comply with the relevant building regulations.

Unigate carried out design and supervisory activities 'in-house'. Relations on site were excellent and the dairy was completed in time, within budget and opened in June 1983. By the time of the fire on 7 March 1995, Unigate Western was not in occupation.

Unigate UK instituted proceedings against Turners for breach of duty and negligence. The allegation against them was that the fire compartment walls, which separated the first floor laboratory and office area, had not been properly constructed and had allowed the passage of fire and smoke across the fire compartment wall on gridline 2, causing damage to the laboratory and office areas. The fire then spread across the fire compartment wall on gridline 3 and had caused widespread damage in the production area. In short the allegation was not defective design but *poor workmanship*. On this preliminary matter, Turners were held liable only in respect of the damage caused by the fire and smoke to the equipment installed in the dairy building. Because their claim against the architects was statute-barred, Turners joined the architects in part 20 proceedings, seeking a contribution under s. 1 of the Civil Liability (Contribution) Act 1978. The judge dismissed the claim and Turners appealed against the findings claiming:

(a) the judge was wrong in law to conclude that the architect's duties in regard to fire protection was limited by their retainer to a purely responsive nature (i.e., producing drawings); and
(b) the judge should have held that an architect producing designs to satisfy the local authority that it conforms to Building Regulations owes a duty to subsequent owners. The duty was to ensure that the design was sufficiently complete, so that the building could be constructed in a manner which would safeguard the health and safety of those who subsequently occupy it.

In the Court of Appeal Potter LJ outlined the broad principles of the scope of the architect's duties:

1 The architect may in certain circumstances owe a duty of care and be liable to subsequent owners of the building which he has designed or the construction which he has supervised in respect of latent defects of which there was no reasonable possibility of inspection.
2 Whether he does depends upon the original design and/or supervisory duties placed upon him. No duty of care will be owed by the architect for a latent defect where he had no design or supervisory responsibility.

3   If a dangerous defect arises due to a negligent omission on the part of the architect, the fact that he delegated the duty to design to a third party will not
excuse him, unless of course he has obtained the permission of his employer to
do so.

4   The detailed duties of the architect in relation to his design duties function
depends on:

   (a)   the application of the general principles outlined above; and

   (b)   evidence of experts as what competent experienced architects would do in
such circumstances.

He concluded that the function of Watkins as architects was limited to producing
drawings and/or instructions necessary to demonstrate the final result required by
Turner and its sub-contractors. As regards the undertaking to the local authority
to provide a sufficiently safe design, he dismissed any suggestion that they had
enlarged their duty. At para. 70 he said, 'again one comes back to the position that
Watkins were not supervising or inspecting architects; they were working with an
experienced contractor who had, so far as the evidence showed, given Watkins no
reason to think that their design intentions were not being carried out.'

The contractual response

Clearly rights to sue for defects were of great importance to all parties to construction
and property-related contracts. These cases reinforced the *distinction* between
contract and tort and the importance of contractual rights. Contractual rights were
created by the use of *collateral warranties*. They flourished in the new climate ranging
from simple 'duty of care' letters to 40-page formal contracts. The basis of a collateral
warranty is that it created an ancillary contract alongside the main contracts. One
party to these contracts agrees that if the contract were transferred to a third party,
they would be owed the same duties. The essential purpose of a collateral warranty
is to enable third parties to sue for economic loss (i.e., the costs of repairs resulting
from a latent defect found after taking possession after a sale or lease of the property).
It replaces the rights previously enjoyed under tort. Since the mid-1980s the use of
collateral warranties has been growing in the UK. Its growth was accelerated in the
wake of the decisions of the House of Lords in *D & F Estates* and later *Murphy*.

What is a collateral warranty?

It is a contract which exists alongside the main contract, sub-contract or design
contract. An example would be a JCT 98 contract between an employer and
contractor. The invitation to tender would require the successful bidder to enter
into another agreement, in the form of a collateral warranty with another party
(e.g., the funding institution). A draft agreement would be contained in the invitation to tender and would thus form a contractual term if the tenderer were
successful in its bid. It is possible to obtain a collateral warranty after the award of
the contract (if no agreement was included in the invitation to tender). There is,
however, no way of insisting that the contractor do so after the award of the
contract. The contractor can, however, agree provided the employer funds the

insurance cover or where the employer has the commercial muscle to insist on one. The invitation to tender should also contain a provision that a collateral warranty is obtained from all sub-contractors and suppliers.

The consequences of a collateral warranty

Below are listed the consequences of a collateral warranty:

1 A contractual link is created between parties who would not otherwise have one. Examples are the contractor and funding institution, or the tenant and the designer of the building.
2 The funding institution could sue the contractor/designer directly for breaches of contract.
3 It bypasses the doctrine of privity of contract that only parties to a contract can enjoy the benefits of and be subject to the liabilities of a contract.
4 Rights to sue in tort are no longer as important as before.

The rise of collateral warranties

The demand for collateral warranties arose out of a number of disparate sources, as outlined below:

1 Developments in property finance created new classes of lenders. This brought about new attitudes of mind and approaches. One of its governing maxims was that everything should be done to protect the lender and provide the necessary tools to realise his security if necessary.
2 The lesson learnt during the property slump of the mid-1970s was applied. What was recognised was that for a lender faced with a borrower's default over a property development, the better option was to 'build out' of the difficulties rather than sell for whatever return the lender could get for it in its uncompleted state.
3 There was an increasing use of sophisticated and complicated leases with the object of obtaining a 'clear rent'. To do this the lease has to allocate liability to the tenant by ensuring the rent will be paid without any liability for the cost of repairs, maintenance or rebuilding. Often this was achieved by placing liability directly on the tenant by a suitably worded repairing covenant or by the use of service charges. There is a substantial body of law on the interpretation of repairing covenants in leases. Quite often it is hard to decide whether the defect was an inherent design fault, a construction fault or merely repairs required to maintain the building in its original state.
4 Developments in the law of tort. Certainly *Junior Books* (1982) can be regarded as a high point of the advantages suing for the cost of repairs in tort. Subsequent developments showed this 'protection' to be illusory.

Who benefits from a collateral warranty?

The list of those seeking to benefit includes anyone likely to have any form of interest in the completed development The financier is one of best known of the prospective beneficiaries. He appears in various guises:

- providing finance to the developer
- mortgagee of the ultimate tenant
- mortgagee of the purchaser of the freehold.

Tenants, sub-tenants and other end users (such as licensees) represent another class of beneficiaries.

Who provides collateral warranties?

Accepting that the source of any defect or deficiency in a buildings may be (i) the product of faulty design, (ii) materials used or poor workmanship or, more likely, (iii) a combination of these, the class of providers can be wide. The building *contractor* is likely to be first in the queue. A design and build contract is likely to contain more onerous warranties than a traditional contractual arrangement using, say, the *JCT 98*. Sub-contractors are likely to carry their own direct responsibilities, especially nominated specialist sub-contractors with design responsibilities. Members of the professional team are other obvious providers. Their significance is increased greatly by the fact that they carry professional indemnity policies. The list may extend to suppliers, especially those providing items of critical importance such as lifts or escalators.

Provisions in collateral warranties

The contents of collateral warranties vary greatly and the documents can consist of one page or more than 40. There are a number of standard forms produced but lawyers for specific projects draft their own. Typical provisions are given below:

1 The provider gives various assurances as to the performance of his obligations in relation to the development. The wording may vary but the purpose is to give the beneficiary a contractual right against the provider if the promise is broken.
2 In warranties given by the contractors and some consultants there may be assurances that certain deleterious materials will not be specified or used in the development. A list will often be given in the warranty.
3 Other provisions have nothing to do with collateral warranties as such. Thus a financier may want a licence from members of the professional team to use their drawings and calculations if he wishes to continue the development in his own name following the developer's default.
4 A tenant or a purchaser of the completed development may also want the right to use consultants' drawings and calculations for the purpose of future maintenance or even for the building of an extension.
5 The collateral warranty may also have a requirement for appropriate insurance policies to be entered into by the provider, and for these to be maintained for the period of the warranty. If giving a warranty, the provider must check against his existing insurance policy. Entering into it may be a 'material' chance, thus accidentally making the policy void.

Some problems

Everyone who enters a collateral warranty naturally wants to ensure that his liability is restricted to his 'contribution' to the defect or deficiency giving rise to the claim:

1 Each provider will therefore want to ensure that the liabilities he takes on do not go beyond the liabilities imposed on him by his original contract.
2 Beneficiaries will of course want an indemnity against 'all losses' suffered as a result of the defect or deficiency. This may open the door to claims for economic or consequential loss.
3 Beneficiaries will also request warranties that the development will be fit for a particular purpose. This poses a special problem for the providers at the time of the giving of the warranty, since they have no idea how in practice the building may be used.

Assignment or novation of collateral warranties

Assignment

The general principle is that the *benefit* of a contract can be assigned. From the view of the contractor the benefit is to receive payment for the work. The right to receive the contract sum can be assigned *without* obtaining the permission of the employer. Where the contractor assigns either (i) the right to receive sums due or (ii) retention monies, the employer's permission is not needed. The contractor gives notice to the employer that this has been done. If the employer ignores the assignment and pays the contractor, it is liable to the assignee. If the contract prohibits such an assignment, the notice is invalid and the assignee cannot enforce a claim for such monies against the employer.

An assignment can be legal or equitable. Where it is legal the assignee can sue in its own name. If equitable the assignee sues in the name of the assignor as third party. Section 136 (1) of the Law of Property Act 1925 requires an assignment to be absolute and in writing. Express notice in writing of the assignment must be given to the debtor (the employer under the contract). The assignment of retention money or instalments due at the date of the assignment is absolute and within the section. For an equitable assignment there must be a clear intention to assign the debt but the formal requirement of s. 136 has not been met.

The contractor assignment cannot assign the *burden* of the contract. The contractor's liability to compete the work cannot be assigned without the *permission* of the employer. Contractors often sub-contract work. This is called vicarious performance of the work and the contractor remains liable for the sub-contractor's performance. Standard form contracts contain complex provisions prohibiting assignment and sub-contracting without the permission of the employer.

Most collateral warranties allow for at least one assignment. The real question for the drafter is what damages would be available to the party who received the transfer. It is a basic rule of assignment that the assignee of a warranty could only recover the same damages that the assignor would receive. This rule was tested in a number of cases that followed in the 1990s. These cases have in themselves little to do with the assignment of collateral warranties. However, they *clarified* the nature of the *damages* that could be recovered following the assignment of contractual rights. A number of rules emerged, as shown below:

1 Contracts and contractual rights can be assigned. A building can be assigned and, with it, rights to sue under the contract under which it was built. *Linden Gardens Trust Ltd* v. *Lanesta Sludge Disposals Ltd*; *St Martin's Property Corporation Ltd and another* v. *Sir Robert McAlpine* [1993] 3 All ER 417 decided that the *employer* could assign both a right to *future* performance of a building contract and the right to claim damages for *past* breaches of it. The assignment can take place after the transfer of the building. Note that these two cases were heard at the same time by the House of Lords.

2 A provision under a clause in a contract requiring permission to assign is effective: see *Linden Gardens*. Failure to seek permission is a breach of contract. The clause in this case was clause 19 of JC 80, which read: 'neither the employer nor the contractor can assign this contract without permission'.

3 However, in *St Martin's*, the House of Lords held that where the assignment was ineffective, the employer could sue the contractor on behalf of the new owner despite suffering no loss. There was an *exception* in the law, otherwise the person with the defective building would have no remedy. The difficulty, of course, was, how could they be made to sue?

4 Where there is valid assignment, the assignee is entitled to be placed in the same position as the assignor. This leads to the *no loss* argument. The 'no loss' argument is that where the building was sold for its full value, the assignor had suffered no loss and hence the assignee could not recover any damages either. This 'technical objection' was rejected in *St Martin's* by the House of Lords. An *exception* was created. Where parties to a building contract expect the building to be transferred (and the assignment is invalid), justice requires the original employer to sue on behalf of the third party.

5 How far does this extend? In *Darlington* v. *Wiltshier Northern Ltd* (1994) 69 BLR 1 this question was answered. When the employer refuses to sue, what can the third party do? The employer becomes *constructive trustee* of the right to claim damages. As trustee the employer *must* enforce those rights and is liable to a third party if it refuses.

6 It is important to realise that in *St Martin's* the companies involved in the transaction were sister companies, and therefore the prospect of asking one to sue on behalf of the other created no real problems for the parties concerned. But what if the contract specifically excluded any responsibility for defects discovered: how then to make the assignor sue? This was the situation in *Darlington*. In fact the assignor had expressly provided in its contract that it would have no liability for any defects arising. The Court of Appeal resolved it by the *constructive trustee* device above. This effectively sidesteps the whole concept of privity of contract, a view supported by Brownsword who regards *Darlington* as another departure from the strictness of the privity of contract rule (2000, p.133).

7 The decision in *Linden Gardens* was based on two grounds:

(a) the narrow ground: the exception to the rule that the plaintiff can only recover its own losses. The House adopted *Dunlop* v. *Lambert*, a carriage of goods by sea case, where the property was in transit and ownership transferred to a third party.

(b) the wide ground: Lord Griffith wanted to do away with privity of contract altogether where contracts were made for the benefit of third parties.

8   In *Alfred McAlpine Construction Ltd* v. *Panatown Ltd* [2001] 1 AC 518, the House of Lords considered both grounds. The contractor undertook the construction of an office and multi-storey car park under a design and build contract. Panatown was the employer under the building contract. In order to save money on VAT, its associated company, Unex Investment Properties Ltd (UIPL), was the owner and developer of the site. As part of the contractual arrangement McAlpine entered into a collateral warranty with UIPL. Panatown alleged serious defects that would require the demolition and rebuilding of the building. McApine's defence was that Panatown had never been the owner of the site and hence had suffered no loss. UIPL had suffered the loss but were not parties to the building contract. The House of Lords held:

(a) The *Albezero* exception recognised in *Linden Gardens* did not apply because on the facts. The employer under the building contract did not need to sue on behalf of UIPL because it already had a direct right of action under the collateral warranty.

(b) Panatown had also argued it should be allowed to sue under the wide ground. A party who contracts to receive certain services should be able to recover substantial damages if the services provided fall short of it. (For its 'performance interests' see the speeches of Lord Goff and Lord Millet.) This approach was rejected by the majority of the House.

Note that the law of assignment of leases is very different from that of assignment of contractual rights. Lord Browne-Wilkinson in *St Martin's* commented on this in saying:

'The law became settled that an assignment in breach of covenant gave rise to forfeiture, but pending forfeiture the term vested in the assignee. In contrast, the development of the law affecting the assignment of contractual rights was wholly different. It started from exactly the opposite position *viz.* contractual rights were personal and not assignable. Only gradually did the law permitting assignment develop. In is therefore not surprising that the law applicable to the assignment of contractual rights differs from that applicable to the assignment of leases.'

The assignment of collateral warranties

This issue came to the fore in *Allied Carpets Group Plc* v. *The Whicheloe Macfarlane Partnership* (*a Firm*) NCN [2002] EWHC 115 (TCC). The claimant, Allied, had purchased the lease to a warehouse from the liquidator Harris Queensway, a large national chain. It found many faults with the warehouse and wanted to sue the builders and the professionals involved in the design and construction. Collateral warranties under a deed had been entered into but the firm of engineers and structural designers denied that they were under any liability since it was never assigned to the claimant. Allied claimed to be successors in title to Harris as owners of the benefit of the warranty. The problem for them was that there had been no express assignment of the benefit of the warranty. In the sale and purchase agreement, a list of warranties had been attached to schedules listing the collateral warranties. HHJ Bowsher QC decided that there had been no express assignment of the warranty, and the handing over of the document containing it did not act as an assignment of the warranty.

The judge decided that in the light of his findings, he need not consider the effect of an assignment made in breach of warranty. *Lanesta Sludge* would have applied if there had been (i) a covenant against the assignment and (ii) an assignment in breach of that covenant. The effect would have been that the assignment was ineffective. As to the commercial effect of such warranties and provisions for assignment, he said at para. 44:

> 'The extent of such warranties is a matter for negotiation. In the absence of consideration, the warranty is normally required to be made by deed and hence has a twelve-year limitation period. The warrantor firstly has no commercial interest in giving a warranty at all. If the commercial pressures (such as getting the work or not getting it) persuade him to give a warranty the warrantor's commercial interest is in favour of making it non-assignable. On the other hand, the commercial interest of the beneficiary of first warranty is in favour of the warranty being assignable because that will add to the sale value of the property. These conflicting commercial interests are resolved by negotiation.'

In *Northern & Shell Plc* v. *John Laing Construction Ltd* [2003] EWCA 1035 a claim under a deed of warranty was held to be statute barred. The arguments centred on whether the relevant date was (i) the date of practical completion, in which case it was statute barred or (ii) whether it ran from the date when it was executed, and in that case it was not. On the proper construction of the contract, it was held that it commenced on the practical completion. The tenant was unable to recover the costs of repairing the building defects. The question of limitation was at issue too in *Aldi Stores Ltd* v. *Holmes Building Ltd* [2003] EWCA Civ 1882. Holmes, the design and build contractor, entered into a contract with a developer to complete two super-stores. These it leased to Aldi and B&Q on 25 year leases. As a condition of the contract, the contractor was required to enter collateral warranties by deed with both tenants. The contractor employed a structural engineer to advise it on its preparatory work and in the submission of its tender. After completion, severe cracking occurred in the buildings. This was caused by differential settlement, resulting from the stores being constructed on four lagoons infilled with chemical sludge. The sludge was separated by bunds and these it appears made settlement inevitable.

Aldi claimed against the contractor under its warranty and the contractor served a Part 20 claim on the structural engineer for an indemnity equal to its damages to Aldi. It claimed that negligent advice had been given in relation to the ground improvement work. Relying on that advice, the contractor had designed the scheme, which was then incorporated into the contract. Almost a year later, B&Q also issued proceedings against the contractor, making in substance, the same claims that Aldi had. The contractor in turned served a Part 20 claim on the structural engineer, but this time the engineer pleaded a limitation defence. Whereas the contractor was under a collateral warranty with a 12 year limitation period, the engineer was limited to a 6 year period in contract and a limitation period in tort depending on the date when it became liable.

Section 14A of the Limitation Act 1980 (inserted by the Latent Damage Act 1986) states that the limitation period is whichever is the later of (i) six years from the accrual of the action or (ii) three years from the date of knowledge. The issue in the High Court and subsequently in the Court of Appeal was whether the contractor could amend its Part 20 claim in the Aldi action. It wanted to add claims made against it made by B&Q and the assignee of the benefit of the collateral warranty (the current landlord), as additional heads of damage in the Aldi claim. As Auld LJ put it in the Court of Appeal 'appeal raises a point as to the circumstances in which party A should be permitted to amend its claim against party B to introduce a claim which was already being made by A against B in other pending proceedings, where . . . it is said the amendment will or may deprive B of a limitation defence available to it'. In allowing the appeal, the court held that the amendment would not add new cause of action and did not add new claims against the engineer.

The original criticism of *Anns* type of negligent liability was that it was distortion of the limitation periods. Parties enter contracts on the basis that their liability is limited by statute to either 6 or 12 years and part of that risk will be reflected in the overall contract price. If they knew that that period could be extended for an unlimited period they might well want a much higher price for the proposed work. A consequence of applying the law of tort to building defects was to extend liability for much longer periods. *Pirelli* led a clarification of the limitation periods for latent defects in buildings and structures. *D&F Estates* and then *Murphy* ended the right to sue for the cost of building defects using the law of tort. Collateral warranties now provide the protection against those defects that third parties require. But it seems that limitation periods are still a major problem for parties to construction contracts.

The Contracts (Benefit of Third Parties) Act 1999

The Act provides that a third party may enforce a term in a contract if:

(a) it expressly states that the third party may enforce the term; or
(b) the term purports to enforce a benefit on him.

The Law Commission in its report outlining the argument for reform of the third party rule identified the particular difficulties for construction contracts: see (Law Commission, www.lawcom.gov.uk/library/lc242/part-3.htm). It stated at 3.18: 'Our proposed reform would enable contracting parties to avoid the need for collateral warranties by simply laying down third party rights in the main contract. Moreover, our proposed reforms would enable the contracting parties to mirror the terms in existing collateral warranties.' The industry has not taken up this invitation and the response has been to expressly exclude the application of the Contracts (Rights of Third Parties) Act 1999. The Act enables parties to a contract to avoid the restriction of privity rules if they choose to do so.

Section 1 (1):
> 'Subject to the provisions of this Act, a person who is not a party to a contract (a third party) may in his own right enforce a term in the contract if –
> (a) the contract provides that he may, or
> (b) subject to subsection (2) below, the term purports to confer a benefit on him.'

Section 1 (2):
> 'Subsection (1) (b) does not apply 'if on the proper construction of contract it appears that the parties did not intend the term to be enforceable by the third party.'

To take advantage of the Act, the parties need to insert in their contract a provision which makes a *class* of persons or beneficiaries a party to the contract. For example a clause allowing prospective purchasers to take advantage of the clause might read: 'The terms of this contract, which are enforceable by [employer] may also be enforced by its purchasers, tenants, sub-tenants, assignees, successors etc.' This will satisfy s.1 (3), which provides: 'The third party must be expressly identified in the contract by name, as a member of a class or as answering to a particular description but need not be in existence when the contract is entered into.' To avoid this result the *ICE 7th* states in clause 3 (2): 'Nothing in this contract confers or purports to confer on a third party any benefit or right to enforce any term of the contract.' The guidance notes merely state that by this clause 'the acquisition of rights by third parties are excluded'. Eggleston (2001) supports this exclusion since it avoids the possibility of:

*   contractors suing the employers professional team (and the reverse)
*   employers suing sub-contractors (and the reverse)
*   sub-contractors suing each other
*   future owners of the works suing firms and individuals involved in the construction process.

So perhaps it is not goodbye to collateral warranties after all as it may be best to keep to the known, using collateral warranties rather than risking the uncertainties of the Act. Eggleston's views probably reflect the commonly held views of the construction industry. See also amendment 4:2002 to the *JCT 98* that also excludes the application of the Act. Furmston (2001) by contrast welcomes the Act, although he notes that to date its practical effect has been a proliferation of clauses designed by cautious lawyers to exclude it. He has, however, no doubt that, in the long run, it will have major effect on how English lawyers view contract law.

The main uncertainties

'Purport'

It is unclear what the words 'purport to confer a benefit' mean. The Law Commission said that the drafting of the section was 'not intended to 'recast the decision in *White* v. *Jones* [1995] 2 AC 307, by bringing the claims of disappointed beneficiaries under negligently drafted wills into our proposed act'. Furmston (2001) observes that this has happened under the American version of the common law. Successful actions have been brought against lawyers in similar circumstances. Doubts such as these means that parties will not want to leave their contractual provisions to the uncertainty of later case law.

Variation and cancellation

Construction contracts make provision for the variations. They also make provision for the termination or cancellation of the contact. See ICE clause 40, where a suspension of the works for more than three months can trigger the contractor's right to terminate its employment under the contract. The Act takes an intermediate position between (i) the rights of the parties once they have created rights, to take them away and (ii) that once the contract is made, third party rights should not be *disturbed* as described below in Section 2(1) which provides:

'Subject to the provisions of this section, where a third party has a right under section 1 to enforce a term of the contract, the parties to the contract may not, by agreement, rescind the contract, or vary it in such a way as to extinguish or alter his entitlement under that right, without his consent, if –
(a) the third party has communicated his assent to the term to the promisor,
(b) the promisor is aware that the third party has relied on the term, or
(c) the promisor can reasonably be expected to have foreseen that the third party would rely on the term and the third party has in fact relied on it.'

The view of the Law Commission was that the rights of the third party crystallise and the third party can refuse to have his rights removed. To avoid crystallisation the parties need express words in their contract to allow them to rescind or vary the contract without the consent of third parties.

Defences

Section 1(6) provides that where a contract term excluded or limits liability the third party rights are limited in the same way. This section provides a shield to the original party on being sued by a third party.

Novation

This is a procedure where both the benefit and burden of the warranty is passed on to another party. To do this the agreement of all three parties are required. In *Darlington* the parties could have avoided all the subsequent litigation had they chosen to novate the contract instead of assigning it. Novation be can used where two companies merge and they wish to novate existing contracts to the new company. Design contracts are sometimes novated where the client wishes to the main contract to be a 'design and build' contract: see *Blyth* in chapter 13 for the difficulties this may cause.

The European dimension

The major problem in creating a common market lies in the harmonisation of laws in Europe. To what extent do national laws inhibit the creation of a level playing field? Post-completion liabilities reflect this dilemma in a very practical manner.

English law views a contract as a *bargain* between the parties. Part of that bargain is that the limitation period is 6 years for a simple contract and 12 years for a contract under seal. If parties wish to stipulate a longer period they should do so. The development of the remedy in tort was simply a distortion of the fundamental rules of a contract under English law.

The rules across the European Union are equally diverse. Only France has a comprehensive system of post-completion liability. In 1988 the European Parliament initiated a study for the unification of liabilities in the European construction industry. Research into post-completion liabilities was undertaken by the Groupe des Associations des Professionelles Européennes de la construction, or GAIPEC.

Their recommendations were as follows:

1 A formal process of 'acceptance' would replace 'practical completion' of the work. Both the employer and the contractor would have the right to initiate the process. The contractor also has a *right* to do any repairs required.
2 The limitation period would be reduced to either 5 or 10 years.
3 A compulsory warranty scheme would come into operation on acceptance underlain by a compulsory insurance policy. The policy would operate on proof of material damage. It would not depend on fault.
4 The warranty would be automatically assigned on sale of the building.

To implement the GAIPEC proposals a *Directive* would be needed. This would be followed in England by a bill to implement it. So far the recommendations have been shelved because further research is needed.

Summary

1 Remedies for defectively constructed buildings post-completion are complicated by recent developments in the case law. There are three different periods.
2 Period 1, pre-1988. The decision of the House of Lords in *Anns* created the means of suing for the costs of repairs in tort using negligence based on *Donoghue* v. *Stevenson*.
3 Period 2, post 1988. The decisions in *D & F Estates* and *Murphy* overruled *Anns* and held that the builder was only responsible for injury to persons and damage to other property. The costs of repairs were classed as economic loss that could only be recovered in contract.

4 Period 3, post-2000. The Contracts (Rights of Third Parties) Act 1999 allows contracting parties to expressly provide that a third party can enforce the contact.
5 *Murphy* extended the principle of *Donoghue* from chattels to buildings. The contractor is responsible for the damage caused by the defective building. This principle has been extended to construction professionals,
6 The construction industry has rejected the Contracts (Rights of Third Parties) Act 1999 in favour of continuing the use of collateral warranties.
7 A collateral warranty provides contractual rights to third parties and allows for the recovery of the cost of repairs post-completion.
8 Contractual rights can also be assigned. In *Linden Gardens* the House of Lords created an exemption to the rule that where the assignment is invalid, the original employer could recover substantial damages on behalf of the assignee.
9 It is simpler to novate a contract where liabilities can be specifically assigned.
10 It is likely that in time construction contracts will adopt the Contracts (Benefit of Third Parties) Act 1999, since its application is likely to prove easier and cheaper than collateral warranties.

15 Determination and Damages

Breach and damages

Standard forms of contract make provision for the paying of damages where one party is in breach. This may be a provision for the payment of LAD for delay in completing the work. In *Phillips Hong Kong Ltd* v. *The Attorney General of Hong Kong* (1993) 61 BLR 41, the Privy Council observed that 'in building contracts . . . parties should know with reasonable certainty what their liability *is* under the contract'. In turn the employer may undertake to pay for any loss and expense arising from its own delay. Despite these types of provision the usual remedy for breach of contract is the payment of *damages*, which is monetary compensation intended to put the injured party *in the position he would have been in*, had the contract been carried out. There are situations where damages would not be an *adequate* remedy. In such a case the court may order the equitable remedies of an order for *specific performance* or grant an *injunction*. Such remedies are *discretionary* compared to the common law remedy of damages. Injunctions are not easily granted in construction contracts: see for example, *Hounslow BC* v. *Twickenham Gardens Development Ltd* [1970] 3 All ER 326. The contractor refused to vacate the site on the determination of its employment under the contract by the employer. The court in turn refused an application for the grant of an injunction ordering the contractor to do so.

Rescission

A further remedy which the innocent party has is a right to *rescind* the contract ('to cut off, take away, remove, sever, annul, to abrogate', according to the *SOED*). *Rescission* is the right to bring the contract to an end *in addition* to claiming damages. The effect of *rescission* is, as far it is possible, to restore the parties to their original position. This is not likely to be possible in a construction contract since once work has started it may be impossible to restore the parties to their original position. In any case, rescission is now recognised as being a *restitutionary* remedy.

   Rescission can also be distinguished from:

(a) a right to *repudiate* the contract at common law; and
(b) the right to effect the determination of the contractor's employment under a provision of the contract.

Such determination clauses are also commonly known as *forfeiture clauses* as they entitle one party to bring the other's employment under the contract to an end in certain defined circumstances. They may also provide for the employer to use the contractor's equipment to complete the work.

Repudiation

Lawyers and textbook writers *use rescission in two senses*. The first sense of rescind is to bring the contract to an end as from the very beginning. The contract is rescinded *ab initio*. In those cases the contract is treated in law as never having come into existence at all. The parties must be restored to their original positions before the contract was entered into. Burrows (1993) calls this a self-help remedy but stresses that the innocent party must give notice to the other party that he is wiping away the contract. In *Car and Universal Finance Co Ltd* v. *Caldwell* [1965] 1 QB 525, a cheque for the sale of a car was dishonoured the next day. It was held that immediately notifying the police and also asking a motoring organisation for help in tracing the car was sufficient to effect rescission.

There are several *bars* or limits to the *right to rescind*. The innocent party loses the right if it has affirmed the contract (indicating that it will honour the contract in spite of the breach), a lapse of time, if the parties cannot be restored to their original positions or if third parties have acquired rights under the contract.

Rescission has been described as effecting the *unwinding or unscrambling* of a contract that is prima facie binding. Such a contract has been *defectively formed*. Contracts can be unwound for negligent or innocent misrepresentation, fraudulent misrepresentation, mistake or undue influence (McMeel, 1996).

The second sense of rescission is where there is 'an accepted repudiatory breach of a contract that has come into existence but [this] has been put to an end, or discharged': see *Johnson* v. *Agnew* [1979] 1 All ER 883. McMeel (1996) claims that this is a purely *contractual remedy* of termination for breach. The innocent party can put an end to the contract because of a fundamental failure or breach of a fundamental term by the other party. This remedy is retrospective only and is combined with a right to damages. Termination is purely concerned with *defective* performance of a contract.

Furmston by contrast states that the correct sense of rescission is the retrospective cancellation of contract *ab initio* where, for example, one party is guilty of fraudulent misrepresentation. In such a case the contract is destroyed as if it never existed. Discharge by breach never impinges on upon rights and obligations that have already matured. He prefers in this case to use the word termination or discharge rather than rescission (1996).

The discussion above clearly illustrates the complications involved in the terminology for discharge of contracts. Standard forms of construction contracts provide *express rights* to terminate as part of the contract. These may be much wider than the position at common law. An example in the *JCT 98* is clause 27 which gives power to the contractor whereby it may bring its employment to an end 'if the employer does not pay a certificate within 14 days'. This is called a *determination* clause. The contract also makes express provision for what is to happen when the contractor's employment is so determined under the contract. In providing expressly for what is to happen in certain events, there is no need to distinguish between whether a term in a contract is (i) a condition or (ii) a warranty.

Conditions and warranties

The modern approach to distinguishing conditions and warranties is to ask 'Has the innocent party been *substantially deprived* of the whole benefit of the contract?' Note that lawyers use the word warranty in a number of senses. The *SOED* defines a warranty (the act of guaranteeing) as follows:

(a) a covenant (promise in a contract) in the conveyance of land that the seller warrants the title to the land conveyed;
(b) an undertaking by one party that he/she will be answerable for the truth of some statement incidental to the contract: especially an assurance by the seller of goods that he will be answerable for some quality possessed by them;
(c) in a contract of insurance, the engagement (formal promise) of the assured that certain statements are true.

Furmston (1996) considers that the law takes two approaches to the circumstances in which a party may terminate a contract:

1 By classifying the term broken into conditions and warranties. This is the approach taken in contracts for the sales of goods.
2 In analysing how *serious* the breach of contract was. Two possibilities exists:
   (a) one party behaves in such away that it is clear that the contract is being *repudiated*;
   (b) the work is so defectively performed it can be treated as repudiation (sometimes called fundamental breach of contract).

Fundamental breach

At one time the courts classified terms in this manner to escape the effect of exclusion clauses. In *Photo Production*, the Court of Appeal held that there had been a fundamental breach of contract by Securicor and that they could not rely on the exclusion clause. The House of Lords disagreed; deciding that whether the breach was fundamental depended on whether it went to the root of the contract.

The following are examples:

1 In a contract for the delivery of goods by instalments, defects in one delivery would not be fundamental.
2 Failure by the employer to pay an instalment would not normally justify rescission. It follows that in the absence of an express clause giving the contractor the right to suspend work, the contractor cannot stop work because no payment has been made. Such express rights are now contained in the Housing Grants Construction and Regeneration Act 1996.
3 *Lubenham Fidelities* v. *South Pembrokeshire DC* (1986) 33 BLR 39 illustrates clearly the difference between determination clauses and the right to repudiate at common law. The contractor abandoned the contract because of what it

considered an under-valuation of an interim certificate of payment. The employer determined the employment of the contractor under an express provision. In leaving the site it was in breach of the obligation to proceed with work regularly and diligently.

Held: No right to suspend for underpayment existed at common law. The contractor had repudiated the contract and was liable in damages to the employer. The employer in following the procedures laid down by the contract had lawfully determined the contractor's employment.

4 Delays in performance or the presence of defects are not fundamental unless made expressly so by the contract. The employer would only be justified in rescinding if it can be shown that the contractor cannot or will not comply with the contract. See *Sutcliffe*. The work was so defectively done and the delays so comprehensive that the employer lost confidence in the contractor's ability to complete the contract.

Treitel (MLR vol. 30, p.139) claims the complexity of this area of law is simply a result of the discussion of the problem being widely scattered in the books. This creates different solutions to problems which appear to be basically similar. He states that 'the conceptual apparatus that has been built up is both *formidable and confusing*.' Thus it is said that:

(a) the effects of a breach depends (at least sometimes) on whether it is 'funda-mental'; *or*
(b) 'goes to the root of the contract'; *or*
(c) whether it 'substantially' deprives a party of what he has contracted for; *or*
(d) on whether promises are 'independent' or 'concurrent'; *or*
(e) whether the performance of one party is a 'condition precedent' to the liabil-ity of the other; *or*
(f) whether the breach is one of a 'condition' or only of a 'warranty'.

Remoteness of damage

A claim for damages has two different aspects: first, the remoteness of the damage, which is being claimed, must be assessed and, second, separated from the mone-tary assessment of that compensation. English law uses the tests formulated in *Hadley* v. *Baxendale* [1854] to decide whether a particular loss in the circumstances of the case is *too remote* to be recovered There are two limbs to *the rule in Hadley and Baxendale*; damages are either:

(a) 'fairly and reasonably be considered as either *arising naturally* (i.e., from the usual course of things) from the breach; *or*
(b) such as may *be reasonably be supposed to have been in contemplation of* both parties at the time they made the contract as the probable result of breach of it.'

See also *Victoria Laundry (Windsor)* v. *Newsman's Industries* [1949] 1 All ER 997, where there was a failure to deliver a boiler which the laundry intended to use in a particularly lucrative contract.

Held: Damages for loss of normal profit was recoverable but for loss of the exceptional profit that could have been made.

In the Court of Appeal it was stated that the knowledge possessed by the parties was of two kinds: imputed or actual knowledge. This aspect was considered in *Balfour Beatty Construction (Scotland) Ltd* v. *Scottish Power Plc* (1994) 71 BLR 70. The House of Lords applied the rules in a construction setting. Attempts to construct a watertight aquaduct failed. This was due to the rupture of the fuses in a temporary supply of electricity to a site batching plant. This interrupted the required continuous pour of concrete needed to ensure the aquaduct was watertight. A claim for damages resulting from the demolition of the aquaduct and its replacement was held to be too remote under the second limb. The House of Lords held that: (i) what one party *knew* about the other's business was a *question of fact* and (ii) the demolition and reconstruction of the aquaduct was not in the contemplation of the respondent.

Note that damages were awarded under the first limb of *Hadley* v. *Baxendale*. These were the damages that arose directly from the effect of the fuses failing.

The measure of damages

The object of damages is to place the innocent party (if the damage is not too remote) in the position he would have been in had the contract been carried out. Damages recovered must relate to the loss that was made by the claimant, not the profit made by the defendant. In *Surrey County Council* v. *Bredero Homes* [1993] 1 WLR 109, surplus land was sold to the developers. They covenanted that they would carry out the scheme in accordance with the planning permission and the approved scheme. Later, they obtained fresh planning permission to build more homes than had initially been planned.

The council sued to recover the profit the developers made out of the sale of the extra houses. The Court of Appeal held that although they had broken the covenant the plaintiffs could only recover nominal damages. (The Council had in fact not lost anything).

The difficult question of what the plaintiff has lost is decided on one of two bases and the plaintiff is free to choose. The first basis is the *expectation of loss* that the plaintiff has suffered. Expectation is what the plaintiff would have received and includes the profit it would have made. Since profit may often be speculative, the plaintiff may choose instead to recover its wasted expenditure, (i.e., its *reliance loss*). An example is *Anglia Television* v. *Reed* [1971] 3 All ER 690, where the defendant, an American actor, repudiated a contract to appear in a film. Since the producers could not find a suitable replacement they had to abandon the project. The making of films is usually a speculative venture with profit difficult to predict. Anglia claimed, and was awarded, the cost of preparation such as hiring other actors, hiring a scriptwriter, scouting for suitable locations, and so on.

Equitable remedies

Injunctions

An injunction is an *order* of a court ordering the party to whom it is addressed *to do an act* or to *refrain from doing an act*. It is *an equitable remedy*. There are several types of injunction, as explained below:.

Perpetual injunction

This is granted at the trial of the action. It stops the defendant from *ever doing the act* complained of *again*. Two principles govern its issue: (i) *damages* must be an *inadequate* remedy, and (ii) the claimant must show his *rights* have been *infringed*.

Mandatory/prohibitory injunction

The principles are the same as those of prohibitory injunctions except the court must *weigh the balance of convenience* between the parties and refuse an injunction if *constant supervision* is required. In addition, the courts will only grant a mandatory injunction if it is sure the defendant knows exactly what he has to do.

Interlocutory injunction

These are granted in the *period between* the action complained of and the case coming to *trial*. *American Cyanamid Co* v. *Ethicon Ltd* [1975] AC 396 laid down the principles to be applied by the court in deciding whether an injunction should be granted. The House of Lords laid down that the correct approach was to ask the following questions:

1 Has the plaintiff shown a *serious issue* to be tried?
2 The plaintiff must show that damages would not be an adequate remedy.
3 The plaintiff must show that damages would adequately compensate the defendant for any loss suffered if the injunction were granted.
4 The injunction should be granted on the *balance of convenience*: who would be likely to suffer the greatest loss?

Specific performance

This is also an equitable remedy. In common with other equitable remedies, failure to obey an order of specific performance can be punished by imprisonment for contempt of court. Specific performance is an order of the court requiring the person to whom it is addressed to perform a contract.

The principles which govern the granting of a decree of specific performance are listed below:

1 There must be an enforceable contract.
2 Damages must not be an inadequate remedy (e.g., all land is unique, and thus damages will always be an inadequate remedy).
3 Specific performance will not be granted where performance has become illegal, impossible or futile: see *Wroth* v. *Tyler* [1983] 2 WLR 405.

4 Specific performance will not normally be granted where constant supervision is required.

Specific performance will not usually be awarded against a builder. *Chitty* (para. 28–025) suggests that there are three reasons for this:

1 Damages are an adequate remedy since another builder can be employed to carry out the work.
2 The fact that the work required has not been adequately described makes it impossible to enforce.
3 Carrying out the work would require supervision by the court.

In *Wheatley* v. *Westminster Brymco Coal Co* (1869) LR 9 Eq. 538 it was said 'this court will not undertake either the construction of a railway, the management of a brewery, or the management of a colliery'. Specific performance of contracts to *build* were ordered in *Wolverhampton Corp.* v. *Eammons* [1901] 1 QB 515 *and Carpenters Estate* v. *Davies* [1940] 1 All ER 13. In *Carpenters* the defendant sold land to the plaintiff, retaining the adjoining land and covenanting to build roads, mains and sewers on the adjoining land. This he failed to do and the plaintiff's land became worthless for development. The sale price had reflected the development value, so damages were not an adequate remedy.

Defective and incomplete work

Where the work being constructed is carried out defectively, the question that often arises is what the quantification of the loss should be. Hudson (1995 at p.1046) states that there are three possible bases for assessing damages for defective or incomplete work:

1 the cost of reinstatement;
2 the difference in cost to the builder of the actual work done and the work specified;
3 the diminution in value of the work carried out due to the breach of contract.

A standard form of contract may make provision for the difference between specified work and its costs to the builder: see Clause 8 of JCT 98. In certain situations the work which is not in accordance with the specification can be kept on site, but the price is reduced to reflect this. In many situations there will in practice be little difference between reinstatement (what it will cost to cure the defect) and the loss in value resulting from the breach. There are however, situations where there can be a significant difference between the two measures. Such an occasion was the case of *Ruxley Electronics and Construction Ltd* v. *Forsyth* (1996) 73 BLR 1 which has clarified the law relating to damages for building defects.

Late in 1986 Mr Forsyth, entered into a contract with Ruxley Electronics and Construction (Ruxley) to build a swimming pool next to his house in Kent. The contract price with extras was £18,800 (approximately) including VAT. The depth of the pool was to be 6½ feet at the deep end. Mr Forsyth then changed his mind and wanted the pool to be 7½ feet at the deep end. Mr Hall, who controlled Ruxley, agreed to do this without extra charges. At no time was there any mention of a diving board and no provision in the contract for it. Work started in June 1987 and was carried out by a sub-contractor who did the work poorly, resulting in cracks to the bottom of the pool. Mr Hall agreed to remove the existing pool and replace it free of charge. He also agreed to pay Mr Forsyth's professional charges resulting from the defective work. The new pool was finished by the end of June 1987. In November, Ruxley submitted their invoice to Mr Forsyth who insisted in a reduction of £10,000 as compensation for the disturbance suffered during the rebuilding of the pool. Mr Hall agreed reluctantly, but Mr Forsyth did not pay the balance outstanding. During March 1989 Mr Forsyth discovered that the pool was only 6½ feet at the deep end. He also complained of corrosion. In response, Mr Hall did some extra work but Forsyth still made no payment for the work. On 19 January 1990, the plaintiffs started proceedings claiming £10,300 as the balance of their account. The trial started before HHJ Diamond QC on 14 March 1993.

Mr Forsyth's defence was as set out below:

1　Since the contract was an entire one, the rule in *Cutter* v. *Powell* applied. This meant no payment was done until all the work had been done properly. *Hoenig* modified the rule. In a lump sum contract, if the contractor has substantially completed the work, he is entitled to the lump sum, less a deduction for reme-dying any defects. In deciding whether there has been substantial completion, the nature of the defects and the difference between the contract price and the costs of rectifying the defects, must be taken into account: see *Bolton*. The result there was that the contractor could not claim the contract price less the costs of making good the work because the work itself was of no use to the employer.
2　He counter-claimed for the sum of £33,800 plus VAT, which were the estimated costs of rebuilding the pool to its correct depth.

The Judge rejected these arguments. He found, as a matter of fact, that the pool was perfectly safe for diving. As a result, he was prepared to award general damages for loss of amenity (for not having 7½ feet pool to dive in), which he put at £2,500. He also awarded Mr. Forsyth £750 for general inconvenience and disturbance instead of the £10,000 that he had originally negotiated. Mr. Forsyth appealed the judgement. In the Court of Appeal Staughton and Mann LJJ disagreed with the judge's approach to the award of damages, while Dillon LJ dissented. The approach of the various judges were as follows:

1　Staughton LJ held that Mr Forsyth was entitled to the cost of reinstatement, however expensive, since there was no other way of giving him what he had contracted for. Whether he intended to rebuild the pool with his damages was irrelevant.
2　Mann LJ agreed, saying that Mr Forsyth's 'personal preference' was for a 7½ feet pool and the only way that could be satisfied was to rebuild the pool. He

conceded that there were times when it would be unreasonable to award the costs of rectifying a failed project.

The court went on to award the full costs of reinstatement and set aside the award of £2,500. Dillon LJ dissented from the approach of the majority. He considered the pool as constructed was deep enough to dive in safely. Since the pool was safe for use there was no loss in value. It was not reasonable to spend £21,560 on reinstating it. He said: 'I see no reason, therefore, why if there is no loss in value, he (Mr Ruxley) should automatically become entitled to the cost of reinstatement however high. That would be a wholly unreasonable conclusion in law'.

Ruxley appealed to the House of Lords. They held, in allowing the appeal that 'where the expenditure was out of all proportion to the benefit to be obtained, the . . . measure of damages was the diminution in value caused by the breach to the work, not the cost of reinstatement'. The Judge's award of £2,500 should be restored.

The reasoning of the House of Lords is given below:

1 Lord Lloyd of Berwick said: '. . . in the Court of Appeal, Mann LJ described the question in this case as a simple one, but one, which had, nevertheless, attracted arguments which went to the foundation of the measure of damages for breach of a contract. It is surprising, and perhaps somewhat disconcerting, that at this stage of the development of the law of damages, such a simple question should have caused such a wide diversity of judicial opinion'.
2 In *Robinson* v. *Harman* (1848) 1 Ex. 850, Parke B. said at p. 855: '[the innocent party] . . . as far as money can do it, is to be placed in the same situation . . . as if the contract had been performed'.
3 In building cases, this money loss is measured in two ways. Damages were either the difference in value of the work done or the cost of reinstatement. Where the costs of restatement are less than the difference in value, the measure of damages will invariably be the cost of reinstatement.
4 But it is not the only measure of damages. See *Jacob and Young's* v. *Kent* (1921) 129 N.E. 899, a judgment of the New York Court of Appeal. The building owner specified that the plumbing should be carried out with galvanised piping of 'Reading manufacture'. Due to a mistake, the builder used piping of a different manufacture. The builder sued for the balance of his account. The building owner counter-claimed for the cost of replacing the pipe work even though it would have meant demolishing most of the completed structure at great expense. The court held that in the circumstances of the case, the measure of damages was not the cost of replacement, which would be great, but the difference in value, which would be either nominal or nothing.

This case establishes two principles. First, the cost of reinstatement is not the appropriate measure of damages if the expenditure would be out of proportion to the good to be obtained. Second, the appropriate measure of damages in such a case is the difference in value, even though it would result in a nominal award.

These principles have been recognised in the leading English case of *East Ham Corporation* v. *Bernard Sunley and Sons Ltd* [1966] AC 406. The House of Lords, accepting Hudson's formulation as the basis for awarding damages. These rules have been followed in a number of subsequent cases including *Atkins Ltd* v. *Scott* (1980) 7 Const LJ 215.

Referring to Mann LJ 'personal preference' approach in the Court of Appeal, the House of Lords stated that the judge, at first instance, had taken this into account in his judgement. Although there was an argument for saying Ruxley was not an ordinary commercial contract, but a contract for a personal preference, such cases should not be elevated into a special category with special rules. For support of this approach see *Channel Island Ferries Limited* v. *Cenargo Navigation Limited ('the Rozel')* [1994] 2 LLR 161, in which Phillips J used it to distinguish *Ruxley* (a Court of Appeal decision) by which he was bound.

In *Ruxley* Lord Lloyd at p.388 considered the question of damages for loss of amenity in a different situation from the position in *Ruxley*:

'Suppose [a builder constructs] a new house. The building does not conform in some minor respect to the contract, as, for example where there is a difference in level between 2 rooms necessitating the building of a step. Suppose there is no measurable difference in value, and the cost of reinstatement is prohibitive. Is there any reason why the court should not award by way of damages some modest sum, not based on difference in value, but solely to compensate the buyer for his disappointed expectations? Is the law of damages so inflexible . . . *that it cannot find some middle ground in such a case*? I do not find the final answer to that question in the present case.' (emphasis added).

Application of the principles

Ruxley and the statement made by Lord Lloyd were considered in *Farley* v. *Skinner* [2001] BLR 1, HL. At issue was the recovery of non-pecuniary damages. The occupier claimed damages against a surveyor for failing to warn him about aircraft noise. A country house he wished to buy was 15 miles from Gatwick Airport and he instructed the surveyor to check the risk of aircraft noise. Relying on the advice that the house was unaffected by such noise, he went ahead with the purchase and spent a substantial amount on refurbishing it. Unfortunately the stacking beacon for flights waiting to land at the airport was directly over his house. Not wishing to move, he sued the surveyor for damages for the inconvenience and nuisance he suffered as a result of the noise. Lord Scott of Foscote said in his judgement at p.62 that this was a case with simple facts and, in his respectful opinion, with a simple answer. At p.74 he said:

'The reason why such an apparently straightforward issue has caused such division of opinion is because it has been represented as raising the question whether and when contractual damages for mental distress are available. It is highly desirable that your Lordships should resolve the present angst on this subject and avoid the need in future for relatively simple claims, such as Mr. Farley's, to travel to the appellate courts for a ruling'.

The legal basis of the recovery of damages for physical inconvenience and mental suffering arising from the breach of a contract has always been difficult. In *Watts* v. *Morrow* (1991) 54 BLR 86 the Court of Appeal sought to reverse the trend of increasing awards under this head of damage. Fenwick Elliot summed up the position before *Watts* (1993, p.200). *Jarvis* v. *Swan Tours* [1973] 1 QB 233 had established the principle that where a person is unable to live in his house, or was caused inconvenience or mental suffering in his home by reason of a breach of contract by another, damage for distress may be awarded. *Perry* v. *Sidney Phillips* [1982] 3 All ER 705 CA, confirmed that the amount should be modest. It was the escalation of the amounts that the court of Appeal sought to contain in *Watts*.

Examples

(a) £250 awarded where there was a landslip causing danger to the plaintiff's house: *Batty* v. *Metropolitan Property Realisations* [1978] 2 All ER 445.
(b) £1,000 awarded following failure of the plaintiff's bedroom floor: *Haig* v. *London Borough of Hillingdon* (1980) 19 BLR 143.
(c) £500 for delay resulting from the delay in completing the plaintiff's holiday home. He lost three weekends in the house and spent Christmas in a hotel: *Franks & Collingwood* v. *Gates* (1983) 1 CLR 21.
(d) £ 600 where the plaintiff had to put up with dust and scaffolding during building works in his house which lasted for 6 months: *Hooberman* v. *Slater Rex* (1984) 1 CLR 63.
(e) £1,500 following a negligent survey the plaintiff's had to move into his mother-in-law's house whilst repairs were carried out: *Roberts* v. *Hampson* [1989] 2 All ER 504.
(f) £2,000 awarded when rising damp repairs costing £14,700 were carried out: *Cloggre* v. *Sovro* (Scottish case, unreported).
(g) £8,000. Remedial work in a farmhouse totally disrupted family life for two years. Only two bedrooms were habitable: *Syrett* v. *Carr and Neave* (1990) CILL 619

Watts was itself a claim by the owners for damages arising out of having their holiday home turned into a building site for eight months. By the time of the trial the plaintiff's marriage had broken down, and at first instance they were awarded £4,000 damages each. The Court of Appeal reduced the sum to £750 each and made some observations as to when damages should be awarded. They emphasised that damages should be modest.

In *Farley* v. *Skinner* the House of Lords held the following:

1 The broad principles laid down by Bingham LJ in *Watts* relating to the recovery of non-pecuniary damages from a contract-breaker were confirmed.
2 Whilst 'a contract-breaker is not generally liable for any distress, frustration, anxiety, displeasure, vexation, tension or aggravation which his breach of contract may cause to the innocent party', there was an exception to this where it was a major or important object of the contract to give pleasure, relaxation and peace of mind.

3 A further exception to this general principle is that general damages can also be recovered for physical inconvenience and discomfort caused by a breach of contract.

4 Aircraft noise capable of causing physical inconvenience and stress may give rise to a claim for general damages. Noise which significantly interferes with a person's enjoyment of a property is something for which general damages can be awarded. If the cause of the inconvenience is a sensory experience (sight, touch, hearing, smell, etc.) damages can be recovered.

5 General damages under both exceptions are recoverable where there has been a contractual failure to use reasonable care and skill; it is not necessary that the contract guaranteed the achievement of a result.

6 Mr Farley did not forfeit a right to general damages by his failure to move out of the property; in deciding to stay he had acted reasonably.

7 Whilst the award of £10,000 and general damages appear high, the case was very unusual because the inconvenience and discomfort would continue. However, awards in this area should be restrained and modest.

8 *Knott* v. *Bolton* (1995) 11 Const LJ 375, was wrongly decided.

Comment

Bingham LJ in *Watts* identified two bases for awarding damages for distress and inconvenience:

1 Where the very object of the contract was to provide pleasure, relaxation, peace of mind or freedom from molestation and failed to do so.

2 Where physical inconvenience caused by the breach and mental distress directly related to that inconvenience and discomfort are foreseeable and are suffered during a period when defects are repaired.

In summary, the difference between the claims in *Ruxley* and *Watts* can be described as follows. *Ruxley* can be considered to be a case where the claimant was deprived of a contractual benefit (i.e., something of value). Whereas *Watts* was about whether or when contractual damages can be awarded for inconvenience and discomfort. Lord Lloyd discussing a minor defect in a house in *Ruxley* asked 'whether the law of damages was so inflexible . . . that it cannot find some middle ground in such a case?' *Farley* supports the proposition that an award could be made either on the basis of *Ruxley* or of *Watts*, but the award should be a modest one. As far as commercial contracts are concerned, in *Farley* Lord Hutton at para. 47 observed that the principle in *Watts* that a 'contract-breaker is not generally liable for any distress, frustration, anxiety, displeasure, vexation, tension or aggravation which his breach of contract' had particular application. He referred with approval to the observation of Lord Thornton at p.108 in *Johnson* v. *Gore Wood & Co* [2001] 2 WLR 72: 'Contract-breaking is treated as an incident of commercial life which players in the game are expected to meet with mental fortitude'. Lord Scott in *Farley* also pointed out that the distinction between commercial and other contracts was too imprecise to be satisfactory.

Ruxley was considered and applied by the Court of Session in *McClaren Murdoch*

& *Hamilton Ltd* v. *The Abercromby Motor Group Ltd* [2002] www.ballii.org. Lord Drummond Young gave the opinion in the Outer House. An architect brought a claim for outstanding fees. The defenders counter-claimed, alleging breach of contract and negligence. Under a professional contract for services, the architects had designed and supervised the construction of four car showrooms and an associated workshop. The counter-claim made was for the costs of replacing the heating system designed by the architects. The parties agreed that it had been designed negligently but disputed the loss arising from the negligent design. The heating system consisted of electrical underfloor heating acting as storage heaters. At night, using cheap rate electricity, it absorbed heating from electrical elements running through the floor. During the day the heat stored in the concrete floor was released.

The system had two flaws. Heat released during the day could not be replenished so that it was impossible to control. The system works in buildings with thick walls and small windows. Here the buildings were of lightweight construction and included large amounts of glazing. In the workshop area large doors were regularly opened so that heat was constantly being lost through them. The defenders replaced the system with a gas-fired system. Its claim included the cost of installing a mains gas supply. The architects disputed that the whole system needed replacement. The decision to do so was unreasonable since they could have obtained the same benefit by providing additional heating at a lower cost. Therefore they were not entitled to their re-instatement costs. The pursuers disagreed, relying on *Ruxley*. Their damages should be measured by the costs of reinstatement of the heating system to provide a proper one. Giving his judgement Lord Drummond Young agreed with the arguments of the pursuers. At para. 28 he held the following:

1 The pursuers were entitled to the cost of reinstating the heating system.
2 Where the contract was one to use reasonable care and skill, the innocent party was entitled to be placed in the position it would have been had the party in breach exercised reasonable care and skill: *Farley* applied.
3 In construction contracts [including professional contracts] the loss was measured in two ways:
    (a) either the cost of making the work conform to contract; or
    (b) the difference in value of the works as built and the value of the works as they should have been (*Ruxley* applied).
4 In some cases the reinstatement would not be the appropriate measure of loss: see *Ruxley*. In this case he held replacing the heating was not disproportionate to the benefit that could be obtained.

Mitigating damages

The purpose of the mitigating rule is to prevent the waste of scarce resources. The innocent party cannot claim for losses that he could otherwise have mitigated by taking reasonable steps to reduce his loss. In construction contracts the law assumes that the claimant will have the defective or incomplete work corrected or completed. Reasonableness of substitute performance was considered in *East Ham Corporation* v. *Bernard Sunley & Sons* [1966] AC 406. See also *Atkins Ltd* v. *Scott* (1980) 7 Const LJ 215 (CA). *Chitty* considers the factors 'relevant to the issue of reasonableness to include:

(a) the claimant has actually had the work done; or
(b) he undertakes to have the work done (but such an undertaking will not, on its own, make it reasonable for the claimant to have it done); or
(c) he shows a "sufficient intention" to have the work done if he receives damages on this basis: the claimant's subjective intention is irrelevant' (Beale, 1999, p.1279).

Contractual provisions for damages

Standard forms of contracts usually make provision for the contractor to be paid extra money for certain events occurring. Its purpose is to spread the risk for these events from the contractor to the employer. The employer saves costs because the contractor does not have to price these risks in advance. If they do not occur, the cost saving is made to the employer instead. Such a right to *loss and/or expense* is given by *JCT 98*, clause 26. The clause states that the contractor can make written application to the architect stating that it is likely to incur *loss and/or expense* for which it is not likely to be reimbursed under any other provisions of the contract. On receiving such application the architect shall from time to time thereafter *ascertain*, or instruct the quantity surveyor to *ascertain*, the amount of such loss and/or expense, and so on.

The phrase *loss and/or expense* has been held to mean the same as damages at common law – see *Wraight Ltd* v. *PH & T (Holdings) Ltd* (1968) 13 BLR 26. Damages at common law are there to put the innocent party in the same position it would have been had the contract been carried out. The limits of damages are contained in *Hadley* v. *Baxendale*. Loss and/or expense are usually regarded as confined to the first limb of *Hadley* v. *Baxendale*.

Matters that may entitle the contractor to direct loss and/or expense are detailed in clause 26.2, examples of which are:

(a) failure to give necessary instructions;
(b) architect's instructions issued under clauses 2.3, 13.2, 23.2;
(c) delay caused by persons engaged by the employer;
(d) failure to give ingress or egress to site.

Note the insertion of a new sub-clause inserted by amendment 18 to JCT 80. Clause 26.2.10 provides an express right to the contractor to *suspend performance* of his *obligations* under the contract. The main stipulation is that the suspension is *not frivolous or vexatious*. The meaning of these words were considered in *John Jarvis* v. *Rochdale Housing Association* (1986) 36 BLR 48, CA. It was held that 'vexatiously' meant ulterior motive, to oppress, harasses or annoy.

In *Photo Production Ltd* v. *Securicor Transport Ltd* [1980] AC 827, the House of

Lords distinguished the primary and secondary obligations of the parties. A failure to perform those primary obligations (i.e., the performance obligations in the contract) is a breach of contract. Lord Diplock said: 'The secondary obligation on the part of the contract breaker . . . is to pay monetary compensation to the other party for the loss sustained by him as a consequence of the breach.' The parties are free to allocate those *secondary obligations*. The primary obligations of the contractor are to carry out and complete the work in accordance with the conditions of the contract. In addition the contractor remains wholly responsible for carrying out and completing the works in all respects in accordance with the conditions: see, for example, clauses 1.4 and 2.1 of JCT 98. The primary obligations of the employer are to pay sums due 14 days after the issue of an interim certificate: see clause 30.1 of JCT 98. When a breach of their primary obligations occurs, a secondary obligation to pay damages arises. However, the parties may decide beforehand what is to happen in the event of a breach of their primary obligations. Such clauses can include matters which the common law would not regard as primary obligations. Examples of such provisions are found in clauses 27, 28 and 28a of *JCT 98*, called 'Provisions for Determination of contractor's employment under the contract before Practical Completion'. The present provisions were altered by amendments 11 (and 18) to JCT 80.

A number of difficulties have been created by termination clauses. See *Hounslow BC* v. *Twickenham Gardens Development Ltd* [1970] 3 All ER 326. Labour troubles and an eight-month strike resulted in delay in completing a housing development. When work resumed the architect complained of the slow rate of progress. After giving the required notices, the employer terminated the contract under the contractual provisions. The contractor refused to accept what it called the 'repudiation' of the contract, would not vacate the site and elected to carry on with the work.

Held: Refusing the employer an injunction for trespass, (a) there was an implied negative obligation on the council not to revoke the contractor's licence to the site and (b) the contract had not been validly determined since the parties disputed the facts which had given rise to the determination.

Hounslow BC has not been followed elsewhere in the Commonwealth. In *Mayfield Holdings Limited* v. *Moana Reef Limited* 1 NZLR 309 the New Zealand Supreme Court refused to follow it in a case on similar facts. On giving a valid notice the employer was entitled to possession of the site and entitled to employ others to complete the works. The court dismissed the concept of a negative implied obligation not to revoke the contractor's licence to the site. To recognise such a concept was to allow the contractor specific performance of the contract. Such a remedy was one that the court would not grant the contractor. The general rule is that specific performance will not be granted where *constant supervision* is required: see *Wheatley* v. *Westminster* (1869).

Notices of determination

Notices given under clauses 27 and 28 of (*JCT 98*), if sent by *registered* or *recorded post*, are deemed to have been received *48 hours after posting*.

Defaults by contractor which entitle employer to give notice to determine

Defaults are of two kinds. The first group deals with the actual carrying out of the work and concerns the *conduct* of the contractor in so doing. This is provided for in clause 27.2. The other kind of default deals with the practical consequences of contractor insolvency.

Clause 27.2 stipulates a number of grounds for the issuing of a notice to determine. The contractor is given 14 days to rectify these defaults. Failure to do so means that within 10 days from the expiry of the 14-day period, the employer is entitled to invoke the provisions. Should the contractor end the specified default or the employer fail to give a notice, the matter ends there. Should, however, the contractor make further default, the employer is entitled after a reasonable time has passed to determine the contractor's employment under the contract. The grounds providing for this are as shown below:

1 Where the contractor wholly or substantially suspends the carrying out of the work. Note that *Lubenham* (1986), above, was an example where the contract was terminated when the contractor suspended the work. This right should not be confused with the right of the contractor to suspend the work for non-payment in accordance with the HGCRA 96 and amendment 18.
2 Failure to proceed regularly or diligently with the work. The meaning of the phrase was considered in *West Faulkner Associates* v. *London Borough of Newham* (1994) 71 BLR 1. It was held that the phrase meant orderly planned progress towards the completion date.
3 Failure to comply with architect's instructions to remove defective work. See the code of practice for dealing with the removal of defective works in Chapter 6.
4 Breach of undertakings contained in clause 19.1 and 19.2 not to assign or sublet without consent. See further in Chapter 12 for limitations on this right.
5 A failure to comply with the Construction Design and Management regulations 1984.

The second group of defaults concerns the financial consequences of the contractor's default on financial grounds. These are detailed in Clause 27.3: before Amendment 11 to the 1980 contract, if the contractor was in financial difficulties and appointed a receiver, a provisional liquidator, or made a voluntary arrangement with its creditors, the employment of the contractor under the contract was *automatically determined*. Now, the contractor must immediately inform the employer in writing if any of these events occur. The appointment of a provisional liquidator or a winding-up order determines the employment of the contractor automatically.

Determination by the contractor

Default by the employer, which entitles the contractor to determine, is set out in clause 28.2.

1 Where the employer does not pay by the final date for payment the amount properly due to the contractor in respect of any certificate (inserted by amendment 18 to reflect s. 110 of the Housing Grants Construction and Regeneration Act 1996).

2 Interference with the issue of certificates. Since in carrying out its certifying duties the certifier has to act independently in making decisions, any interference by the employer is a serious matter. In *Hony Huat Development (Pte) Ltd* v. *Hiap Hong & Co* (2000) CILL 1787, it was held that a persistent pattern of late interim certificates does not necessarily imply that the employer is aware of this.

3 Assignment of the contract prohibited by 19.1.1

4 Suspension of the work for one month (unless specified in the Appendix) due to:

  i   failure to give necessary instructions, etc;

  ii   architect's instructions issued under clauses 2.3, 13.2, 23.2;

  iii   delay caused by persons engaged by the employer;

  iv   failure to give ingress or egress.

Note that these grounds involve issues that are at the heart of any construction contract and will always prove contentious. The contractor has to give notice to the employer specifying the events, and the employer has 14 days to end the specified defaults.

Where the employer makes a composition with its debtors, appoints a provisional liquidator or passes a winding-up resolution, the contractor may by notice determine its employment under the contract. Note that the provisions in clauses 27 and 28 make detailed arrangements for what is to happen in the event of a determination of the employment of the contractor by either party.

Determination by either employer or contractor

The grounds are set out in clause 28A and cover events which are not the fault of either party. These include *force majeure*, damage by specified perils, civil commotion, breaches by statutory undertakers, hostilities (whether war is declared or not) and terrorist activities.

The differences between the clauses

The provisions in clause 27 (with the exception of financial ones) are difficult for employer to prove, whereas the provisions in clause 28 go to the heart of the construction contract in that they cover matters that will always create dispute.

Summary

1 Construction contracts make provision for what is to happen should one party be in breach of contract.

2 These include provisions for the payment of a fixed sum for delay caused by the contractor or sums not fixed but payable by the employer in the event of specified matters delaying the contractor.

3 The terminology used by English law for dealing with breach is complex and confusing.

4 Rescission is a restitutionary remedy and is used to unwind defectively formed contracts.

5 Repudiation of the contract is a self-help remedy available to the innocent party where the other indicates it will not honour its agreement. It is concerned with the defective performance of a contract.

6 Breach of a contract entitles the innocent party to monetary compensation in order to put itself in the position it would have been in, had the contract been carried out.

7 In a construction contract this may be cost of cure or the difference in value.

8 No damages will be payable if the costs are unreasonable compared to the benefit to be gained.

9 In such a case the innocent party may be entitled to compensation because it has been deprived of a contractual benefit or suffered inconvenience; these sums will be modest.

10 Construction contracts provide termination clauses to deal expressly with defaults occurring during the work.

16 Methods of Dispute Resolution

Background

The introduction of the Housing Grants Construction and Regeneration Act 1966 Act 1996, part II (HGCRA 96) has transformed the resolving of construction disputes. Disputes arise in construction contracts for a variety of reasons. In many instances the dispute arises when one party makes an application for payment and the other disputes the value of the work carried out. Amongst the recommendations of the Latham Report was that tenders should be evaluated on quality and price. It also identified the following examples as areas of likely conflict:

Item 3.10: the lack of co-ordination between design and construction
Item 4.18: a mismatch between reasonable care and skill and fitness for purpose
Item 4.20: nomination of specialist contractors
Item 8.2: unfair practices and a lack of teamwork on site
Item 8.9: the use of unfair terms in contracts
Item 9.10: dissatisfaction with current methods of dispute resolution

The nature of disputes

Disputes arise during a construction project for a number of reasons. These include the standard of workmanship and the quality of materials, applications for extensions of time not being granted, claims for direct loss and/or expense being rejected, contractor delay and subsequent deduction of LAD or applications for payment and sometimes the meaning of contractual terms. The variety and scope of disputes in adjudication has been wider than this but disputes are primarily triggered by applications for payment.

Dealing with disputes

When a dispute arises the parties quite often spend a great deal of time in negotiations in an attempt to settle the matter. Initially, recourse to lawyers or starting legal proceedings is the least favoured option. Quite often a spirit of give and take prevails and the matter is settled. But, as the Latham Report found, many disputes are left to be settled at the end of the project. One result is that relationships on site deteriorate further after each unresolved dispute.

To deal with these disputes the Latham Report recommended the introduction of a system of *statutory adjudication* (a quick and easy method for resolving disputes during the construction process). Amongst the concerns expressed by many eminent commentators was the 'unavoidable and permanent downgrading of the role of the professional contract administrator, necessarily affecting the status of all professional architects and engineers'. This was a rather special bit of pleading seeing that those very professionals were often acting as contract administrators on contracts where there was little protection of the rights of sub-contractors.

Disputes before the introduction of adjudication

Adjudication existed as a method of dispute resolution before the introduction of statutory adjudication. The sub-contract form DOM/1 introduced it in 1976 to deal with sub-contractor concerns about the misuse of rights of set-off as excuses for non-payment. Its major problem was the status of the award made by the adjudicator, the courts refusing to recognise it as having the same status as an arbitrator's award.

Arbitration was the preferred method of dispute resolution in the standard forms of contract, especially after the decision in *Crouch* (only an arbitrator could open up and revise certificates issued by the architect or contract administrator). Litigation, before the introduction of the Arbitration Act 1996, was often seen as proving to be a quicker and cheaper alternative, especially bearing in mind the experience and expertise of the (then) Official Referees. The growth of alternative dispute resolution (ADR) was due to dissatisfaction with both arbitration and litigation. ADR however has always suffered from the fact the outcome was not legally binding on the parties.

Housing Grants Construction and Regeneration Act 1996

The Act provides for *statutory adjudication* and came into force in May 1998. The provision implementing the Latham Report's recommendations for its introduction are found in s.108 and provides that:

1 'A party to a construction contract has a right to refer a dispute arising under the contract for adjudication under a procedure complying with this section. For this purpose "dispute" includes any difference.
2 The contract shall
   (a) enable a party to give notice at any time of his intention to refer a dispute to adjudication;
   (b) provide a timetable with the object of securing the appointment of an adjudicator and referral of the dispute to him with 7 days of such notice;
   (c) provide a timetable with the object of securing the appointment of an adjudicator and referral of the dispute to him with 7 days of such notice;
   (d) require the adjudicator to reach a decision within 28 days of referral or such longer period as is agreed by the parties after the dispute had been referred;
   (e) allow the arbitrator to extend the period of 28 days by up to 14 days, with the consent of the party by whom the dispute was referred;
   (f) impose a duty on the arbitrator to act impartially; and enable the adjudicator to take the initiative in ascertaining the facts and the law.
3 The contract shall provide that the decision of the adjudicator is binding until the dispute is finally determined by legal proceedings, by arbitration (if the

contract provides for arbitration or the parties otherwise agree to arbitration) or by agreement. The parties may agree to accept the decision of the adjudicator as finally determining the dispute.

4 The contract shall also provide that the adjudicator is not liable for anything done or omitted in the discharge or purported discharge of his functions as adjudicator unless the act or omission is in bad faith, and that any employee or agent of the adjudicator is similarly protected from liability.

5 If the contract does not comply with the requirements of subsection (1) to (4), the adjudication provisions of the Scheme for Construction Contracts apply.

6 For England and Wales, the Scheme may apply the provisions of the Arbitration Act 1996 with such adaptations and modification as appear to the Minister making the scheme appropriate. For Scotland, the Scheme may include provisions conferring power on the courts in relation to the enforcement of the adjudicator' award.'

Note the provisions of s.114 (4) of the HGCRA 96: 'Where any of the provisions of the Scheme apply by virtue of this part in default of contractual agreement by the parties, the Scheme shall have effect as an implied term of the contract concerned.'

What is a construction contract?

Section 104 of the HGCRA 96 defines a construction contract as including:

- all normal building and civil engineering works, including operations such as scaffolding, site clearance and painting and decorating
- consultants' agreements on construction operations
- labour-only contracts
- contracts of any value

Certain contracts are excluded from the operation of the Act. These are:

- supply and fixing of plant in process industries (including supporting steelwork)
- off-site manufacture
- contracts with residential occupiers
- contracts not in writing, but 'writing' is widely defined, so word-of-mouth agreements referring to a written document or an exchange of letters is sufficient to bring it into the section (but see below)
- private finance initiative contracts
- finance agreements

What is meant by a contract in writing

The meaning of the phrase was considered in *RJT Consulting Engineers Ltd* v. *DM Engineering (Northern Ireland) Ltd* (2002) 18 Const LJ 425 CA. RJR accepted an invitation from sub-contractors to carry out design work on their behalf. The same consulting engineers had previously been advisers to the employers and had carried out on their behalf negotiations with the sub-contractors with regard to the scope of work, design requirements and price. Subsequently they recommended the sub-contractors for mechanical and engineering work to the main contractor. A dispute arose out of the consultancy contract between the parties. The sub-contractors alleged serous failures in performance of the consultants and sought a referral to adjudication. The consultants challenged this on the basis that the Act did not apply, as the contract was *not in writing*. After the appointment of an adjudicator, but before his decision, they sought a declaration in the High Court to the effect that the HGCRA 96 did not apply to the parties' contract. The judge refused the declaration but gave permission to appeal. Hearing the appeal, Ward LJ commented that the court had to decide what was meant by the time-hallowed words, an agreement which is *'evidenced in writing'*. As a result the Court of Appeal had to consider the effect of s. 107 of the Act.

Section 107 (1) 'The provisions of this Part apply only where the construction contract is in writing, and any other agreement between the parties as to any matter is effective for the purposes of this Part only if in writing.

The expressions 'agreement', 'agree' and 'agreed' shall be construed accordingly:

(2) There is an agreement in writing –
   (a) if the agreement is in writing (whether or not it is signed by the parties),
   (b) if the agreement is made by exchange of communications in writing, or
   (c) if the agreement is evidenced in writing.
(3) Where the parties agree otherwise than in writing by reference to terms which are in writing, they make an agreement in writing.
(4) An agreement is evidenced in writing if an agreement made otherwise than in writing is recorded by one of the parties, or by a third party, with the authority of the parties to the agreement.
(5) An exchange of written submissions in adjudication proceedings or in arbitral or legal proceedings in which the existence of an agreement otherwise than in writing is alleged by one party against another party and not denied by the other party in his response constitutes as between those parties an agreement in writing to the effect alleged.
(6) References in this Part to anything being written or in writing include its being recorded by any means.'

Counsel for the *appellant* made two submissions: (i) the judge had confused the documents proving there was a contract with documents recording the oral contract and (ii) the whole agreement had to be evidenced in writing to give certainty to the decision of the adjudicator in the short time frame allowed.

Counsel for the *respondent* argued that (i) all that had to be in writing was evidence of a contract and (ii) the purpose of the Act was to spread the benefits of the Act as widely as possible.

The judgment

Ward LJ in construing s. 107 (1) said that the requirement for writing was part of the earlier attempt to force the construction industry to submit to a standard form of contract. Although that did not succeed, writing lends certainty. In the short time that the adjudicator has, certainty as to the terms of the contract was very important.

Section 107 (1) created three categories of agreements in writing:

(a) an agreement in a written document recording the agreement though not signed by the parties;
(b) communications in writing containing all that is needed about the agreement;
(c) where the evidence in writing supports the existence of the contract.

Section 107 (3) he construed as meaning that 'where the parties agree by reference to terms that are in writing, the legislature are envisaging that all the *material terms* are *in writing* and that oral agreement refers to that written record' (emphasis added). Section 107 (4) allows the agreement to be evidence in writing if it is recorded by one of the parties (or a third party) with the authority of the parties to the agreement. Section 107 (6) contemplates that this record can be in writing or recorded but it has to be a record of the whole agreement.

Section 107 (5) provides that where the parties exchange written submissions in adjudication proceedings, and one party alleges an agreement otherwise than in writing, if the *other party does not deny* this, then the exchange constitutes an agreement in writing *to the effect alleged*. Ward LJ said:

> 'The last few words are important. The exchange constitutes an agreement in writing which does more than evidence the existence of the agreement. It also evidences the effect of the agreement alleged, and that must mean such terms which may be material to allege for the purpose of that particular adjudication.'

He also adopted the observations of HHJ Bowsher QC in *Grovedeck Ltd* v. *Capital Demolition Ltd* [2000] BLR 181. An oral agreement was entered into for demolition work on two sites. The parties went to adjudication under the Scheme for Construction Contracts despite the protests of one side that the Act did not apply. HHJ Bowsher refused to enforce the adjudicator's award as there was no written contract. The adjudicator had no jurisdiction to make the award. In that case, he said:

> 'Disputes as to the terms, express or implied, of *oral* construction contracts are surprisingly common and are not readily susceptible of resolution by summary procedure such as adjudication. It is not surprising that Parliament should have intended such disputes should not be determined by adjudicators under the Act.'

Ward LJ agreed with those sentiments. Writing was essential. The written record of the agreement provided the adjudicator with certainty as to the terms of the agreement. All three judges interpreted s. 107 to mean that the *terms of the agreement* had to be in writing, not the agreement itself. This was a surprising finding since Ward LJ said at 19 that: 'Here we have a comparatively simple oral agreement about the terms of which there may be very little, if any, dispute.' Note that the only term that could be implied into the consulting contract was whether the engineers had to use reasonable care and skill in carrying out the work and whether they could be sued in contract and tort. Terms are readily implied into oral contracts for professional services.

Adjudication and formation of contracts

In *VHE Construction* in 1997 HHJ Bowster OR remarked that the issue in this case was (whether) 'there was a contract between the parties, and if so what were the terms'? He then described how remarkable it was that it 'was probably the most frequently raised issue in the construction industry'. That in contracts worth thousands and sometimes millions of pounds, the first issue, whenever a dispute arose, was whether (i) there was a contract and (ii) if there was, what were the terms? The reasons for this situation are explained in chapter 3. What however, is not surprising is that the requirement for writing in the HGCRA 96 should highlight the formation problem.

Prior to the introduction of the Act it was clear to many commentators that the status of a letter of intent was always likely to create difficulties. This reservation has been reflected in a number of appeals from the awards made by adjudicators. *Shimsu Europe Ltd* v. *LBJ Fabrications Ltd* [2003] EWHC 1229 involved a contract made by the issue of a letter of intent. The letter stated that the sum to be paid was fixed and was subject to the DOM/1 standard form of contract and other non-standard conditions. The fixed sum plus VAT was subject to 'claim or set off which we might make against you for breach of contract'. One of the additional clauses, 21.2.4 held that payment was not due until the defendant delivered to the claimant a VAT invoice or an authenticated VAT receipt. The adjudicator found that as the sum was not due for payment until this was done, the defendant could issue a withholding certificate against it. It could also do so against the adjudicator's award.

A further issue was whether the adjudicator had jurisdiction to decide the contractual position. The judge decided that the parties had accepted that there was a contract based on the letter of intent. It was not necessary for him to consider whether there had been a breach of natural justice by the adjudicator, in not giving the parties an opportunity to deal with the matter. However, he went on to do so and concluded that the issue of whether there was a contract or not, was outside the adjudicator's jurisdiction.

May LJ recalled what he described as the long discredited decision of 'this court' in *Dawnays* in *Tally Wiejl (UK) Ltd* v. *Pegram Shopfitters Ltd* [2003] EWCA Civ 1750. This was reference to Lord Denning's view that cash flow was the lifeblood of the industry and that interim certificates should be honoured: see chapter 10. *Tally Wiejl* was an appeal against the decision of the first instance judge to enforce the adjudicator's award summarily. The respondent (Pegram) disputed the enforcement of the award on two grounds. One was that there was no construction contract in existence and two that if there was one, it was different to the one found by the adjudicator. The formation of the contract involved the traditional battle of the forms.

Work started on site without contract documents being prepared and without any tendering process. The project architect introduced the shopfitters to the client, who needed the work of fitting out of its Oxford Street shop in London, to be carried out at once and with haste. The parties agreed at a meeting that the contract would be a prime costs contract and that a letter of intent would be issued. Instead of a letter of intent, the client wrote instructing the carrying out of the work and proposing the JCT Standard form of Prime Costs contract 1998. The claimant responded by declining these terms and suggested instead his own standard terms.

Once it started to employ sub-contractors, the claimant became concerned about the lack of any contractual basis for carrying out the work. It sent a fax seeking assurance about payment and raised the matter at a subsequent site meeting. The day after the meeting, the claimant sent the client four marked-up plans showing the partitions with a budget price included. It was later signed by both parties and formed the basis of the submission by the claimant that a construction contract was formed. It also relied on its standard terms which contained a term that the acceptance of any quotation was an agreement to the terms. In response, the client sent a letter of appointment which sought to incorporate the JCT 98 prime costs conditions. The letter requested confirmation of acceptance by requesting the return of the enclosed copy, signed where indicated. This was not done and no further communications passed between the parties. Court of Appeal concluded that there was no contract between the parties. It decided that the judge, in deciding to enforce the award, was wrong, (as was the parties and the adjudicator) in assuming a valid construction contract existed under the HGCRA 96. May LJ at para. 9, stressed that while the court should be vigilant in examining technical defences to the enforcement of the award, it supported the policy of the legislation which was 'pay now, argue later'.

However, there were situations where legal principles prevailed over the broad-brush approach and this was such a case. What this case also illustrates is the complex negotiations that often take place during the formation of a construction contract. Quite often too the contractual position is unclear and it may well be the case that the parties have no contract after all. This difficulty is further illustrated by the another Court of Appeal case. *Thomas-Frederick's (Construction) Ltd* v. *Keith Wilson* [2003] EWCA Civ 1494 was an appeal against an order for summary judgement in favour of the claimant in an adjudication. The basis of the appeal was that the adjudicator had no jurisdiction to make the award as the contract was not 'evidenced in writing'. An additional ground of appeal was that the adjudication had been directed at the wrong person. Mr Wilson maintained that the construction contract was made with a company called 'Groversand Ltd' and not with him personally. The only evidence in writing was a letter confirming details of the contract arrangements. The Court of Appeal applied *RJT Consulting* held that there was not contract in writing and the adjudicator had no jurisdiction. The appellant had contested the bringing of adjudication proceedings on the ground that he was not a party to the contract. The respondent argued that in taking part in the adjudication he had waived his objections. The court concluded that the respondent had not submitted to the jurisdiction of the adjudicator and dismissed the application for summary judgement.

Galliford Try Construction Ltd v. *Michael Heal Associates Ltd* [2003] EWHC 2888 (TCC) also concerned the question as to whether the adjudicator had jurisdiction. Again, the issue was whether there was (i) a contract in writing and (ii) the Scheme for Construction contracts was incorporated. These issues came before HHJ Seymour in an application by Galliford to enforce the award of the adjudicator. In deciding whether there was a contract between the parties he said that the legal principles he had to apply were clear. He then referred to the well-known judgement of Lloyd LJ in *Pagnan Spa*: see chapter 3 for further details. See also the case *Co-operative Society (CWS) Ltd (formerly Co-operative Wholesale Society Ltd)* v. *International Computer Ltd* [2003] EWHC 1 (TCC) where the same judge gave a most incisive, complete and up-to-date summary of the present state of the law on contract formation at para. 40–53. He decided that no contract had been made either as a result of negotiations or by conduct. That conclusions made it unnecessary to decide whether it was a contract in writing. He concluded that it was not, by applying *RJT Consulting*. If there was a binding contract arising from applying the principles in *Percy Trentham* (where both parties of completed their obligations, there may be no need for offer and acceptance) the terms of such a contract had to be evidenced in writing.

The conclusion which can be drawn from these cases, is that the particular problem of whether there was a contract (and its terms) identified in *VHE Construction* is still very important. The requirement that the contract be in writing has complicated the issues leading to abortive adjudication and enforcement proceedings and still requires resolution. This situation is no great surprise, since the formation of the contract is an ever present problem in construction contracts. The need to get the work done while at the same time negotiating the contract, has been further complicated by the requirement to 'get it in writing' in order to comply with the legislation. It is instructive that there have been three cases on this issue alone in the Court of Appeal.

Adjudication and residential occupiers

The HGCRA excludes contracts with residential owners. However, standard forms have introduced contractual adjudication and payment provisions mirroring the Act; see for example *Rupert Morgan* where the written contact was one prepared by the Architecture and Surveying Institute. In *Picardi (Trading as Picardi Associates)* v. *Mr & Mrs Cuniberti* [2003] EWHC 2923 (TCC) issue between the parties was whether the RIBA Conditions of Engagement/99 was incorporated into the agreement. The conditions allowed for the appointment of an adjudicator by the parties and in default, by the president of the RICS. The occupiers contested the adjudicator's award on the following grounds:

1 There was no concluded contract that incorporated the RIBA conditions.
2 If there was such a contract the provision for adjudication in clause 9.2 was unenforceable because insufficient notice of onerous conditions were given.

3 Clause 9.2 was ineffective under the under the Unfair Terms in Consumer Contracts Regulations 1999.

The court concluded that the defendants did not agree the RIBA terms, nor were they incorporated into the contract by conduct. Hence the adjudication was invalid because the parties had not agreed to it. Although it was not strictly necessary to deal with the other issues, the judge made some further findings in relation to the Regulations. As the terms of the RIBA contract were unusual, they had to be brought properly and fairly to the attention of the other party. This, the architect had failed to do. The RIBA Guidelines specifically draw attention to the need to do so. Under the Regulations a provision for adjudication can also cause a serious imbalance in the rights and obligations of the parties: see *Director of Fair Trading* since the consumer, in prosecuting or defending it will be caused irrecoverable expenditure. This needed to be properly explained to the consumer, together with the fact that the adjudicator, despite being appointed by the architect's professional body, would be neutral. A failure to do so may create to the consumer the appearance of unfairness. Although the dispute in *Lovell Projects Ltd* v. *Legg and Carver* [2003] CILL 2019 was similar to *Picardi*, the outcome was very different. The parties entered into a contract using the JCT Agreement for Minor Works (1998 edition). Since the contract was with residential owners the HGCRA 96 did not apply. However, the contract with amendments MW1-11 provided for adjudication. The builder commenced adjudication for non-payment of certified sums, a claim for an extension of time and damages for wrongful determination. In the adjudication it was awarded approximately £86,000 and applied for summary judgement to enforce the award.

It was held that the terms were not unfair and did not cause a significant imbalance in the relationship between the parties. The judge considered the Unfair Terms in Consumer Contracts Regulations 1999, and decided adjudication provided a speedy settlement of disputes and either party could use it. In addition, the occupiers had the benefit of professional advice and had been advised by their architect. There were no hidden traps and the builder had not taken advantage of them. Furthermore, following *Levolux* there was no right of set-off against money they claimed to have paid to the joinery sub-contractor.

What is a dispute or difference?

The words *dispute* or *difference* have the same meaning. Since the right to refer the subject matter of a claim to an adjudicator only arises when a dispute or difference arises, the first defence to such a request is likely to be 'What dispute? We haven't got one.' *Fastrack Contractors Ltd* v. *Morrison Construction Ltd* (2000) 16 Const LJ 273 decided that a dispute only arose when a claim had been notified and rejected. A rejection could occur when an opposing party refused to answer the claim. In doing so it adopted the reasoning in *Halki Shipping Corp.* v. *Sopex Oil Ltd* (1998) 1 WLR 726. There, in relation to a reference to *arbitration*, the Court of Appeal stated that the word 'dispute' should be given 'its ordinary meaning and in order for there to be a dispute the matter would first have to raised and then denied or ignored by the other party.'

Who is an adjudicator?

Adjudicators were initially drawn from the ranks of practising arbitrators and construction lawyers. An adjudicator must be a natural person not a company, must not be an employee of either party, and must declare any interest in the matter, financial or otherwise (Scheme Pt 1 para. 4).

Who appoints the adjudicator?

There are 16 *Adjudication Nominating Bodies* (ANBs). An ANB is any body which holds itself out publicly as such (Scheme Pt 1 para. 2(3)). Nominating bodies include the Royal Institute of Chartered Surveyors (RICS), the Chartered Institute of Builders (CIOB), and the Chartered Institute of Arbitrators (CIA).

Contractual provisions: the JCT 98

Article 5

'If a dispute or difference arises under this Contract either Party may refer it to adjudication in accordance with clause 41A.' This wording narrows the Act since adjudication under the Act is a right, not an obligation.

Contractual provisions: ICE 6th (but see also 7th)

The scheme of dispute resolution differs in the ICE contracts from that contained in the JCT; see Clause 66.

1 A dispute arises only after the time for giving a decision by the engineer on a matter of dissatisfaction has expired: see clause 66 (3).
2 The parties may then refer the dispute to the Institution of Civil Engineers' Conciliation Procedure: clause 66 (5).
3 The Employer and the Contractor each have the right to refer a dispute about a matter under the Contract for adjudication: clause 66 (6).

What is the adjudicator's role?

Parliament's aim in implementing the HGCRA 96 was to provide a speedy mechanism for resolving construction disputes. The award is made on an interim basis and can be reviewed after practical completion of the project. A particular mischief the Act was intended to conteract the abuse of sub-contractors, whether due to the absence of effective payment provisions, onerous contractual terms, the absence of arbitration clauses or clauses permitting arbitration only after the main contract work had been completed. Section 108 (2) enables a party to give notice at any time of his intention to refer a dispute to adjudication.

Time limits

Once a party has given notice of a reference to arbitration – which it can give at any time – the arbitrator must be appointed and the dispute referred to him within 7 days. Once appointed the adjudicator has 28 days to reach a decision subject to the power exercised by the referring party to extend by up to 14 days.

The duty of the adjudicator

The judges in the Court of Appeal in *Bouygues (UK) Ltd* v. *Dahl-Jensen (UK) Ltd* [2000] BLR 522 (CA) had the opportunity to disagree with the robust approach of HHJ Dyson QC in the TCC but chose not to do so. Giving judgment at first instance, he said: 'Inherent in the Scheme [i.e., for adjudication] is that injustices will occur, because from time to time adjudicators will make mistakes. Sometimes these will be glaringly obvious and disastrous in their consequences for the losing parties.' The Adjudicator awarded the sum of £207,741 to Dahl-Jensen the sub-contractor. He arrived at the figure by taking a gross sum (which *included retention*) and deducting from it the sums that had been paid. In doing so he released the retention money early. Had he taken a net sum excluding the retention, Bouygues instead of being liable for £207,741, would have been *entitled to receive* £141,254.

The Court of Appeal upheld the judgment of HHJ Dyson QC that the purpose of adjudication was to provide a speedy method of resolving construction disputes. Dahl-Jensen was now *in liquidation*. The court held that where the receiving party was in liquidation and the paying party had a viable defence or a cross-claim, *Rule 4.90* of the Insolvency Rules applies. In such a situation the court can take an account between the parties and make only the balance either way payable. In other words, if the matter had been raised before the judge summary judgment *would not* have been granted.

Chadwick LJ said at pp. 528–9:

> 'Part 24, rule 2 of the Civil procedure rules enables the court to give summary judgement on the whole of the claim, or on a particular issue, if it considers that the defendant had no real prospect of successfully defending the claim or there is no reason why the case or issue should be disposed of at trial. In the circumstances such as the present, where there are latent claims and cross-claims between the parties, one of which is in liquidation, it seems to me that there is a compelling reason to refuse summary judgement on a claim arising out of an adjudication which is necessarily provisional. All claims and cross-claims should be resolved in liquidation in which a full account can be taken and a balance struck. That is what rule 4.90 of the Insolvency Rules 1986 requires.'

In *Rainford House Ltd* v. *Cadogan Ltd* (2001) BLR 416 a related issue arose. What was the position where the receiving party was in *administrative receivership*? The claimant had been awarded £75,1333.78 by the adjudicator. The losing party resisted the claim for summary judgment, and made the following claims.

1   It had served a defence and counter-claim arising out of a breach of contract in failing to proceed regularly and diligently in carrying out the work.
2   The claimant had been overpaid for work it had carried out.
3   The defendant had paid various sub-contractors in order to induce them to stay on-site and continue working. These sums amounted to £79,950.66.

Held: Summary judgment was stayed. Where there was a serious doubt about the ability of the claimant to repay the monies awarded once the matter was finally determined, the court would not enforce the adjudicator's award. Cadogan, being in administrative receivership, was such a case.

Powers under the scheme

It is instructive to look at the powers given to the adjudicator under the Scheme. Para. 12 states that the adjudicator shall:

(a) act impartially in carrying out his duties and shall do so in accordance with any relevant terms of the contract and shall reach his decision in accordance with the applicable law in relation to the contract; and
(b) avoid incurring any unnecessary expense.

Para. 13 allows the adjudicator to take the initiative in ascertaining the facts and the law necessary to determine the dispute, and to decide on the procedure to be followed in the adjudication. Note that many ANBs have produced their own rules and procedures for use when an adjudicator is appointed by them.

The enforcement of the adjudicator's award

At the time of the introduction of the HGCRA 96 the main stumbling block fore-seen to the enforcement of the award were that:

(a) Parliament had not provided a mechanism to do so; and
(b) the availability of summary judgment of the award would be caught by s. 9 of the Arbitration Act 1996 which made a stay of court proceedings mandatory where the contract contained an arbitration clause.

This argument has been dismissed by the courts in favour of a policy of enforcing the intention of Parliament. Using their inherent jurisdiction the courts have enforced the adjudicator's award by means of summary judgment whether or not there has been an irregularity or error in the award. See *Macob Civil Engineering* v. *Morrison Construction* (1999) 3 BLR 93, and see also *Bouygues*, above.

In a paper given to the Society of Construction Law, HHJ Forbes QC described the TCC's approach to the enforcement of adjudicators' awards. He explained that it has developed procedures to enable applications for summary judgment to be heard quickly. Applications can be heard within 10–14 days of the award of the adjudica-tor. Where there is a *triable issue*, it too can be heard quickly, perhaps in a matter of weeks or a few months, depending on whether that is practical for the parties.

The general position

Provided the adjudicator acts within his jurisdiction and answers the question put in the *Notice of Adjudication* then, right or wrong, his decision is enforceable (see

Bouygues). 'If he had answered the right question in the wrong way, his decision will be binding. If he has answered the wrong question, his decision will be a nullity': Know J in *Nikko Hotels (UK) Ltd* v. *MEPC plc* [1991] 2 EGLR 103 at 108B, approved by Buxton LJ in *Bouygues*. *Jurisdiction* means that the adjudicator has the power to decide the particular dispute referred to him. His authority comes from the contract between the parties (Forbes J) 'whether the scheme applies or not. So it is that, unless the parties otherwise agree – an adjudicator – unlike a judge – has no inherent jurisdiction to make a final decision as to whether he does have jurisdiction in a particular case.' The adjudicator thus cannot decide on his own *jurisdiction*. When a challenge is made that he does not have it, he must consider it. HHJ Thornton QC provided valuable guidance on the way an adjudicator should approach this task in *Christiani & Nielsen Ltd* v. *The Lowery Centre Development Company Ltd* (2000) unreported, when he said this:

> 'It is clearly, indeed desirable, for an adjudicator faced with a challenge which is not a frivolous one to investigate his own jurisdiction and to reach his own non-binding conclusions as to that challenge. An adjudicator would find it hard to comply with the statutory duty of impartiality if he or she ignored such a challenge.'

A clear illustration of a dispute where an adjudicator's error that did not deprive him of jurisdiction is *C & B Scene Concept Design Ltd* v. *Isobars* [2002] BLR 93 CA. This was an appeal against the dismissal of the claimant's application for summary judgment. The appellant had been awarded payments by the adjudicator under the Scheme because their design and build contract had not made adequate provisions for payment. The parties had *failed* to adopt either of the alternatives provided by the JCT Standard Form of Building Contract with contractors Design, 1998 edition. Clause 30 of the form requires the parties to adopt the provisions of either:

(a) Alternative 'A' – interim payments to be made in accordance with pre-determined stages in the work; or
(b) Alternative 'B' – interim payments to be made according to the lapse of pre-determined periods of time.

The adjudicator awarded the claimant three interim payments amounting to £115,996.33 plus VAT and interest. The defendant challenged the award in resisting an application for summary judgment. Its grounds were that in applying clause 30 of JCT 98, the adjudicator acted outside his jurisdiction. He failed to appreciate that the contractual provisions of clause 30 were superseded by the Scheme. He was thus acting outside his jurisdiction by addressing the wrong question. The adjudicator should have applied the provisions of the Scheme. The Scheme allowed the defendant to challenge the correctness of the claim whilst clause 30 did not. That clause did not allow challenges to the correctness of the sum where notice had not been served within 5 days. The Scheme does allow the defendant to assert in an adjudication that the sum claimed was not 'due under the contract' for the reasons it wished to rely on.

The Recorder, Mr Moxon-Browne QC, accepted that by failing to select alternative A or B, the contractual provisions fell away and the Scheme applied. He refused summary judgment and gave the defendant permission to defend. The claimant appealed.

The Court of Appeal adopted with the formulation made in *Northern Developments (Cumbria) Ltd* v. *J & J Nichols* [2000] BLR 759. There HHJ Bowsher QC cited with approval the following formulation of principles stated by HHJ Thornton QC in *Sherwood and Casson Ltd* v. *McKenzie Engineering Ltd* (2000) 2 TCLR 418:

1 A decision of an adjudicator whose validity is challenged as to its factual and legal conclusions or procedural error remains a decision that is both enforceable and should be enforced.
2 A decision that is erroneous, even if the error is disclosed by the reasons, will still not ordinarily be capable of being challenged and should, ordinarily, be enforced.
3 A decision may be challenged on the grounds that the adjudicator was not empowered by the Act to make a decision, because there was no underlying construction contract between the parties or because he had gone outside his terms of reference.
4 The adjudication is intended to be a speedy process in which mistakes will inevitably occur. Thus, the court should guard against characterising a mistaken answer to an issue which is within an adjudicator's jurisdiction as being an excess of jurisdiction.
5 An issue as to whether a construction contract ever came into existence, which is one challenging the jurisdiction of the adjudicator, so long as it is reasonable and clearly made, must be determined by the Court on the balance of probabilities with, if necessary, oral and documentary evidence.

The Court of Appeal decided that he did not exceed his jurisdiction, since he had decided the matter referred to him. The parties had agreed the scope of the dispute: the employer's obligation to make payment and the contractor's entitlement to receive payment following the receipt by the employer of the contractor's application for payment.

In giving judgment Sir Murray Stuart – Smith said that:

'It is important that the enforcement of an adjudicator's decision by summary judgement should not be prevented by arguments that the adjudicator made errors of law in reaching his decision, unless the adjudicator has purported to decide matters that are not referred to him. He must decide as a matter of construction the referral, and therefore as a matter of law, what he has to decide. If he erroneously decides that the dispute referred to him is wider than it is, in so far as he exceeds his jurisdiction, his decision cannot be enforced. But in the present case there was an entire agreement as to the scope of the dispute, and the adjudicator's decision, albeit he may have made errors of law as to the relevant contractual provisions, is still binding and enforceable until the matter is corrected in the final determination.' (Rix and Porter LLJ both agreed with these conclusions)

In *RJT Consulting Engineers Ltd* all three judges in the Court of Appeal expressed concern at the spread of 'legal wrangling' arising out of jurisdictional challenges. They declined, however, to criticise the consulting engineers for taking a legal point which was open to them.

More jurisdiction difficulties

Hurst Stores and Interiors Ltd v. *M.L Europe Property Ltd* [2003] EWHC 1650 (TCC) An application to enforce the decision of the adjudicator was refused for two reasons. The issue between the parties was whether there had been compromise agreement concerning the final account. It was held that the project manager had had no authority whether actual or ostensible to make the agreement. Second, the document that was claimed to be a compromise agreement was unenforceable because the project manger was acting under a unilateral mistake when entering into it. As such it could be rectified to reflect the mistake.

RSL (South West) Ltd v. *Stansell Ltd* [2003] EWCA 1390 (TCC) The adjudicator was held to be in breach of natural justice and the award overturned. A dispute rose over the final account in a contract incorporating the DOM/2 1981 conditions of contract. With the agreement of both parties, the adjudicator employed the assistance of a colleague to complete a programming report. Stansell asked for a copy and time to comment on it. This was not done and the adjudicator was held to be in breach of the rules of natural justice.

Orange EBS Ltd v. *ABB Ltd* [2003] EWHC 1187 (TCC) confirmed that the award will be enforced despite one party 'ambushing' the other by initiating adjudication proceedings during the Christmas holiday period. The judge commented that holidays at any time of the year present practical problems and companies must deal with them.

Procedural unfairness

The claimant entered into a contract with the defendant for the design, manufacture and erection of steel balconies at Davy House, Harrow, London, in *Discain Project Service Ltd* v. *Opecprime Developments Ltd* [2001] BLR 402. A dispute arose over payment and the claimant served a notice of adjudication. No provision for it had been made and the Scheme applied. After the award was made, the claimant brought an action to enforce the award. The defendant disputed the validity of the award and the judge set down a trial of the action. The claimant's case was that the adjudicator had been in breach of the rules of natural justice and that his award should not be enforced. At the heart of the defendant's case were two issues: (i) he had taken into account matters which he had not disclosed to one party and (ii) he had entered into discussions by telephone of material facts without informing the defendant.

In deciding whether there had been bias on the part of the adjudicator, the judge applied the tests set out in *Director General of Fair Trading* v. *Proprietary*

Association of Great Britain [2000] All ER (D) 2425. HHJ Lloyd QC had reviewed the authorities in *Glencot Developments and Design Co Ltd* v. *Ben Barret & Sons (Contractor) Ltd* [2001] BLR 207 and concluded, in refusing summary judgment of the award: 'It is accepted that the adjudicator has to conduct the proceedings with the rules of natural justice or as fairly as the limitations imposed by Parliament permit.'

The claimant in *Balfour Beatty Construction Ltd* v. *The Mayor and Burgesses of the London Borough of Lambeth* [2002] EWHC 597 (TCC) sought to enforce the award of the adjudicator. The dispute concerned claims for extensions of time and for damages deducted for delayed completion. The claimants' referral to adjudication did not include a critical path analysis or a reliable as-built programme. The adjudicator, with the assistance of his colleagues checked the as-built information to produce his own critical path analysis. Using this information he computed that the claimant was entitled to an extension of time of 35 weeks and one day. He awarded the repayment of liquidated and ascertained damages of £283,997.14 plus interest. Lambeth resisted enforcement and challenged the application on two grounds:

1 In reaching his conclusion the adjudicator had not acted impartially. He had failed to give them an opportunity to deal with arguments that had not been advanced by either party.
2 He had in 'breach of contract and without jurisdiction' employed others to do work beyond that which the parties had agreed he could employ assistants to do.

In reaching his decision HHJ Lloyd QC commented at para. 33:

'Thus in my judgment, the adjudicator not only took the initiative in ascertaining the facts but also applied his knowledge and experience to an appreciation of them and thus in effect did [Balfour Beatty's] work for it . . . If an adjudicator intends to use a method which was not agreed and has not been put forward as an appropriate method by either party he ought to inform the parties and obtain their views as it is his choice as how the dispute might be decided.'

In failing to do this he had exceeded his jurisdiction by making good fundamental deficiencies in the claimant's case. In not informing Lambeth of his approach he failed to act impartially and was in breach of the rules of natural justice. On the facts he found that the use of assistance had been with the knowledge of the parties.

Assistance resulted in a challenge on grounds of lack of jurisdiction and procedural unfairness in *Try Construction Ltd* v. *Eton Town House Group Ltd* [2003] EWHC 60 (TCC). At issue were claims for an extension of time and the repayment of liquidated damages. Two separate notices of adjudication were issued and the same adjudicator appointed to resolve them. With the agreement of the parties a programming expert was appointed to assist the adjudicator. At a further meeting it was agreed that the expert could independently contact the programming experts of the parties. HHJ Wilcox found that both parties had been given an opportunity to deal with the analysis produced by the expert and relied on by the adjudicator. At para. 62 he described it as 'a transparent process sensibility and pragmatically agreed by the parties'. He concluded that there was no unfairness in the adjudication proceedings.

Statutory demand

This is one approach to the enforcement of the adjudicator's award that has emerged. Where the losing party has refused to pay the award, a statutory demand is served instead (section 268 (1) (a) of the Insolvency Act 1986), and the Insolvency Rules 1986 permits a claim for a debt to be enforced in this manner in the county court.

In *Oakley and anor* v. *Airclear Environmental Ltd and anor* (2001) the parties intended to enter into a nominated sub-contract under the sub-contract conditions NAM/SC 1998 edition issued by the JCT Ltd. No formal contract was concluded but work was carried out. The main contractor made stage payments and from these made deductions from the certified work. The deductions included retention money, main contractor discount and various contra-charges. The sub-contractor disputed these on two grounds: one was that there was no contact and thus they were entitled to a *quantum meruit* at a fair commercial rate with no deductions; second, if there was a contract it was under the NAM/SC. The main contractor applied to the President of the RICS for the appointment of an arbitrator under clause 35B of NAM/T. The sub-contractor disputed this on the basis that the proper nominating body was the RIBA not the RICS. At a meeting between the parties the sub-contractor indicated that it would go to adjudication over the disputed payment.

The main contractor applied to the RIBA for the appointment of an arbitrator. The sub-contractor in turn applied to the RICS for the appointment of an adjudicator. The main contractor refused to take part in the adjudication. The adjudicator awarded the sub-contractor the monies deducted in his award. No payments were made and a statutory demand made for £25,870.69 as a debt. The county court judge refused to set aside the statutory demand and the main contractor appealed to the Chancery Division of the High Court.

The appeal

In its appeal, the main contractor raised the following argument:

1 There was no contract between Oakley and Airclear and no contract incorporating of the NAM/SC and NAM/T conditions. The adjudicator had *no jurisdiction* and his decision was a nullity.
2 If there was contract it was *not in writing* within s. 107 and hence no provision for adjudication.
3 They were entitled by the Insolvency Rules to make a *counter-claim* (for a counter-claim, set-off or cross-demands) that could extinguish the debt specified in the statutory demand.

4 The claim under the Insolvency Rules should be set aside on the ground that the bankruptcy procedure was an *inappropriate route* for the enforcement of an adjudicator's award.

The findings

The judge held that the learned judge in the county court was entitled to find on the facts, that at the time the application was made for the appointment of an arbitrator, there was a common assumption by both parties that their contractual relations were governed by the NAM/T and NAM/SC, including in particular their dispute resolution provisions (at 48). At 49 he said:

> 'In my judgement, the learned judge plainly had material on which he could properly come to a conclusion of a common assumption, sufficient to *found an estoppel by convention*, that, at the time the application for a adjudicator, the provisions of NAM/T and NAM/SC were binding on the parties.'

He also found that there was nothing unconscionable in Oakley resiling from that commonly held position so they were not bound by the estoppel. They had done so at the earliest possible time so that little detriment was suffered by the other party as a result. The contract was not in writing so the adjudicator was not validly appointed under Ss. 107 and 108. The award was a nullity. No statutory demand could stand as there was *no legal basis* for the debt demanded.

Estoppel was also raised in *Christiani & Nielsen Ltd* v. *The Lowery Centre Development Company Limited* (2000 unreported). This dispute involved the deduction of liquidated damages. That depended on whether the deed entered into should have contained a contract period of 57 weeks as opposed to 81. At issue was whether the adjudicator had the power to rectify the deed. Also in dispute was whether the claimant was estopped by its pre-agreement that the HGRCRA 96 did not apply to the deed.

Held: 'The adjudicator did not have ad hoc to decide his own jurisdiction, but the decision that he had jurisdiction to decide the particular dispute was correct.' The construction contract was subject to the HGCRA 96 since it was entered into after 1 May 1998. Even if the claimant was potentially estopped by its 'pre-agreement understanding' that the Act was *not to apply* to the deed:

(a) the respondent had waived its entitlement to rely on the estoppel; and
(b) the parties could not in any event, by estoppel, prevent the claimant from relying on the HGCRA 96 since its terms are mandatory and cannot be contracted out of.

The adjudicator had jurisdiction to decide the dispute in question even though the decision involved considering whether the respondent was entitled to claim rectification of a deed.

Total M and E Services v. *ABB Building Technologies Ltd* (2002) CILL 1857–1861 was a lump sum contract for labour only provided by the sub-contractor for the installation of mechanical and electrical work. Additional work was ordered orally. The defendants deducted certain sums which were disputed and the parties went to adjudication. As their contract made no provision for it the Scheme applied. The defendants argued was that the extra work was not part of the reference since the adjudicator had no jurisdiction over it. Each piece of work was a separate oral contract and the HGCRA 96 did not apply. The contract contained no mechanism for variation in any case.

In enforcing the award the court dismissed these claims. What the parties had was a contract for labour only, which was enlarged by the ordering of extra work. An oral variation of a written contract fell within s.107 (3) and the adjudicator had jurisdiction to deal with it. No set-off against the sums could be allowed because the defendant had failed to give notice of deduction 7 days before making payment. He dismissed a further claim for the costs of the adjudication as a head of damage for breach of contract. The scheme made no provision for it since it envisaged that both parties could go to adjudication and incur costs they could not recover. Further arguments that the claimants would not be able repay the money if later an arbitrator found against them was rejected by the judge. When considering an application to stay summary judgment the court has to consider the overriding considerations of Part 1 of the CPR in ensuring justice and fairness between the parties: 'In considering what is just and fair in an application for a stay of execution of summary judgement in circumstances such as these the court must be careful not to reallocate the commercial risks accepted by the parties who engage in a construction contract.' In this case, although the winning party had few assets, paid-up capital of £10,000 and had not filed accounts since 1998, the defendant had continued to do substantial business with them. No evidence had been produced that showed the claimant was likely to become insolvent.

Postscript

In *Ferson Contractors Ltd* v. *Levolux AT Ltd* [2002] EWCA Civ 11, the Court of Appeal had to consider whether the adjudicator's award should be enforced where the right to suspend the work under the Act conflicted with provisions for termination of the contract for wrongly suspending the work (the facts are given in Chapter 10). In upholding the adjudicator's award and finding that the work had been validly suspended despite those contractual provisions, the court made a number of observations on the nature of adjudication:

1 HHJ Dyson QC had explained the scheme of s.108 in *Macob Civil Engineering Ltd* v. *Morrison Construction Ltd* (1999) at para. 24. It introduced a speedy mechanism for settling disputes in construction contracts and required adjudicator awards to be enforced pending final determination in arbitration or litigation.
2 The timetable for adjudication is tight and the adjudicator has a free hand as far as the procedure is concerned. He may act inquisitorially or he may invite representations from the parties. Its purpose is 'to enable a quick and interim, but enforceable, award to be made in advance of what is likely to be complex and expensive disputes' (per Buxton LJ in *Bouygues*).

3 The character of s.108 is draconian. The process is swift and an effective means of resolving disputes and, where the adjudicator is wrong, the matter can be corrected in subsequent arbitration or litigation.
4 *Bouygues* was considered to be a good illustration of the scheme in practice. There the Court of Appeal allowed the award to stand even though the adjudicator had a made an obvious and fundamental error. He had not exceeded his jurisdiction but had simply arrived at an erroneous conclusion.

What is arbitration?

Arbitration is the settlement of a dispute by *technically competent* experts. The principles of arbitration are set out by s.1 of the Arbitration Act 1996, Part I:

(a) 'the object of arbitration is obtain the fair resolution of disputes by an impartial tribunal without unnecessary delay or expense;
(b) the parties should be free to agree how their disputes are to be resolved, subject only to such safeguards as are necessary to safeguard the public interest;
(c) in matters governed by this Part the court should not intervene except as provided by this part.'

Arbitrators have a duty under s.33 to act 'fairly and impartially as between the parties giving each party a reasonable opportunity of putting his case and having it dealt with by his opponent'. They also have to adopt procedures suitable to the circumstances of the particular case, avoiding unnecessary delay or expense, so as to provide 'a fair means of resolution' of the dispute. The parties also have duties imposed on them. Section 40 requires them to do 'all things necessary for the proper and expeditious conduct' of the proceedings.

Requirements

Section 6 defines an 'arbitration agreement' as an agreement to submit to arbitration both present and future disputes (whether they are contractual or not). Such an agreement must be in writing or evidenced in writing (see s.5). The definition of writing includes any means by which the agreement can be recorded, including electronic transmission and communication such as a sound recording. A written document may include signed and unsigned documents, letters, faxes or a memorandum written by one party or by a duly authorised party. The meaning of dispute includes any difference. The power to appoint an arbitrator arises out of the contract, which must contain such a clause allowing disputes or differences to be referred to arbitration. In addition the contract must contain the procedure for appointing one.

Appointment

The arbitrator may be named in the contract agreed by parties or the contract will contain a means of appointment in the absence of agreement.

Jurisdiction

Under the Arbitration Act 1996 the arbitral tribunal has the power to rule on its own jurisdiction, meaning it has the power to resolve the dispute. The standard form, *JCT 98*, *Article 6* and *Clause 41B* confers wide powers on the arbitrator to resolve any dispute or difference and to open up, renew and revise any certificate, opinion, decision, and so on.

Procedure

In arbitral proceedings the general duty of the Tribunal is laid down by s.33.

1 The tribunal shall
   (a) act *fairly and impartially* as between the parties, giving each party a reasonable opportunity of putting his case and dealing with that of his opponent;
   (b) *adopt procedures suitable* to the circumstances of the particular case, so as to provide a fair means for the resolution of the matters falling to be determined. (emphasis added)
2 The tribunal shall comply with that general duty in conducting the arbitral proceedings in its decisions on matters of procedure and evidence and in the exercise of all other powers conferred on it.

This general duty is a broad statement of the approach of the tribunal in meeting the objective of s.1, a fair resolution of disputes by an impartial tribunal without unnecessary delay or expense. The tribunal is positively required to adopt flexible procedures to meet the individual circumstances of a dispute. It is unnecessary to follow court procedures slavishly.

Preliminary meeting

Once appointed the arbitrator will convene a preliminary meeting. Although this is not mandatory it does give the arbitrator an opportunity to issue instructions on how the arbitration is to be conducted, the issue of pleadings, the time scale and the appointment of experts.

Hearing

A hearing can be a document-only one. The parties submit all relevant documents and the arbitrator makes the award having studied the documents. This is where the dispute is an uncomplicated one. Where the dispute is technically complex the parties may appoint lawyers to represent them. In such a case an oral hearing may be essential.

Award

The parties are free to agree the form the award should take. Where they have not done so the award must be in writing, signed by the arbitrator(s). Such an award shall contain reasons and state the seat of the award (s.52). 'Seat' is defined in s.53 as the place the award was made regardless of where it was signed, dispatched or delivered to any of the parties.

Problems with arbitration

'Contrary to popular belief in the construction industry, arbitration in building contract matters is generally slower, more expensive and less certain than High Court litigation' (Elliott, 1993, p.86). There are many explanations for this situation, including the fact that the use of litigation procedures adds to the costs of arbitration compared to other forms of dispute resolution. Hudson (at p.1,667) makes the following comments on the procedure in arbitration:

1   Arbitration has been popular in the past because the parties had great *confidence* in the tribunal, especially when the dispute was entirely technical.
2   The fact that the tribunal itself was technically competent in the subject matter of the dispute *restrained the parties* and their expert witnesses and made them cautious in their evidence.
3   However, the degree of *specialism* in the construction industry is such that architects and engineers, when acting as arbitrators, may not have the relevant specialist expertise themselves.
4   The relevant *documentation* can be *massive* both on liability and quantum. The inexperience of technical arbitrators in sifting and weighing evidence means that they may be in no position to judge with real confidence between legal submissions and objections in procedural matters.
5   Arbitrators of great note in their field can have real difficulty in analysing the *reason for rules of substantive* or *procedural law*. Failure to do so can give way to impulses of sympathy or compromise produces anomalous and sometimes startling results. The instinct to compromise or to hold no claim invalid is perhaps the greatest fault of non-legal arbitrators.
6   The practice of the High Court of referring building and engineering disputes to the *Official Referees* for trial (now the Technology and Construction Court) has resulted in judges who possess a considerable *knowledge of building and engineering problems and practice*. It will usually be cheaper as well (emphasis added).
7   Nature of construction disputes:
    (a)   lengthy and badly drafted contracts which fail to address everyday issues;
    (b)   lengthy contracts with scope for argument and financial disagreement;
    (c)   massive contemporaneous documentation;
    (d)   very little regard for long-term goodwill.

The difference between adjudication and arbitration

The differences are as outlined below:

1   The timetable for adjudication is fixed at 28 days. Arbitration is flexible, either agreed between the parties or fixed by the arbitrator.
2   The adjudicator's award can be opened up at a later date and a completely new hearing. The only appeal against an arbitrator's award is on the grounds of a procedural irregularity (see s.68).

3 The status of the adjudicator's award is not defined by the HGCRA 96. The enforcement procedures are more complex than an arbitrator's award.
4 The adjudicator, unlike an arbitrator, has no general power to award costs.

Alternatives

The growth of ADR resulted from the dissatisfaction with arbitration as a cost-effective method of solving disputes quickly and cheaply. There are a number of ways in which the various forms of ADR differ from litigation or arbitration, as explained below:

1 ADR is non-binding on the parties. The agreement making a provision for ADR usually provides that either party may withdraw from the process at any time. It is possible to provide a form of wording that makes the process binding on the parties. This, however, can be counter-productive in that it undermines the voluntary nature of the process.
2 Like arbitration and litigation, a third party is involved in facilitating agreement who is called a Neutral or Neutral Adviser. Unlike a judge or arbitrator the role of the third party is not to do justice between the parties but to obtain a resolution of their dispute.
3 There are no restrictions on communication in ADR. The parties are free and encouraged to confide in the Neutral matters they would not disclose to the other side. In formalised sessions these are known as 'caucus sessions'.
4 ADR is free from interference by the court. Although there are now (limited) grounds of appeal in the Arbitration Act of 1996, the courts supervise arbitration.
5 ADR agreements usually provide that each party should bear its own costs and half that of the Neutral.

The attraction of ADR for disputing parties is that there is a range of established techniques from which the parties can choose.

Conciliation

This process requires an impartial third party (the Neutral). The Neutral holds separate meetings with the party until a consensus emerges. It will then chair a meeting between both sides and facilitate agreement.

Quasi-conciliation

The Neutral provides a second opinion as a basis of negotiation.

Mediation

This is more akin to shuttle diplomacy. The mediator shuttles between the parties until a consensus emerges. The mediator then makes recommendations.

Mini-trials

In this type of case the subject matter is taken to executives of the companies in dispute. They must have authority to settle. At a hearing each side outlines its case and lawyers may well represent the parties. The mini-trial takes place before a bench of three: an executive from each side who has not been involved in the dispute, and the Neutral. Following the presentations the three openly discuss and agree the relative strength of each side's case.

This chapter has provided an outline of the process of adjunction. The general position as illustrated by the case law is that the courts will enforce the award provided the adjudicator has asked the right question even if the answer given is wrong. A number of issues have resolved by the courts. The meaning of a contract in writing is now clear – the terms of the contract need to be in writing. The meaning of a 'dispute or difference' has been settled by adopting the approach in arbitration. The courts have enforced the award, in the absence of two issues by summary judgment. Only where the adjudicator did not have jurisdiction, has the proceeding to enforce the award has been unsuccessful. The same applied to absence of procedural fairness. The enforcement proceedings indicate that in many cases adjudication is closely linked to payment provisions. What these proceedings have provided is a much better idea of the nature of construction disputes and their wide variety. It has also illustrated the wide range of legal issues that can arise.

Summary

1 Adjudication is a statutory method of resolving deputes quickly, introduced by s.108 of the HGCRA 96.
2 The adjudicator has 28 days to make an award and it is enforced by summary judgment in the High Court.
3 The award can only be challenged on the ground of procedural unfairness or a breach of natural justice.
4 Where the adjudicator makes a mistake or error, even where it is a glaringly obvious one, provided he has answered the question asked, the award will be enforced.
5 The courts have a taken a robust approach towards securing the intention of Parliament that the method of dispute resolution should be quick.
6 The award can be challenged in either arbitration or litigation after the completion of the project.
7 Arbitration has evolved with the passing of the Arbitration Act 1996. The arbitrator can rule on his own jurisdiction. He can initiate procedures for resolving the dispute with the agreement of the parties.
8 ADR is non-binding and works well when the parties want to resolve their dispute.

Bibliography

Adams, J. and Brownsword, R., *Understanding Contract Law*, 3rd edn (London: Sweet & Maxwell, 2000).

Bartlett, A.G., *Emden's Construction Law* (London: Butterworth, 2001).

Baster, J. *et al.*, *Construction Law Handbook* (Tonbridge: Thomas Telford, 2000).

Beale, H.G. (ed.) *Chitty on Contracts*, 28th edn (London: Sweet & Maxwell, 1999).

Birks, P., *An Introduction to the Law of Restitution* (Oxford: Clarendon Press, 1996).

Brownsword, R., *Contract Law – Themes for the Twenty-First Century* (London: Butterworth, 2000).

Burr, A., *European Construction Contracts*, 4th update (London: Chancery Press, 1966).

Burns A., *The Legal obligations of the Architect*, 1st edn (London: Butterworth, 1994).

Cato, Mark D., *Arbitration law and Procedures*, 2nd edn (London: LLP Ltd, 1998).

Cato, Mark D., *Sanctuary House Case: an Arbitration Work Book* (London: LLP Ltd, 2000).

Cornes, D.L., *Design Liability in the Construction Industry*, 4th edn (London: Blackwell Science, 1994).

Dannemann, G., *An Introduction to German Civil and Commercial Law* (London: British Institute of International and Comparative Law, 1993).

Duncan-Wallace, I. N., *Hudson's Building and Engineering Contracts*, 11th edn (London: Sweet and Maxwell, 1995).

Duncan-Wallace I. N., *Construction Contracts: Principles and Policies in Tort and Contract*, (London: Sweet & Maxwell 1986).

Eggleston, B., *The ICE Conditions of Contract 7th edn*, 2nd edn (Oxford: Blackwell, 2001).

Foster, N., *German Legal System and Laws*, 2nd edn (London: Blackstone, 1996).

Friedman, C., *Contract as Promise – A Theory of Contractual Obligation* (Cambridge, Mass: Harvard University Press, 1985).

Furmston, M., *Building Contract Casebook*, 3rd edn (London: Blackstone, 2000).

Furmston M., *Cheshire, Fifoot & Furmston's Law of Contract*, 14th edn (London: Butterworths, 2001).

Furst, S., and Ramsey, V., (eds), *Keating on Building Contracts*, 7th edn (London: Sweet & Maxwell, 2001).

Gray, J., *Lawyers' Latin – A vade-mecum*, 1st edn (London: Robert Hale, 2002).

Grubb, A., *Law of Contract*, 1st edn (London: Butterworth Common Law Series, 1999).

Haanappel, P., and Mackaay, E., *New Netherlands Civil Code Patrimonial Law* (Deventer: Kluwer, 1990).

Harkamp, A., *et al.* (eds), *Towards a European Civil Code* (The Hague: Kluwer International, 1998).

Kotz, H., and Flessner A., *European Contract Law*, Volume 1, translated by T. Weir (Oxford: Clarendon Press, 1997).

Loots, P., *Construction Law and Related Issues*, 1st edn (Kenwyn, South Africa: Juta, 1995).

Markesinis, B.S., *et al.* *The Law of Contracts and Restitution: A Comparative Introduction* (Oxford: Clarendon Press, 1997).

Marsh, P. D. V., *Comparative Contract Law*, 1st edn (Aldershot: Gower, 1993).

McKendrick, E., *Contract Law*, 4th edn (Basingstoke: Palgrave, 2000).

McMeel, G., *Casebook on Restitution* (London: Blackstone, 1996).

McMeel, G., *The Modern Law of Restitution* (London: Blackstone, 2000).

Merryman, J. H., *The Civil Law Tradition*, 1st edn (Stanford, Cal: Stanford University Press, 1969).

Milsom, S., *Historical Foundations of the Common Law*, 2nd edn (London: Butterworth, 1981).

Murdoch, J., and Hughes, W., *Building Contract Law* (London: Longman, 1992).
Murdoch, J., and Hughes, W., *Construction Contracts Law and Management*, 3rd edn (London: Spon, 2000).
Nicholas, B., *The French Law of Contract*, 2nd edn (Oxford: Oxford University Press, 1992).
Odams, M., (ed.) *Comparative Studies in Construction Law: the Sweet lectures* (London: Construction Law Press, 1995).
Pike, A., *Engineering Tenders, Sales and Contracts* (London: Spon, 1982).
Rogers, W.V.H., *Winfield & Jolowicz on Tort*, 7th edn (London: Sweet & Maxwell, 2002).
Samuel, G., and Rinkes, J., *Law of Obligations and Legal Remedies* (London: Cavendish, 1996).
Samuel, G., *Sourcebook on Obligations and Legal Remedies*, 2nd edn (London: Cavendish, 2000).
Sealy, L.S., and Hooley, R.J.A., *Commercial Law: Text, Cases and Materials*, 3rd edn (London: LexisNexis 2003).
Simpson, A.W.B., *The History of the Common Law of Contract: The Rise of the Action of Assumpsit* (Oxford: Clarendon Press 1966).
Smith J., *The Law of Contract*, 4th edn (London: Sweet & Maxwell, 2002)
Uff, J., (ed.) *Contemporary Issues in Construction Law* (London: King's College Press, 1997).
Uff, J., *Construction Law*, 7th edn (London: Sweet & Maxwell, 1999).
Uff, J., and Lavers, A., *Legal Obligations in Construction* (London: Centre of Construction Law and Management, King's College, 1992).
Uff, J., and de Zylva, Odams (eds), *New Horizons in Construction Law* (London: Construction Press, 1998).
Zimmerman, R., *The Law of Obligations: Roman Foundations of the Civilian Tradition* (Oxford: Clarendon Press, 1996).
Zweigert, K., and Kotz, H., *An Introduction to Comparative Law*, 2nd edn, translated by T. Weir (Oxford: Clarendon Press, 1992).

Reports

Latham, M., *Constructing the Team* (London: HMSO 1996)
Law Commission, Section B arguments for reform www.lawcom.gov.uk/library/lc242/part-3.htm
Law Commission, Limitations of Actions, item 2 of the Seventh Programme of Law Refoms (Law Com no 270) www.lawcom.gov.uk
Principles of European Contract Law, Parts I and II www.ufsia.ac.be/storme/pecl2en.html

Journals

Cornes, D. L., 'The Second Edition of the New Engineering Contract' [1996] ICLR p.97.
Duncan-Wallace, I. A., 'Not what the RIBA/JCT meant' (1995) *Construction Law Journal*, p. 188.
Duncan-Wallace, I. A., ' RIP *Crouch*: An Extraordinary Experience' [1998] ICLR p.513.
Duncan-Wallace, I. A., 'The HGCRA: A Critical Lacuna' (2000) 18 *Construction Law Journal*, p.117.
Giaquinto, V. W., *Construction Law Journal*, October 1999 p.2.
Harrison, R. and Jansen, C., 'Good Faith in Construction Law: The Inaugural King's College Construction Law Lecture' (1999) 15 *Construction Law Journal*, p.346
Jones, N., 'Set-off in the Construction Industry', *Construction Law Journal*, 1999, p.84.
Lavers, A., 'Final certificate revisited', www.scl.org.uk.
Lenihan, M., 'Private Finance Initiative/Private Public Partnership: A Contractor's Perspective' 2002, www.scl.org.
Speaight, A., 'I Have No Direct Contract with the Wrongdoer – who can I sue in 2003'?, www.scl.org.uk.

Rice, N. (1984), 'Producing a Standard Form', *Construction Law Journal*, p.255.
Trotman, T., *'Pacific Associates* v. *Baxter*: Time for a Reconsideration? (1995) 15 *Construction Law Journal*, p.449.
Treitel, G. (1967) 'Some Problems of Breach of Contract', *Modern Law Review*, vol.30, p.139.
Uff, J., 'Standard Contract Terms and the Common Law' (1993) 9 *Construction Law Journal*, p.108.
Uff, J., 'Alwyn Walters Memorial Lecture: The Making of a Standard Form' (1995) *Construction Law Journal*, p.170.
Wharton Construction Law Journal, 11, 1995, vol. 6.
Windward R., 'Do Construction Contracts Benefit Parties or Just Their Lawyers?' 1992 Fourth Annual Construction Conference.

Statutory Instruments

Construction Contracts (England and Wales) Exclusion Order 1998 (SI 1998, No. 648).
Insolvency Rules 1986 (SI 1986, No. 1925).
Public Works Contracts Regulations 1991 (SI 1991, No. 2680).
Scheme for Construction Contracts (England and Wales) Regulations 1998 (SI 1998, No. 649).
Unfair Terms in Consumer Contracts Regulations 1999 (SI 1999, No. 2083).

Index